METHODS OF PROTEIN ANALYSIS

ELLIS HORWOOD SERIES IN ANALYTICAL CHEMISTRY

EDITORS: Dr. R. A. Chalmers and Dr. Mary Masson, University of Aberdeen

"I recommend that this Series be used as reference material. Its Authors are among the most respected in Europe". *J. Chemical Ed.,* New York.

Application of Ion-selective Membrane Electrodes in Organic Analysis
G. BAIULESCU and V. V. COŞOFREŢ, Polytechnic Institute, Bucharest
Industrial Methods of Microanalysis
S. BANCE, May and Baker Research Laboratories, Dagenham
Inorganic Reaction Chemistry
Volume 1: Systematic Chemical Separation
D. T. BURNS, Queen's University, Belfast, A. TOWNSHEND, University of Hull, and A. H. CARTER, North Staffordshire Polytechnic
Handbook of Process Stream Analysis
K. J. CLEVETT, Crest Engineering (U. K.) Inc.
Automatic Methods in Chemical Analysis
J. K. FOREMAN and P. B. STOCKWELL, Laboratory of the Government Chemist, London
Fundamentals of Electrochemical Analysis
Z. GALUS, Warsaw University
Laboratory Handbook of Thin Layer and Paper Chromatography
J. GASPERIČ, Charles University, Hradec Kralove
J. CHURÁČEK, University, of Chemical Technology, Pardubice
Handbook of Analytical Control of Iron and Steel Production
T. S. HARRISON, Group Chemical Laboratories, British Steel Corporation
Handbook of Organic Reagents in Inorganic Analysis
Z. HOLZBECHER et al., Institute of Chemical Technology, Prague
Analytical Applications of Complex Equilibria
J. INCZÉDY, University of Chemical Engineering, Veszprém
Particle Size Analysis
Z. K. JELÍNEK, Organic Synthesis Research Institute, Pardubice
Operational Amplifiers in Chemical Instrumentation
R. KALVODA, J. Heyrovský Institute of Physical Chemistry and Electrochemistry, Prague
Atlas of Metal-ligand Equilibria in Aqueous Solution
J. KRAGTEN, University of Amsterdam
Gradient Liquid Chromatography
C. LITEANU and S. GOCAN, University of Cluj
Titrimetric Analysis
C. LITEANU and E. HOPÎRTEAN, University of Cluj
Statistical Methods in Trace Analysis
C. LITEANU and I. RÎCĂ, University of Cluj
Spectrophotometric Determination of Elements
Z. MARCZENKO, Warsaw Technical University
Separation and Enrichment Methods of Trace Analysis
J. MINCZEWSKI et al., Institute of Nuclear Research, Warsaw
Handbook of Analysis of Organic Solvents
V. ŠEDIVEC and J. FLEK, Institute of Hygiene and Epidemiology, Prague
Handbook of Analysis of Synthetic Polymers and Plastics
J. URBANSKI et al., Warsaw Technical University
Analysis with Ion-selective Electrodes
J. VESELÝ and D. WEISS, Geological Survey, Prague
K. ŠTULÍK, Charles University, Prague
Electrochemical Stripping Analysis
F. VYDRA, J. Heyrovský Institute of Physical Chemistry and Electrochemistry, Prague
K. ŠTULÍK, Charles University, Prague
B. JULÁKOVÁ, The State Institute for Control of Drugs, Prague

METHODS OF PROTEIN ANALYSIS

Edited by

ISTVÁN KERESE

Formerly Senior Lecturer
Department of Biochemistry
József Attila University
Szeged, Hungary

Translation Editor:

R. A. CHALMERS

Department of Chemistry
Aberdeen University

ELLIS HORWOOD LIMITED
Publishers · Chichester

Halsted Press: a division of
JOHN WILEY & SONS
New York · Brisbane · Chichester · Ontario

This work was originally published as
Fehérjevizsgálati módszerek
by Műszaki Könyvkiadó, Budapest

Translated by
Dr. D. A. DURHAM

First published in English in a revised and updated form in 1984 by

ELLIS HORWOOD LIMITED, PUBLISHERS
Market Cross House, Cooper Street, Chichester, West Sussex, England
and
AKADÉMIAI KIADÓ, Budapest

Distributors:

Australia, New Zealand, South-east Asia:
Jacaranda-Wiley Ltd., Jacaranda Press,
JOHN WILEY & SONS INC
G.P.O. Box 859, Brisbane, Queensland 4001, Australia
Canada:
JOHN WILEY & SONS CANADA LIMITED
22 Worcester Road, Rexdale, Ontario, Canada.

Europe, Africa:
JOHN WILEY & SONS LIMITED
Baffins Lane, Chichester, West Sussex, England

East European Countries, China, People's Republic of Vietnam, Korean People's Republic, Cuba and Mongolia:
AKADÉMAI KIADÓ, Budapest

North and South America and the rest of the world:
Halsted Press: a division of
JOHN WILEY & SONS
605 Third Avenue, New York, N. Y. 10016, U.S.A.

© **Akadémiai Kiadó, Budapest 1984**

British Library Cataloguing in Publication Data
Methods of protein analysis.—(Ellis Horwood series in analytical chemistry)
1. Proteins—Analysis
I. Kerese, István II. Fehérjevizsgálati módszerek *English*
547.7'5 QD431

ISBN 0-85312-176-1 (Ellis Horwood Limited)
ISBN 0-470-27497-2 (Halsted Press)
LIBRARY OF CONGRESS Card No. 83-18476

Printed in Hungary

COPYRIGHT NOTICE:
All Rights Reserved. No part of this publication may be reproduced, stored in a retrieval system, or transmitted, in any form or by any means, electronic, mechanical, photocopying, recording or otherwise, without the permission of the copyright owners.

To the memory of
Prof. W. Grassmann

Contributors

DR. LÁSZLÓ GÁSPÁR, C. Sc.
Agricultural Research Institute
of the Hungarian Academy of Sciences,
Martonvásár, Hungary

DR. HUBA KALÁSZ, C. Sc.
Department of Pharmacology,
Semmelweis University Medical School,
Budapest, Hungary

DR. ISTVÁN KERESE, C. Sc.
Department of Biochemistry,
József Attila University,
Szeged, Hungary

DR. ÖDÖN TAKÁCS
Department of Biochemistry,
University Medical School,
Szeged, Hungary

DR. ERNŐ TYIHÁK, C. Sc.
Research Institute for
Plant Protection of the
Hungarian Academy of
Sciences, Budapest, Hungary

Contents

Introduction

Protein chemistry has not been exempt from the ever-accelerating development of the natural sciences. Progress in protein chemistry was made possible by the advances in the protein-analysis methods. At the same time, the number of chemists dealing with the study of proteins has also increased since there is a constant need for the investigation and evaluation of protein-containing products in all areas of production, both in industry and in agriculture. It may not be the protein-analyst who through his work directly saves from want the large number of people suffering from protein-starvation, as this is essentially a question of economic investment, but protein-analysis work is indispensable when establishing the economic trends.

Like other methods, those of protein analysis are not immune to rapid obsolescence. This has been taken into consideration in the writing of this book, which has accordingly been compiled in such a way that the material presented will not become out-dated too quickly, even with the further development of the methods described.

This book is intended to serve a practical purpose, and this may have influenced the selection of methods. We have attempted to present methods that can be utilized with the equipment generally available in the average laboratory. Accordingly, those methods have not been discussed which, although they may be the finest procedures in protein analysis, are employed only in leading institutes specializing in research on the structures of proteins, nor have those methods which are the characteristic investigation procedures of some specialized branch of industry.

We set out to achieve a comprehensive and uniform systematization of the separation techniques generally used in protein analysis. As its basis we chose those physical properties through which the separation effects are manifested: the net charges and the sorption characteristics of proteins. Consequently, the methods discussed fall into two large categories: electrophoresis and chromatography. The further sub-divisions within these take into account the particular physico-chemical basis of the separations, and the technical solutions found.

The differences in the set-up of certain chapters are due to the characteristics of the technical performance of the individual methods. In the chapter on laboratory practice, the individual methods are given as parts of an interrelated experimental series; the technical details provided are such that, on their basis, any similar task can be carried out if the aim is carefully considered and a good laboratory technique and scientific outlook are applied.

In the description of the methods, mention of the producer of a product or of the proprietary name of a preparation does not mean any exclusive recommendation or preference over some other similarly utilizable product or apparatus; it is justified only by the more faithful presentation of the procedure under discussion.

The abbreviations in the text are in accordance with IUPAC nomenclature. When the words 'proteins' and 'peptides' are mentioned separately in various places, or are emphasized side by side, then by the word 'peptide' we refer to a fragment obtained from the cleavage of a protein molecule, although its molecular weight may perhaps be larger (e.g. a high molecular-weight peptide from collagen cleaved with CNBr) than that of an original, naturally-occurring molecule (e.g. insulin).

Our aims are complex in nature: to give the theoretical basis and correlations of methods that frequently are technically very different but serve the same object, to provide a guide for selecting the method appropriate for the solution of a given task, and to describe the practical procedure of the methods. There will therefore naturally be certain failings, and hence the authors would be grateful for any comments, advice or criticism regarding theoretical, practical or methodological deficiencies and errors.

The authors would like to take this opportunity to express their thanks to Ellis Horwood Ltd and Akadémiai Kiadó for collaborating in this publication.

CHAPTER 1

General aspects of protein investigations

I. Kerese

Technical development, especially the dramatically rapid development in instrumentation during the past 30 years, has led to a massive expansion of biochemical research, and consequently biochemical knowledge. The progress achieved in protein chemistry since the turn of the century may perhaps be illustrated most strikingly in connection with the determination of the amino-acid composition of proteins: at the time of E. Fischer a quantity of 100 g was necessary for study of the amino-acid composition and the analysis required one month; nowadays, from 10–20 μg of a protein hydrolysate a high-pressure amino-acid analyser will give the result within 70 minutes, with a limiting error of $\pm 1\%$.

These new methods have enabled the rapid isolation of the most varied proteins and the establishment of their structures. Choice of the best method to use depends on the properties of the proteins, and on the aims of the work. In planning protein-analysis work it must always be remembered that different methods may give very different answers (an example is given in Chapter 7, in connection with various methods of separating egg-white).

It would be impossible to systematize the methods for protein isolation and analysis on the basis of classification of the proteins, because there is no such classification of general validity. Whatever the basis used—the constituents, the biochemical and physiological roles in the organism, the solubility, etc.—there is always some degree of overlap between groups. It follows that there can also be no generalized guide to the methods for preparation and investigation of the proteins within the different groups. The methods used to isolate given proteins may also differ according to the origin of the proteins. For instance, because of the different consistencies of the

Table 1.1

Material for examination	Characteristic data, features	Preparative processing		Analytical examinations applicable
		Aim	Manner	
Native plants and animal organs	Plant and animal tissues with high water content	Extraction and possibly isolation of component proteins	Application of homogenization, dissolution, concentration and separation methods	Nitrogen or protein determination, amino-acid determination, examination of homogeneity with electrophoresis and chromatographic methods, molecular weight determination
Physiological solutions, tissue fluids	Solutions with low protein and free amino-acid concentrations	Detection, extraction and possibly isolation of component proteins, determination of free amino-acids	Application of concentration and separation methods	Nitrogen or protein determination, amino-acid determination, free amino-acid determination, examination of homogeneity with electrophoresis and chromatographic methods, molecular weight determination

Fermentation industry products	Fermentation fluids containing dissolved and insoluble proteins, suspensions, fermented feeds	Obtaining each component protein, examination of utilization of partially decomposed proteins, detection and determination of free amino-acids and amino-acid destruction	Dependent on the technological objectives	Nitrogen or protein determination, amino-acid determination, free amino-acid determination, examination of components with electrophoresis and chromatographic methods, molecular weight determination
Food and feed proteins	Agricultural and industrial products	Determination of protein content and amino-acid composition, examination of solubility and tendency to undergo lysis	Generally standard prescriptions for the individual preparations, depending on the practical use	Nitrogen and amino-acid determination, possibly solubility study, examination of composition of component proteins, and of molecular weight proportions
(a) Isolated proteins, hydrolysis and reaction products (b) Synthetic peptides	Biochemical products Products of preparative chemical synthesis	Chemical and biological identification, establishment of homogeneity, structural examination in the case of research products	Examination of solubility and of stability and physical properties of solution, with electrophoresis and chromatography, examination of homogeneity	Determination of physico-chemical characteristics, determination of nitrogen and amino-acid contents, partial hydrolyses, determination of terminal amino-acids and amino-acid sequence

starting materials, in the cases of soybean, pancreas and egg different methods must already be employed in the first step (the extraction) in the isolation of a particular protein such as trypsin inhibitor.

In protein analysis, the object is not usually establishment of the chemical structure. In the study of plant breeding, for example, if the object is the knowledge of only one of the components of the seeds (e.g. total nitrogen content, or a histone, or the nutritionally important methionine), a complete protein analysis is not necessary. The choice of method is therefore influenced jointly by the properties of the raw material and of the proteins to be investigated, and by the purpose of the analysis. Some insight into protein analysis is provided by a scheme which attempts to systematize the most common starting materials on the basis of their origin (Table 1.1).

Since proteins are products of biosyntheses proceeding in plant or animal organisms, the native plant and animal tissues, physiological solutions and tissue fluids form the starting materials for protein examination, which may range in complexity from the simplest determination, that of nitrogen content, to the most complex problem, elucidation of the structures of the proteins and peptides isolated from them. The experimental work will correspondingly be extremely varied. The nitrogen content may be determined with only a few operations, but the result may not reveal very much. In contrast, isolation of a biologically active protein necessitates a whole series of difficult separations.

The last type of materials listed in Table 1.1 consists essentially of the product of these purification and separation operations: an isolated protein or peptide, as proved by purity tests, which is suitable for the determination of its physical, chemical and biological characteristics.

Fermentation industry products, together with food and feed proteins (which deserve special mention for economic reasons), do not usually require a full analytical examination in everyday practice. To establish the enzyme-output of an industrial-scale fermentation and the biological activity of the product, it is generally unnecessary to perform the entire series of preparative separations. Similarly, the nutritional value of food and feedstuffs can be estimated from their nitrogen content, amino-acid composition and readiness to undergo lysis, without extraction of the component proteins.

In research into protein structures, the isolation steps may sometimes be omitted, as shown by the elucidation of the insulin sequence, but then great ingenuity is needed, such as that shown by Sanger in his Nobel Prize work. The isolation of proteins can be divided into two steps: *preparation* and *separation.*

The *preparation* steps (including mechanical disintegration, homogenization, suspension, dissolution, sedimentation, etc.) are determined not only by the analytical objectives, but also by the mechanical, physical and chemical properties of the starting material.

Before the protein isolation, the material homogenized or extracted from the starting material is purified by ultracentrifugation, fractional precipitation, filtration, etc., depending on its nature.

The *separation* and isolation of the individual component proteins (peptides) are done on the more or less purified solutions of the proteins, at the appropriate concentration. The separation may be based on differences in some fundamental physical property of the proteins: the sedimentation rate, mass, size, electric charge, hydrophilic nature or isoelectric point of the protein molecules.

With the development of isoelectric focusing (IEF), the separation techniques have undergone considerable modification. Until very recently, gel-filtration was used to achieve separation on the basis of molecular mass. Isoelectric focusing gives direct knowledge of the isoelectric points of the component proteins and hence indicates the optimum conditions for further separation or isolation, on the basis of the character of individual proteins.

Figure 1.1 outlines a flow-chart for protein purification; it illustrates various possibilities for combining individual techniques.

The following comments may be made on the individual steps:

1. With analytical PAG IEF in the pH interval 3–10, the pI values of the component proteins are obtained, which can serve for identification or confirmation of identity.

2. On the basis of the IEF results from 1, the protein(s) of importance can be obtained by preparative flat-bed IEF with an ampholyte in a narrow pH range corresponding to the pI value(s).

3. By the simplest PAGE, electrophoresis in homogeneous 7.5% PAG, information is obtained on (a) the purity of the fraction; (b) its approximate molecular mass; (c) in the case of several components, the approximate amounts of the individual fractions.

4. Further information is obtained on the molecular mass(es) of the protein(s).

5. From comparison of the SDS-PAGE results and those of the previous two PAGE operations 2 and 3, it may be decided whether the protein is present as a monomer or a polymer.

6. Accompanying proteins (with similar pI values, but different molecular masses) in the band acquired from 2 are separated from the desired material by preparative gel-filtration with the appropriate Sephadex, on the basis of 3, 4 and 5. The success of this step depends on correct selection of the Sephadex (G-75, 100, 150).

7. Repeated purification of the detected fractions on the basis of their pI values, in the narrow pH range used in 2, with a restricted electrode distance, to separate the desired protein from those protein fractions not removed by the gel-filtration.

8. Analytical IEF in PAG with an ampholyte giving a narrow pH range, to confirm the results of 7. Initially the longest possible gel-field is used, then a restricted electrode distance. If isolation

Fig. 1.1

① Analytical PAG IEF, pH 3–10
② Preparative flat bed IEF, within the optimum interval for the fractions to be separated

Pre-examinations for further purification:
③ Homogeneous analytical PAGE
④ Gradient analytical PAGE
⑤ SDS analytical PAGE

Further purification on the basis of the net charge of the proteins

⑦ Preparative flat-bed IEF on the necessary fractions

⑧ Restricted analytical IEF in PAG, in the optimum pH interval

⑨ Special isolation methods (e.g. affinity chromatography)

Further purification on the basis of molecular mass and charge

⑥ Preparative column gel-filtration

⑩ Ion-exchange preparative chromatography of charged proteins

⑪ Uncharged proteins

⑫ Cation-exchange chromatography of positively-charged proteins

⑬ Anion-exchange chromatography of negatively-charged proteins

⑭ Adsorption chromatography

⑮ Partition chromatography

⑯ Restricted analytical IEF in PAG, in the optimum pH interval, or
⑰ Affinity chromatography

Abbreviations: IEF = isoelectric focusing
PAG = polyacrylamide gel
PAGE = polyacrylamide gel electrophoresis
SDS = sodium dodecylsulphate

is successful, the desired material is eluted from the flat-bed used in 7 and the ampholyte removed by gel-filtration.

10, 12, and 13. Preparative ion-exchange chromatography on the basis of the charge of the protein, which depends on the pH of the solvent; a cation-exchanger is used for positively-charged proteins dissolved at a pH below the isoelectric point, and an anion-exchanger for negatively-charged proteins dissolved at a pH above their isoelectric points.

11, 14, and 15. Preparative chromatography of uncharged molecules by partition or adsorption chromatography.

16. As for 8.

17. With affinity chromatography it is possible to separate the substances according to their biological function.

The analytical methods used are employed both for identification and determination, and should preferably be applied after each major separation step to assess the efficiency. Examination of the starting material yields useful information on the choice of the amounts and proportions to be used. The efficiency of dissolution can be checked by determination of the amount of protein dissolved, e.g. by nitrogen determination, the Lowry reaction, the biuret reaction, or some specific reaction of the solute, such as an enzymic assay.

From the protein and biological activity determined in the course of extraction of biologically active substances, the purity achieved can be assessed. It is useful to measure the following quantities and tabulate them:

Protein concentration, mg/ml
Volume, ml
Total protein, mg
Total activity units
Specific activity, units/mg
Yield, %
Purity, %

One of the most important tasks is to establish whether a fraction exhibiting a single chromatographic peak or electrophoretic band corresponds to a single component or an unresolved mixture. This can be done by using essentially the same techniques as for the preparative separation: sedimentation, electrophoresis, gel filtration and chromatography.

Different authors have different ideas about characterization and structural examination of isolated proteins. We will illustrate this from the proposals of two authors.

Jollès [1] classifies structural methods into two main groups. The first contains the classical structure research methods in ten steps.

1. Determination of the amino-acid composition.

2. Determination of the molecular weight.

3. Qualitative and quantitative determination of the *N*-terminal amino-acid and sequence.

4. Qualitative and quantitative determination of the *C*-terminal amino-acid and sequence.

5. Separation of the peptide chains, if the molecule consists of several.

6. Specific cleavage of every peptide chain into smaller peptide fragments, preferably with trypsin.

7. Purification and separation of the fragments.

8. Determination of the structures of the fragments.

9. Compilation of the fragments.

10. Establishment of the positions of the disulphide bridges.

The second group contains the three methods for direct establishment of the structure.

1. X-ray structural analysis, with which the stereostructure of the protein can be derived from knowing the primary structure.

2. Totally automated Edman decomposition.

3. Mass spectrometry, for identification of peptides.

Rothfus [2] gives the following characterization of proteins: titration [3], hydrogen exchange [4], enzymic cleavage [5], reactions with bifunctional reagents [6], dye-binding [7–9], sedimentation [10], light scattering [10], viscosity measurement [10], osmometry [11], dialysis and diffusion [12], ultraviolet spectroscopy [13–15], infrared spectroscopy [14], optical rotatory dispersion [14, 16, 17], circular dichroism [18], fluorescence measurements [14, 19–22], electron paramagnetic resonance spectrometry [23–26], nuclear magnetic resonance spectrometry [27–29], differential thermal analysis [30, 31], X-ray crystallography [32–34], sequence analysis [35].

The listings by both authors are subjective, and neither can lay claim to exclusive validity. In any event they provide a picture of the complexity of characterizing proteins and show that at this level a significant proportion of protein analysis no longer belongs only to the sphere of biochemistry.

In this book we are confined to presentation of only the possibilities of application of protein analysis, and a fundamental technical description of the individual methods. This is supplemented in the appropriate chapters by a list of those monographs and technical books which may satisfy the needs of those interested in the theoretical basis of the methods, and in the more complex aspects of protein research [36–53].

REFERENCES

[1] Jollès, P.: *Chimia*, **25,** 1 (1971).

[2] Rothfus, J. A.: *J. Am. Oil Chem. Soc.*, **47,** 316 (1970).

[3] Nozaky, Y., Tanford, C.: *Examination of Titration Behavior*, in *Methods in Enzymology*, S. P. Colowick and N. O. Kaplan (eds), Vol. XI, Academic Press, New York, 1967.

[4] DiSabato, G., Ottesen, M.: *Hydrogen Exchange*, in *Methods in Enzymology*, S. P. Colowick and N. O. Kaplan (eds), Vol. XI, Academic Press, New York, 1967.

[5] Rupley, J. A.: *Susceptibility to Attack by Proteolytic Enzymes*, in *Methods in Enzymology*, S. P. Colowick and N. O. Kaplan (eds), Vol. XI, Academic Press, New York, 1967.

[6] Wold, F.: *Bifunctional Reagents*, in *Methods in Enzymology*, S. P. Colowick and N. O. Kaplan (eds), Vol. XI, Academic Press, New York, 1967.

[7] Steinhardt, J., Beychok, S.: *Interaction of Proteins with Hydrogen Ions and Other Small Ions and Molecules*, in *The Proteins*, H. Neurath (ed.), 2nd Ed., Vol. II, Academic Press, New York 1964.

[8] Furano, A. V., Bradley, D. F., Childers, L. G.: *Biochemistry*, **5,** 3044 (1966).

[9] Lang, J. H., Lasser, L. C.: *Biochemistry*, **6,** 2403 (1967).

[10] Tanford, C.: *Physical Chemistry of Macromolecules*, Wiley, New York, 1961.

[11] Amstrong, J. L.: *Modern Methods for Determining Number-Average Molecular Weights*, in *International Symposium on Polymer Characterization*, K. A. Boni and F. A. Sliemers (eds), Interscience, New York, 1969.

[12] Craig, L. C.: *Techniques for the Study of Peptides and Proteins by Dialysis and Diffusion*, in *Methods in Enzymology*, S. P. Colowick and N. O. Kaplan (eds), Vol. XI, Academic Press, New York, 1967.

[13] Herskovits, T. T.: *Difference Spectroscopy*, in *Methods in Enzymology*, S. P. Colowick and N. O. Kaplan (eds), Vol. XI, Academic Press, New York, 1967.

[14] Weber, G., Teale, F. W. J.: *Interaction of Proteins with Radiation*, in *The Proteins*, H. Neurath (ed.), 2nd Ed., Vol. III, Academic Press, New York, 1965.

[15] Donovan, J. W.: *J. Biol. Chem.*, **244,** 1961 (1969).

[16] Schellman, J. A., Schellman, C.: *Conformation of Polypeptide Chains in*

Proteins, in *The Proteins*, H. Neurath (ed.), 2nd Ed., Vol. II, Academic Press, New York, 1964.

[17] Urnes, P., Doty, P.: *Adv. Protein Chem.*, **16,** 402 (1961).

[18] Beychok, S.: *Science*, **154,** 1288 (1966).

[19] Weber, G.: *Adv. Protein Chem.*, **8,** 415 (1953).

[20] Brand, L., Witholt, B.: *Fluorescence Measurements*, in *Methods in Enzymology*, S. P. Colowick and N. O. Kaplan (eds), Vol. XI, Academic Press, New York, 1967.

[21] Cowgill, R. W.: *Biochim. Biophys. Acta*, **168,** 417, 431, 439 (1968).

[22] Weber, G., Andersen, S. R.: *Biochemistry*, **8,** 361 (1969).

[23] Ogawa, S., McConnel, H. M.: *Proc. Natl. Acad. Sci. U. S.*, **58,** 19 (1967).

[24] Hsia, J. C., Piette, L. H.: *Arch. Biochem. Biophys.*, **129,** 296 (1969).

[25] Landgraf, W. C., Inesi, G.: *Arch. Biochem. Biophys.*, **130,** 111 (1969).

[26] Beinert, H., Orme-Johnson, W. H.: *Ann. N. Y. Acad. Sci.*, **158,** 336 (1969).

[27] Blears, D. J., Danyluk, S. S.: *Biochim. Biophys. Acta*, **147,** 404 (1967).

[28] Fuller, M. E., Brey, W. S. Jr.: *J. Biol. Chem.*, **243,** 274 (1968).

[29] Takahashi, A., Mandelkern, L., Glick, R. E.: *Biochemistry*, **8,** 1673 (1969).

[30] Green, D. B., Happey, F., Watson, B. M.: *Differential Thermal Analysis Applied to Polypeptides*, in *Conformation of Biopolymers*, Ramachandran (ed.), Vol. 2, Academic Press, New York, 1967.

[31] Chrigton, J. S., Happey, F., Ball, J. T.: *The Study of Some Fibrous Proteins by Differential Thermal Analysis*, in *Conformation of Biopolymers*, Ramachandran (ed.), Vol. 2, Academic Press, New York, 1967.

[32] Phillips, D. C.: *Sci. American*, **215,** 78 (1966).

[33] Perutz, M. F., Muirhead, H., Cox, J. M., Goaman, L. C. G.: *Nature*, **219,** 131 (1968).

[34] Dickerson, R. E.: *X-Ray Analysis and Protein Structure*, in *The Proteins*, H. Neurath (ed.), 2nd Ed., Vol. II, Academic Press, New York, 1964.

[35] Hirs, C. H. W.: *Enzyme Structure*, in *Methods in Enzymology*, S. P. Colowick and N. O. Kaplan (eds), Vol. XI, Academic Press, New York, 1967.

[36] Neurath, H. (ed.): *The Proteins*, 2nd Ed., Academic Press, New York, 1964.

[37] Alexander, P., Block, R. J.: *A Laboratory Manual of Analytical Methods of Protein Chemistry*, Pergamon Press, London, 1960.

[38] Morris, C. J. O. R., Morris, P.: *Separation Methods in Biochemistry*, Pitman, London, 1963.

[39] Lübke, K., Schröder, E., Kloss, G.: *Chemie und Biochemie der Aminosäuren, Peptiden und Proteine*, Thieme Verlag, Stuttgart, 1975.

[40] Eastoe, J. E., Courts, A.: *Practical Analytical Methods for Connective Tissue Proteins*, Spon, London, 1963.
[41] Bailey, J. L.: *Techniques in Protein Chemistry*, 2nd Ed., Elsevier, Amsterdam, 1967.
[42] Dévényi, T., Gergely, J.: *Amino Acids, Peptides, Proteins*, Elsevier, Amsterdam, 1974.
[43] Haschemeyer, R. H. H., Haschemeyer, A. E. V.: *Proteins. A Guide to Study by Physical and Chemical Methods*, Wiley-Interscience, New York, 1973.
[44] Work, T. S., Work, E. (eds): *Laboratory Techniques in Biochemistry and Molecular Biology*, I. Vol. 1, Vol. 2, II. Vol. 2, V. Vol. 2, North-Holland, Amsterdam, 1969, 1970, 1975, 1976.
[45] Bergmeyer, H. U. (ed.): *Methods of Enzymatic Analysis*, Academic Press, New York, 1963.
[46] Needleman, S. B. (ed.): *Protein Sequence Determination*, Springer-Verlag, Berlin, 1975.
[47] Rauen, H. M. (ed.): *Biochemisches Taschenbuch*, Springer-Verlag, Berlin, 1964.
[48] Wold, F.: *Ann. Rev. Biochem.*, **50,** 783 (1981).
[49] Gross, E., Meinehofer, J. (eds): *Peptides, Structure and Biological Function*, Pierce Chemical Co., Rockford, 1979.
[50] Gross, E., Meinehofer, J. (eds): *The Peptides*, Vol. 2, Part A, Academic Press, Inc., New York, 1979.
[51] Srinivasan, R. (ed.): *Biomolecular Structure, Conformation, Function* and *Evolution*, Vol. 1, Pergamon Press, Oxford, 1980.
[52] Whitaker, J. R., Fujimaki, M. (eds): *Chemical Deterioration of Proteins*, Am. Chem. Soc., Washington DC, 1980.
[53] Day, C. E. (ed.): *High-Density Lipoproteins*, Dekker, Basel, 1981.

CHAPTER 2

General laboratory methods

L. Gáspár

2.1 PREPARATION OF THE MATERIAL FOR EXAMINATION

2.1.1 Sampling

Depending on the heterogeneity of the material, the mass of the sample may range from a few grams to several kilograms. In any event it is many times the quantity taken for the actual analytical examination. As an example, agricultural feed (e.g. roughage, hay) may be the most heterogeneous material to be examined in laboratory practice. It must be made as representative as possible, and many 200–250 g samples must be taken (at least 20 samples per 25 tons of unpressed hay, or per 50 tons of pressed hay) and combined to give average samples of about 5 kg. In the course of the laboratory preparation, therefore, there is of necessity a second sampling (we may call this internal sampling), but this must be preceded by rough homogenization of the primary sample, without which the original sampling would lose its significance. The correct procedure then is to bring the whole of the primary sample into a state suitable for homogenization, and to employ laboratory equipment or semi-plant-scale machines to homogenize it in the manner most appropriate for the purpose. For example, solids that can be ground should be reduced to a grain size of at most 1–2 mm.

For umpire, referee or official examination, the entire analysis must be carried out in the prescribed agreed manner. Since these procedures are readily available in official handbooks, it is not necessary to describe them here. We are primarily concerned with the analyses connected with various fields of research and application in agriculture and industry, the methods for which are selected by the specialist, according to the objectives, number of analyses, accuracy required and the technical level of the laboratory.

In practice there are three possibilities for decreasing the scatter in the results of an examination:

(a) increasing the quantity of material taken for analysis;
(b) increasing the number of replicate determinations;
(c) improvement of the homogeneity of the average sample.

Increasing the amount of material taken precludes use of a whole series of modern methods that have been developed for micro and semimicro-scale work, and at the same time increases the costs and possibly the time of the analysis. This approach is therefore used only in the opening-out phase (dissolution, destruction, ashing) and the analysis is subsequently continued on an aliquot. The optimum number of replicates demanded by statistics is usually much larger than the number dictated by practical considerations such as time, equipment, labour-force, costs, etc. Frequently, therefore, the best solution is to improve the homogeneity of the sample. Thus, in the internal sampling, a solid sample must be ground to such a fineness that a quantity of 0.1–1 g will be adequately representative of the whole. According to practical experience, this requires a particle size range of 200–500 μm. The comminution must be done so that the temperature does not exceed 45 °C during the homogenization. It is advisable to homogenize moist materials in two steps, if the starting material is heterogeneous and in the kg range. It is very important to mix the samples again just before portions are taken for analysis, to avoid introduction of heterogeneity by sedimentation.

Table 2.1
Standard deviations of nitrogen analyses as a function of the nitrogen content of the sample [1]

Nitrogen, %	$\bar{R}$, %	S, %	v, %
1	0.036	0.032	3.2
3	0.057	0.051	1.7
5	0.062	0.055	1.1
15	0.086	0.076	0.5

The smallest quantity of sample that should be taken is governed not only by the homogeneity of the material, but also by the sensitivity of the analytical method. Further, the scatter in the results depends not only on the systematic error in the analytical procedure but also on the concentration of the determinand in the sample to be analysed. For example, the estimated standard deviation for duplicate Kjeldahl nitrogen determinations was found to increase with nitrogen content, but the relative standard deviation (coefficient of variation) decreased (Table 2.1) [1].

In Table 2.1, $\bar{R}$ is the mean range (difference between duplicates) for a large number of samples, S is the estimated standard deviation calculated from the relation $S = \bar{R}k_n$ where n is the number of replicates (the values of k_n are given in Table 2.2a), and v is the coefficient of variance given by

$$v = \frac{S}{x} \cdot 100$$

It can be seen that the standard deviation increases with increasing nitrogen content, the coefficient of variation decreases.

Table 2.2a
Values of the coefficients k_n for estimation of the standard deviation, according to Dean and Dixon [2]

n	k_n
2	0.886
3	0.591
4	0.486
5	0.430
6	0.395
7	0.370
8	0.351
9	0.337
10	0.325

If the procedure is particularly difficult or time-consuming there will seldom be more than duplicate analyses performed. If the results agree closely enough the average may be used; otherwise a third analysis must be performed to decide which of the results is more nearly correct. However, the question arises as to what constitutes a close enough agreement. With only two results we cannot proceed according to the strict methods of mathematical statistics; instead, an estimation procedure must be employed. The maximum difference R_{max} between the results of duplicate analyses may be calculated from

$$R_{max} = b\bar{R}$$

where the b values are taken from Table 2.2b, for probability levels $P = 10$, 5 and 1%, and $\bar{R}$ has the same meaning as before.

Thus for the Kjeldahl nitrogen determination, if the nitrogen content of the samples is about 3%, the value of $\bar{R}$ is 0.057%, and for a probability level $P = 5\%$ and $n = 2$ repetitions, we have $b = 2.46$.
Hence

$$R_{max} = 2.46 \times 0.057\% = 0.14\%$$

Suppose that the two results are 3.19 and 2.96%. The difference is 0.23%, which is larger than the permitted $R_{max} = 0.14\%$. One of the results therefore

Table 2.2b
Values of b for calculation of the maximum range permitted for probability levels $P = 10\%$, 5% and 1%, in the case of n replicate analyses [1]

n	b		
	$P = 10\%$	$P = 5\%$	$P = 1\%$
2	2.06	2.46	3.23
3	1.71	1.96	2.43
4	1.57	1.76	2.14
5	1.50	1.66	1.98

involves a gross error, and accordingly a third analysis must be done. Suppose it gives 2.92%. In this case, the difference $2.96 - 2.92 = 0.04\%$ is smaller than the calculated 0.14%, and the two values can therefore be averaged. The value 3.19%, on the other hand, must be rejected, as involving a gross error.

2.1.2 Homogenization of samples

For sample preparation, various types of apparatus may be used in one or more steps to attain the required degree of comminution and homogeneity. This equipment uses percussion, shearing, cutting and friction, which produce various amounts of heat, depending on the resistance of the material, the construction of the apparatus and the rate of the operation. The requirements are similar to those to be met in drying. The temperature must not be so high that the examination material undergoes damage. This can be achieved by appropriate choice of apparatus, correct rate of addition of the sample, and by cooling if necessary.

A further requirement is that the apparatus should be chemically resistant to the materials in the sample and another very important factor is that it should be easy to clean.

Comminution of dry solid materials

Comminution of a large mass of solid samples such as dry plant materials is best begun with slicing machines and continued with a universal mill. The rotor of a universal mill operates at 3000–15 000 rpm and is of various types such as a rotating cross, a swing hammer, a series of hammers or a serrated disc. The stationary part of the mill may be smooth or serrated, and there may be a sieve over a greater or smaller portion of the cylinder jacket. The use of disc grinders cannot be recommended, as these are difficult to clean and it is difficult to avoid the heat-effect associated with the shearing.

Ball and roller mills may be used for the comminution of brittle, relatively inelastic materials containing fibres. In the case of naturally-occurring materials the possibilities of their use are limited. The comminution is slow, the heat-effect is high (although this can be counteracted somewhat by external cooling), it is not easy to take out the sample, and the apparatus is difficult to clean. However, there can be no doubt that a very high degree of homogeneity can be achieved with such equipment, with reduction to an average particle size of 0.5 mm.

The efficiency of comminution of dry solid materials may be checked by classification with a set of sieves shaken mechanically. It must be emphasized that the sieving is intended primarily to be a means of controlling the comminution process. If it is used for grading, it must always be taken into account that the sample (which should be uniform for analysis purposes) will have been resolved into fractions on the basis of particle size, and that these fractions will differ in composition because each component will have its own particle size distribution, and these will all be different. In certain cases, however, e.g. with milling-industry products, the aim may well be to examine the fractions with different particle sizes.

Comminution of moist solid materials

The objective is the production of homogeneous suspensions with various consistencies. Depending on the original moisture content of the material and on the aim and means of the further processing, the suspensions may be prepared with or without the addition of a suspending agent. The suspensions are prepared with homogenizers. The apparatus known as a 'Waring blender' comes in two different types. In the first, the sample is placed in a ribbed glass or metal vessel, various amounts of water or some solution are added, and comminution is effected by two-armed, cross- or star-shaped knives, generally rotating in several planes on a common spindle, at 10 000–25 000 rpm. The ribbing helps to prevent the mixture from flowing together with the knives, and the material is comminuted by the knives as a consequence of the resulting eddy resistance. The homogenizer vessel is usually surrounded by a cooling jacket, to prevent overheating. The cutting knives of the other type rotate in a serrated castellated sheath, the apparatus operating on the principle of centrifugal turbines.

Larger samples may be homogenized in a vessel into which a coarse suspension is fed so that it flows through at a rate corresponding to the efficiency of homogenization.

Homogenizers of 'Potter–Elvehjem' type are suitable for the preparation of cell-free material. The essence of these homogenizers is that a ground-glass or Teflon piston is moved, either manually or mechanically, in a concentric ground-glass tube. The material passing between the two surfaces attains a very fine degree of dispersion as a consequence of the friction, and is homogenized so as to be cell-free. This type of apparatus is suitable only for the homogenization of small samples which have previously been coarsely dispersed in some other homogenizer.

It must be noted that very good homogenization can also be obtained by grinding the material with a mortar and pestle, by the traditional simple method, with the addition of powdered sand or glass.

For preparation of the homogeneous suspension, water or some aqueous solution is used as suspending agent. The suspension medium may be a buffer solution, or salt or sugar solutions of various concentrations. If the aim of the homogenization is to obtain intact cells in the suspension, then solutions of isotonic concentration are necessary. Detergents, chelating agents, anti-oxidants, enzyme-inhibitors, etc. may be employed in the homogenizing solution.

Ultrasonic disintegrators may be used to advantage, mainly for extraction of the proteins of micro-organisms [3]. Instruments manufactured for laboratory use operate in the frequency range around 20 kHz. With these, and a vessel and vibrator of good geometry, an energy concentration of as much as 50 W/cm^2 can be achieved. In the ultrasonic instruments, the output of a high-frequency generator is transformed by a piezoelectric cell into mechanical vibration, which is transmitted to the liquid by a vibrator made from a metal (e.g. titanium) resistant to mechanical and chemical effects. The vibration causes high pressure and suction forces in the liquid and the formation of extremely small bubbles in the interior of which the pressure is very low. This phenomenon is known as cavitation. Since the vibration frequency is very high, the bubbles have a short life-time, less than 10 μsec. As a consequence of the disappearance of the bubbles, very high ultrasonic pressures develop in the liquid. The liquid comes into motion, the velocity of the particles attaining 30 m/s, and their acceleration 10^5 g. This mechanical effect is capable of rupturing the cell wall, and thus the internal content of the cell becomes free. Since the different biomembranes differ in resilience, ultrasonic disintegrators may be used to carry out fractionated cell rupture according to a programme developed by previous investigation with the particular instrument and materials. Though the method is most suitable for obtaining cell proteins, various effects must be reckoned with. Ultrasonic vibration is accompanied by a thermal effect, and without satisfactory cooling the material may be overheated. There is also a depolymerizing effect, and thus aggregated protein molecules are disaggregated. It may readily be observed that the electropherograms of protein solutions display changes after ultrasonic treatment of the solutions: certain fractions disappear, and new ones arise. In liquids, the ultrasonic waves are decelerated over a short distance, and uniform treatment can be attained only in a small volume. A flux regulator can be used to achieve uniformity of treatment.

2.1.3. Concentration and drying of the material

The samples to be examined may be liquids or solids of biological origin, that are heterogeneous to various extents; by virtue of their composition and moisture content, they may be readily liable to change, either in themselves or as a result of microbial processes. The sample preparation therefore has to be considered in conjunction with storage of the samples. Until dried or processed, liquid samples or solids with a high moisture content must be stored in deep-freezers, in which they can be kept at a temperature of −25 °C or below. A temperature of −60 °C is necessary for the complete freezing of animal tissues.

The technical equipment for mild drying consists of vacuum-driers and lyophilizing apparatus.

The vacuum-drying ovens used in the laboratory can be heated in the range 40–150 °C and are pressure-tight; they are cylindrical in shape, and fitted with a glass-windowed door, through which the state of the sample may be inspected. If the water pressure is favourable and constant a water-pump can be used to reduce the pressure, but a rotary oil-pump with gas-ballast valve is generally used. In such equipment, bulkier samples may be dried at 40–60 °C.

Lyophilization, the most expedient means of removing water from thermally labile biological materials, is used in several phases of the work in a protein laboratory. Lyophilization, or freeze-drying, removes water by the sublimation of ice from the frozen sample at a temperature in the range between 0 and −60 °C, at a pressure of between 13.3 and 0.13 Pa (0.1 and 0.001 mmHg). Apart from the low temperature, the method has the great advantage (particularly in the case of colloidal solutions) that dissolution of the lyophilized preparations is much easier than with any other drying procedure.

Concentration and drying by vacuum-evaporation

A frequent task in biochemical laboratory practice is concentration of dilute aqueous solutions, and sometimes they must be evaporated to dryness. In almost all cases this must not be done by simply boiling off the water at atmospheric pressure, especially with protein, peptide and amino-acid solutions, for which a lower temperature is necessary.

Vacuum-evaporation and lyophilization are very suitable for this purpose, vacuum-evaporation being used mainly for peptides and amino-acids, and lyophilization for proteins. The essential difference between the two procedures is whether the water is evaporated from the liquid phase, or sublimed from the solid phase at a temperature below freezing-point.

Table 2.3
Vapour pressure above water and ice

Temperature °C	Pressure Pa (mmHg)	Temperature °C	Pressure Pa (mmHg)
+60	19918 (149.4)	0	613 (4.6)
+50	12333 (92.5)	−10	260 (1.95)
+40	7372 (55.3)	−20	103 (0.77)
+30	4240 (31.8)	−30	31.3 (0.28)
+20	2333 (17.5)	−40	12.9 (0.097)
+10	1227 (9.2)	−50	4.0 (0.030)
0	613 (4.6)	−60	1.1 (0.008)

It is clear from Table 2.3 that water vapour can be sublimed from ice at a temperature lower than 0 °C only at a pressure below 600 Pa (4.5 mmHg). At temperatures higher than 0 °C, water vapour can be removed from the liquid phase at a pressure less than the water vapour pressure corresponding to the given temperature.

The basis of successful vacuum technique is choice of the proper pump for the job.

In simple laboratory operations, water-pumps or rotary oil-pumps are employed for filtration and vacuum-distillation. Water-pumps made of glass or plastic have the advantage that they are also suitable for the removal of corrosive vapours (e.g. hydrogen chloride) without the use of absorbers or cold-traps. Their disadvantages are that the performance is comparatively poor, the water-consumption high, and the attainable pressure-reduction not very high; moreover, if there is a decrease in the water pressure they readily suck back, and a suitable valve must therefore always be employed.

Rotary oil-pumps are most frequently used for laboratory operations requiring a vacuum. In the case of distillation and evaporation, it must be

Table 2.4

Operating data on vacuum-pumps

Type of pump	Operating range, Pa (mmHg)	End-vacuum, Pa (mmHg)	Suction rate
Water-pumps	1600 (12)	Pressure of water vapour at operating temperature. Water requirements: 1 l of water/0.6 l of air sucked off	100 l/h
Rotary oil-pumps	0.13 – 1.3 (10^{-3} – 10^{-2})		
one-stage		0.26 – 13.3 (0.002 – 0.1)	1 – 10^3 m^3/h
two-stage		0.00133 – 0.0133 (10^{-5} – 10^{-4})	
Diffusion-pumps	0.000133 – 1.3 (10^{-6} – 10^{-2})	0.000133 (10^{-6})	0.1 – 10^3 l/s
prevacuum	13.3 (0.1)		

borne in mind that vapours condensing or dissolving in the oil of the pump, or forming an emulsion with it, will lower the efficiency. To remove such vapours, it is advisable to insert a cold-trap between the evacuated vessel and the pump. Even at a pressure of 0.013 Pa (10^{-4} mmHg) water vapour can be condensed in a glass or metal cold-trap placed in a Dewar flask kept at –80 °C with a mixture of 'dry ice' + alcohol. In the case of corrosive gases and vapours, traps containing suitable absorbents must be inserted. If only a medium vacuum is required, the efficiency of a rotary pump can be maintained (in spite of the effect of water vapour on the pump) if the oil is continuously exchanged during operation. The condensed vapour is removed from pumps fitted with a gas-ballast valve. Oil-pumps will give their specified performance only if vacuum-oil of suitable quality is used.

In selecting the size of pump to use it is practical to apply the following rough correlation:

$$t = \frac{6.5V}{S}$$

where t is the suction time necessary to attain a pressure of 133 Pa (1 mmHg), V is the volume to be evacuated, and S is the suction rate of the pump.

The vacuum applied must reduce the pressure to below that of water vapour at the desired evaporation temperature. The use of a vacuum-pump with a performance high enough to remove the vapours formed would not be economical. Accordingly, the vapour formed must be condensed. The efficiency of evaporation will obviously be higher, the more perfect the condensation of the vapour, which can be achieved by increasing the cooling surface and the intensity of cooling. A spiral condenser is generally used, cooled by mains water, but performance can be enhanced by circulation of a refrigerated coolant, e.g. 25% w/w or 23% v/v ethylene glycol—water mixture at —107 °C.

The efficiency of vacuum-evaporation may be raised by increasing the surface area of the liquid being evaporated. This is done in rotary film-evaporators. The principle was originated by Craig *et al.* [4]. A not too large quantity of the liquid to be evaporated is placed in a spherical or pear-shaped flask immersed in a water-bath and rotated at 50–300 rpm.

During recent years, various types of rotary evaporator have been developed. With the normal type, about 2.5 litres of water can be evaporated per hour at a pressure of 2.67–9.33 kPa (20–70 mmHg).

Small amounts of liquids can be evaporated from a test-tube or a beaker in a vacuum desiccator. Phosphorus pentoxide may conveniently be used to bind the water vapour.

In general, whatever evaporation method is employed, if the solution to be evaporated contains a volatile solvent such as alcohol, acetone, ether or a volatile acid, it is advisable to insert an absorption vessel in order to protect the oil-pump from corrosion and from the deleterious effects caused by the solvents dissolving in the oil. The cooling should be regulated so that the water vapour is condensed quantitatively. If this cannot be achieved, then a calcium chloride, silica gel or sulphuric acid water-trap must also be inserted. During the evaporation, the boiling point of the solution to be evaporated rises as a consequence of the concentration; accordingly, it is necessary to increase the vacuum or, if doing so has no harmful effects, the temperature. If the solution undergoing evaporation also contains electrolytes, these will be concentrated together with the material of interest. Any damaging effects of the increased electrolyte concentration must be reckoned with. In the case of evaporation of volatile acids, there will be a change in the ionic equilibrium in the concentrated solution, and hence in the pH. These effects must be taken into consideration in planning the evaporation. It is advisable to remove the excess of electrolytes beforehand by dialysis or gel-filtration.

Drying by lyophilization

Lyophilization is used for removal of the water contents of various water-containing plant and animal tissues and organs and micro-organisms, in the same way as for the evaporation of solutions of thermally labile biochemical substances. It is the mildest drying procedure. Lyophilization of thermally labile solutions may be necessary at almost any stage in an analysis and particularly in preparative work.

Apart from fundamental physico-chemical and biochemical considerations, the lyophilization of substances possessing structure also demands much experience, and in practice the most suitable procedure must be worked out for each material. In this connection it is especially important that the first stage of the operation, the freezing of the material, should be carried out correctly. The freezing of an aqueous solution begins with the crystalline freezing-out of ice (this also holds for tissues). In the following stage, eutectic mixtures crystallize out. The length of this stage is governed mainly by capillary effects and surface forces. The intactness of the fine structure depends on the dimensions of the crystals formed. If the temperature falls slowly, large crystals are obtained. Rapid cooling results in a microcrystalline structure, which damages the structure less.

In addition to resulting in large ice crystals, which mechanically damage the fine structure, slow freezing has the disadvantage that the concentration of solutes in the liquid phase increases because of conversion of part of the water into ice. This solution will be hypertonic relative to the cells and will cause plasmolysis. There will also be a large depression of the freezing point, particularly if the tissue to be frozen contains Ca^{2+} and Cl^- ions in large amounts, since the eutectic temperature for $CaCl_2$ is very low. Generally, tissues may best be maintained in the native state if a temperature of -60 °C is attained as quickly as possible. In the case of protein-containing materials, it must also be considered that the intramolecularly-bound water cannot be frozen out, even at the lowest temperature. The retention of viability of micro-organisms during lyophilization can be enhanced by use of protecting solutions. These usually contain two components: a colloid (this may be 5–20% protein) and 5–10% sugar. The purpose of the sugar is to prevent too rapid and strong a dehydration. With materials sensitive to oxidation it is practical to add to the protecting solution a low concentration of ascorbic acid or other anti-oxidant. The use of protecting solutions is again largely empirical because it is impossible to find a universally applicable system.

Numerous factors must also be taken into account in the freezing of aqueous solutions. Experience shows that at temperatures between 0 and −40 °C the comparatively labile secondary, tertiary and quaternary structures of a protein, which determine the spatial conformation, do not undergo irreversible change. The stability is altered, however, by a high salt content and by a pH value different from the optimum. The presence of organic solvents such as acetone and alcohol in the solution of the protein to be lyophilized is not desirable, since it causes strong depression of the freezing point. Hence, the danger increases that the frozen solution will melt during lyophilization and the protein will coagulate. Solvents with low boiling-points have high partial vapour pressures; this impairs the vacuum, and thus the sublimation proceeds more slowly. Further, the solvent-diluted pump-oil must be exchanged at frequent intervals.

For the reasons already mentioned, it is preferable to apply lyophilization to salt-free solutions, or to select a salt concentration so low that the concentration attains the desired or permitted value only at the completion of the operation. Similarly, it is necessary to plan the pH change, particularly if the buffer mixture contains a volatile component.

The construction of the lyophilization apparatus must also be carefully considered.

If the pressure in the system is less than 1.3 Pa (0.01 mmHg), the mass Q of water evaporated and condensed (related to unit surface area and unit time) may be calculated from the following formula, based on the kinetic gas theory:

$$Q = \frac{4}{3}\left(\frac{2\pi M}{RT}\right)^{1/2} \frac{r^2}{L}(P_a - P_b)$$

where M is the molecular weight of water, R is the universal gas constant, T is the absolute temperature, r is the radius of the cross-section of the vacuum line, L is the path-length from the substance to the pump, P_a is the effective water vapour pressure above the ice, and P_b is the partial water vapour pressure at the pump.

A whole series of various types of lyophilization apparatus is manufactured by the instrument industry; if a pump of suitable performance (0.01 Pa vacuum or better) is available, laboratory lyophilization apparatus may also be constructed from glassware. Two-stage, gas-ballast valve, rotary oil-pumps are satisfactory for this purpose. A higher vacuum can be attained by the use of a diffusion-pump and an appropriate prevacuum-pump.

2.1.4 Preparation of protein solutions

Dissolution of tissue proteins

If the protein is to be obtained from animal or plant tissues, many factors must be taken into consideration, and it is necessary to reckon with the fact that the yield can scarcely be 100%. In this section we shall not touch on the problem of extraction of biologically active proteins; we shall attempt merely to describe those procedures by which protein solutions suitable for analytical purposes may be prepared without damage to the primary structure.

A decisive factor in the effectiveness of all protein-extraction procedures is attainment of the greatest possible degree of disintegration of the animal or plant tissue to be extracted. To break down the cell structure, apart from mechanical comminution, it is advantageous to employ disintegration by freezing or plasmolysis with hypertonic solutions, especially in the case of naturally occurring moisture-containing tissues.

The proteins display very varied solubilities. The scleroproteins do not dissolve in water, dilute salt solutions or other solvents in general. Lipoproteins can generally be dissolved out of the cell structure only after the extraction of the lipids. One group of plant proteins, the prolamines accumulating in the storage tissues of the endosperm of cereal seeds, dissolve only in 60–80% alcohol–water mixtures. The other proteins dissolve to various extents in water or in dilute aqueous salt solutions. The dissolution depends on the nature of the ions in the aqueous solution, their concentrations, the ionic strength, the pH value and the temperature.

At extreme temperature and pH the dissolution is accompanied by denaturing, and at the pH value corresponding to the isoelectric point the proteins are precipitated. Every procedure for obtaining proteins is at the same time a 'fractionation'. Individual solvents may dissolve the different proteins to different extents, as the degree of dissolution depends not only on the composition of the solvent, but also on the structure of the protein. In this respect, the solubility depends primarily on the ratio of the polar hydrophilic and the non-polar hydrophobic groups, and on their arrangement, and consequently on the resulting dipole moment. Depending on the nature of the ions in the aqueous solution, very different quantities and types of proteins may be dissolved out of a given material, as the solubilities of proteins follow the general regularity known from colloid chemistry. The dissolution effect generally increases in accordance with the sequence of the Hofmeister lyotropic series. The peptizing effects of neutral salt and buffer solutions may be increased, for example, by the addition of anionic detergents. The sulphite-reduction of the disulphide bridges causes folding of the peptide chain and so increases the dissolution. The solubility of a protein is similarly increased by complex formation with the Cu^{2+} ion at pH ≥ 8, but this is accompanied by a more profound structural transformation.

If a number of different solvents are used one after another to obtain proteins, then the resulting fraction distribution will depend on the sequence of solvents employed. For instance, if the proteins of cereal seeds are fractionated by the conventional Osborne [5] method, the sequence of solvents is: water, dilute salt solution, 70% alcohol, dilute alkaline solution. If the extraction is begun with a basic solution, this will dissolve out practically all the proteins.

In their procedure for the extraction of maize and other plant seeds, Mertz and Bressani [6] obtained good separation of protein fractions by dialysis after extraction with alkali and cupric sulphate. Kent-Jones and Amos [7] reported a simple procedure for dissolving albumin and globulin out of cereal seeds with a 5% potassium sulphate solution; methanolic alkaline extraction was then performed, and the glutelin was precipitated by adjustment of the pH to 6.4, the gliadin remaining in solution.

The extraction of albumin and globulin from animal tissues is very difficult, for the tissues always contain ions, and these may be extracted together with the albumins, globulins and various pseudoglobulins. The individual fractions may be separated by precipitation, dialysis, gel filtration and other techniques. Helander [8], for example, extracts the plasma protein from muscle with potassium phosphate buffer at pH 7.3.

Pressing-out of tissue proteins

The soluble protein contents of plant or animal tissues may also be obtained by pressing. With laboratory-scale hydraulic presses, a pressure of 300–400 kg/cm^2 may be achieved. The casing of the press is made of stainless steel, with run-off openings at the bottom. The sample thus obtained does not represent the total protein content of the material under examination, but only that part of the soluble protein that is not bound to structural elements and is obtainable by pressing at the given pressure.

2.1.5 Filtration of protein solutions, and their concentration by filtration

Filtration and filters

In laboratory practice the most frequent means of separating solid particles from a dispersion in a liquid phase is filtration on fibrous or porous layers.

Filtration can be done by gravitation, suction or pressure. Various types of apparatus are used: filter funnels, Buchner funnels or sintered-disc filters and pressure filters.

The most important characteristics of filters are the retentivity and permeability (which determine the fineness of the particles which the filter is able to retain or let through), and the filtration rate. The filtration rate is generally taken as the volume of distilled water passed per unit area of filter in unit time under the suction pressure to be used.

Of the fibrous filters, filter papers are used most often in general practice. In the protein-analysis laboratory their use is mainly restricted to qualitative operation. Filter papers are made of short-fibre cotton, consisting of α-cellulose of about 95% purity. For special purposes, filters are also made from glass, PVC, polyester or asbestos fibre. The filtration characteristics can be described only qualitatively.

In protein analysis cellulose filter papers may be used for collection of the coarse dispersed fractions of suspensions, and of proteins coagulated with precipitating agents. For the first purpose, qualitative papers of high permeability are employed. Hardened papers have the advantages that they are resistant to acids and bases up to a given concentration limit, even at 20–50 °C, and their surfaces are smooth, so precipitates can readily be washed off them.

Schleicher and Schüll produce a filter paper with low nitrogen content (N 2095), especially for protein analysis purposes. This is mainly used when, to avoid losses, the filter paper is mineralized together with a precipitate of low nitrogen content in a Kjeldahl nitrogen determination.

The advantage of filters prepared from glass and plastic fibres is that they are able to resist corrosive chemicals at even higher temperatures. Although these properties are rarely utilized in the protein-analysis laboratory, the filters have the added advantage that they do not swell or adsorb in aqueous medium. Glass-fibre paper may be used for prefiltration before use of ultrafilters. Very coarse dispersions are filtered on cheese-cloth or glass-fibre cloth in preparative procedures. Special impregnated filter papers are also utilized in certain cases. The impregnating agent may be an adsorbent, such as active carbon or kieselguhr. If these are used, it is also necessary to reckon with losses arising as a consequence of undesired adsorption. Active-carbon papers are suitable for the removal of dyes. Filter paper impregnated with kieselguhr, with a silica content of about 20%, is able to retain very finely dispersed particles of a semicolloidal nature. Filter papers treated with silicone may be employed to separate aqueous and lipophilic phases.

Porous filters are made of glass or porcelain. For downward suction filtration they are prepared in the form of filter plates, crucibles or funnels, and for upward filtration are in the form of immersed plate filters, filter sticks and filter candles. Buchner and Hartley funnels are used with filter papers; asbestos can be used in a Gooch crucible. Otherwise sintered-discs are used, made of glass or porcelain. Their porosities are well defined, and are denoted by the numbers of the filters (Table 2.5).

Table 2.5
Pore sizes of sintered-glass filters
(sintered-porcelain filters are similar)

Number	Jena filters μm	Pyrex filters μm
G-0	150–200	–
G-1	90–150	100–120
G-2	40–90	40–50
G-3	15–40	20–30
G-4	9–15	5–10
G-5	1–1.7	2

Sintered-glass filters are made thicker or thinner, according to the diameter of the filter plate, so that they can withstand a pressure of 1 atm; filter plates of larger diameter are made slightly convex to increase the bearing strength. Filters 0, 1 and 2 may also be used for gravitational filtration of liquids of low viscosity. With the finer-pored filters, suction is applied with a water-pump or rotary vacuum-pump. A suction pressure of 2700–8000 Pa (20 60 mmHg) is generally sufficient for filtration at a satisfactory rate. The porosity of the filter must be chosen so that the diameter of the dispersed particles is somewhat larger than the maximum pore diameter of the filter, and in this way clogging of the filter plate can be avoided. It must be noted that the nominal pore size and flow rate constantly vary in the course of filtration, because the precipitate held on the filter diminishes the flow rate and makes the porosity finer. The G-5 filters are so fine that they may be used as filters for bacteria. This decelerating effect of the precipitate settled out on the filter may be decreased if auxiliary filtration materials are added to the suspension, though with protein solutions, adsorption must be reckoned with. Kieselguhr is the most satisfactory additive for biochemical work, e.g. 'Hyflo Super Cell', which has a particle size of 5–25 μm. The use of cellulose is not advisable in protein chemistry. Glass filters may be sterilized, but they should not be cleaned with chromic-sulphuric acid, because chromium compounds may be adsorbed on the glass and disturb the biochemical analysis. Perchloric-sulphuric acid is a very good cleaning mixture, however. They may be dried or sterilized by slow heating up to 120–160 °C, but a higher temperature or a sudden change in temperature may give rise to stresses in the filter, which lead to its cracking.

Ultrafiltration

Ultrafilters are based on permeable membranes, which separate dispersed ionic and molecular particles and the dispersing medium (water) from colloidal particles, and thus simultaneously concentrate the macromolecular solution. The wide choice and well-defined porosities of the industrially-produced membrane filters even allow a certain fractionation of the macromolecules according to molecular weight. However, the shapes of the molecules must also be taken into consideration: fibrillar macromolecules pass through an ultrafilter of given porosity in a different manner from globular molecules with the same molecular weight.

Depending on their structure, the filter devices may be either open systems suitable for continuous operation, or closed systems with intermittent operation. The through-flow of the liquid is assisted by suction or pressure. The small-scale laboratory ultrafilters are set up as filter funnels operated by vacuum suction. The intermittent ultrafilters, operating under pressure, are closed vessels in which the liquid to be filtered is placed and forced through under a pressure of 9.81×10^4–98.1×10^4 Pa (1–10 atm) applied from a cylinder of compressed nitrogen. Very small membrane filters operating under mechanical pressure are also made in the form of an injection syringe. Continuous pressure filters, on the pattern of industrial filter presses, are also made on the laboratory scale. They have the advantage that the liquid to be filtered flows parallel to the filter surface and in this way the material sedimenting out does not form a layer obstructing filtration. The solution to be filtered may be transferred to smaller filters from a reservoir by means of nitrogen pressure. At lower and higher pressures it is customary to use a peristaltic-tube pump and a displacement pump, respectively.

The deposition of the coarse dispersed particles on the filter surface may be prevented by using fine glass-fibre prefilters.

Centrifugal ultrafiltration. In a special case of ultrafiltration under pressure, the ions and low molecular-weight substances, together with the dispersing solvent, water, are forced through the filter not by gas or mechanical pressure, but by an increased gravitational force. The method is therefore suitable for the separation of protein solutions from the ionic and molecular-weight dispersed particles, with simultaneous concentration. The 'Centriflo' ultrafilter cones produced by Amicon N. V., The Hague, can be fitted into a special centrifuge tube. Swinging-head centrifuges are best used. With the aid of CF-50 ultrafilter cones at about 1000 *g*, 5 ml of 0.01% albumin solution can be concentrated to a volume of 1 ml. The CF-50 ultrafilter retains substances with a molecular weight of 5×10^4, but has no significant retentivity for molecular weights below 2×10^4.

In current laboratory operation, artificial membranes with well-defined permeabilities are used exclusively. The previously employed membranes of animal origin have lost their importance in laboratory practice. The gelatin filter, which is readily prepared in the gel state but easily damaged, is rarely used in the laboratory. The semipermeable filters employed are solid-structure membranes that may be stored in the dry state; they are produced industrially from various materials, with pore sizes corresponding to the particular purpose. However, since they can be prepared simply in the laboratory, collodion membranes are still used for dialysis.

The membranes may be made of cellulose nitrate, cellulose acetate, regenerated cellulose, pol(vinyl chloride) and polyamide (nylon). For purposes of dialysis they are prepared in the form of tubes and sheaths, and in the form of sheets for ultrafilters and electrodialysers.

Numerous descriptions are to be found in the literature with regard to the preparation of the collodion (dinitrocellulose) membranes employed in the laboratory [9]. A semipermeable membrane may be prepared on some support. Filter paper is immersed in water, laid on a glass plate, and coated with a 4% collodion solution (in alcohol–ether, 1 : 7). The excess of solution is drained off, and after the collodion membrane has hardened (5–10 min) the filter is coated with a further layer. When this layer too has hardened, the filter paper is again immersed in water to dissolve the alcohol out of the layer. After about 30 min the filter is ready for use. In a similar manner a semipermeable membrane can also be layered onto an extractable sheath. A membrane can also be prepared without a supporting layer: a collodion solution (200 ml of 6% collodion solution diluted with a mixture of 200 ml of ether and 500 ml of alcohol) is poured onto a horizontal sheet of glass; after gelation has taken place the glass plate is immersed in water, and after a time the layer may be peeled off. A readily-separable collodion sheet filter is also obtained if collodion is poured onto the surface of mercury. The commercially-available membranes have the advantage that their pore sizes are well defined. The permeabilities of collodion membranes prepared in the laboratory, however, depend on the conditions of gelation (solvent ratio, temperature, etc.). Filters prepared from old collodion solutions have much larger pore sizes than those made from fresh ones. A collodion solution in glacial acetic acid may be aged by heat-treatment at 90–98 °C for 150 min, and thus filters with wider pores may be obtained.

Determination of the pore size is discussed in handbooks on practical colloid chemistry. It can be done by filtration of dye sols of known dispersities, or by measurement of the minimum pressure necessary to press air bubbles through.

The most important characteristic of commercially-available membrane filters is the pore diameter, which varies between 10 and 0.05 μm, depending on the material used and the manufacturing technology. Since the basis for separating the dispersed particles is the sieve effect, i.e. the difference between the dimensions of the dispersed phase and the pores, particles larger than the pores are generally deposited on the surface of the membrane, without any adsorption. The efficiency of filtration is independent of the operating pressure, the stepwise gradient and the electrical potential. These factors affect only the rate of dialysis or ultrafiltration. Electron microscopy shows that the pores occupy about 80% of the volume of a filter, their number being around $10^8/cm^2$.

Filters with the following pore sizes can be prepared from the materials listed:

cellulose nitrate	8–0.01 μm
cellulose acetate	0.5–0.2 μm
regenerated cellulose	0.6–0.2 μm
PVC	*ca.* 0.3 μm
polyamide (nylon)	*ca.* 0.8 μm

Table 2.6

Comparison of various ultrafilter membranes

Pore diameter, nm	Molecular weight excluded	Sartorius	Amicon	Millipore	Kalle
20–10	1×10^5	SM 11533	XM-100		
15–10	5×10^4	SM 12133	XM-50	PSDM	
5	3×10^4	SM 11539	PM-30	PSED	Dialysing tube
5	1×10^4	SM 12136	PM-10		
	1×10^4		UM-10		
	1×10^3		UM-2	PSAC	
	5×10^2		UM-05		

Cellulose-based filters can be sterilized at 130–180 °C. PVC and polyamide cannot endure a higher temperature, but they are highly resistant to corrosive compounds. The thickness of membrane filters lies in the range 50–500 μm. The rate of through-flow of the pure solvent depends on the thickness. The through-flow rate is generally referred to that of water, which at a pressure difference of 93.3 kPa (700 mmHg) is 1 ml · min^{-1} · cm^{-2} at 25 °C.

Depending on the thickness, the through-flow rates vary between 100 and 0.002 ml · min^{-1} · cm^{-2}. Table 2.6 lists the pore sizes and corresponding separating capacities of some of the membrane filters used in the protein-analysis laboratory.

Filtration of low molecular-weight substances by dialysis and electrodialysis

Dialysis. The essence of simple dialysis is the application of diffusion combined with the sieve effect for the separation of macromolecules from ions and compounds with low molecular weights. In its simplest form it requires only the dialysis tube. The liquid to be dialysed is poured into a membrane sheath (a dialysis tube), and this is immersed in pure water. The diffusion of the ions and small molecules into the water begins as a consequence of the concentration gradient. The rate of dialysis depends on the ratio of the volume of the solution to be dialysed and the surface area of the dialysis membrane, on the through-flow rate permitted by the dialysis membrane, on the magnitude of the concentration gradient between the internal and external solutions, and on the temperature. In accordance with Fick's diffusion law, the rate of dialysis is given by the following equation:

$$\frac{dS}{dt} = KA\frac{(C_k - C_b)}{dx}$$

where S is the amount of substance diffusing in time t through a membrane of surface area A, thickness d and diffusion constant K; C_k and C_b are the concentrations of the diffusing substance on the two sides of the membrane.

For a membrane of given quality, the rate of dialysis can be enhanced most simply by increasing the gradient of the external and internal concentrations. At equilibrium the dialysis will come to a stop. The gradient may be maintained by continuous exchange of the external solution, and by mixing, which prevents increase of the concentration at the interface. If ions are being removed, the gradient can also be maintained by adding ion-exchange resin to the external solution, or by circulating the solution through a mixed-bed ion-exchange resin column. The advantage of this is that the substance removed can be recovered if required. If it is desired to concentrate the non-ionic, low molecular-weight constituent, the Hahn apparatus is suitable [10]. The dialysis sheath of this is placed in a closed system fitted with a vacuum-distillation condenser. The condensing vapour flows onto the dialysis sheath and is led by an overflow into the distillation flask, where the low molecular-weight component is concentrated. For maintenance of the concentration gradient, an economic method is recommended by Hospelhorn [11], who winds a thin cord spirally on the dialysis sheath suspended in a cylindrical vessel; the pure dialysing liquid flows throughout on this at a rate of 2 ml/min.

Dialysis may also be used instead of ultrafiltration to concentrate protein solutions by removal of the ions, low molecular-weight compounds and the dispersing medium (water) through the pores of the dialysis sheath under vacuum or pressure. Hence, the high molecular-weight solution is concentrated. A good description of the practical details is given by von Hofsten and Falkbring [12].

Electrodialysis. The final stage of dialysis, when the electrolyte concentration is very low, is extremely long drawn-out. Electrodialysis is mainly used to shorten this stage. A large proportion of the electrolytes is removed by common dialysis, and this is followed by electrodialysis. The efficiency of electrodialysis is increased by the simultaneous application of ionophoresis. The simplest type of electrodialyser is that with three cells. In this apparatus the solution to be dialysed is held between two membrane filters, in contact with the anode and cathode compartments; the wash-liquid, generally tap-water, flows into these two compartments. The rate of dialysis can be regulated by adjusting the flow of wash-water, and also the potential. Depending on the electrolyte concentration, the potential must be regulated so that a power of *ca.* 1–2 J is available per cm^2 of the dialysis membrane. The detrimental effect of increased temperature may be diminished by increasing the flow of the wash-liquid, to act as a coolant. Electrodialysis may be disturbed by a charge on the membrane. Synthetic membranes in general have negative electric charge; they are therefore easily permeable to cations, but hinder the through-flow of anions as a result of the like charge. This effect can be eliminated by the use of a positively-charged membrane on the anode side, and a negatively-charged membrane on the cathode side. Breakdown of the ionic equilibrium as a result of the diffusion of H^+ and OH^- ions at different rates may give rise to a shift in the pH value in the middle (dialysis) cell, and consequently the protein solutions may undergo irreversible damage. By appropriate selection of the membranes, it may be arranged that the anions and cations depart in equivalent amounts. The colloidal particles accumulate and are possibly adsorbed on the surfaces of the polarized dialysis membranes, through which they are not capable of permeating. It is advisable to inhibit this process by stirring the solution to be dialysed, or by reversing the polarity of the supply potential for a short period from time to time, in this way changing the direction of dialysis. In the case of platinum or graphite electrodes this can be done without any problem, but a nickel electrode may only be connected as the cathode. The progress of the dialysis can be checked by conductivity measurement. It is also expedient to check the pH of the solution to be dialysed, with a glass-electrode pH-meter. Cells are often connected together in series to increase the dialysis surface area, and to improve the ratio of the surface area to volume; in these, washing and dialysis cells are connected alternately, separated by dialysis membranes. The wash-liquid flows from one washing cell to another in a labyrinth system. The dialysis cells are similarly interconnected.

A combination of electrodialysis and ion-exchange may be employed for the removal of salts from protein hydrolysates before paper or thin-layer chromatography [13]. The anode and cathode cells of the three-cell electrodialyser are separated by anion-exchange and cation-exchange membranes, respectively, from the hydrolysate to be made salt-free. The ion-exchange membranes have extremely narrow pores, which practically let through only water. The ions diffusing towards the electrodes are exchanged on the ion-exchange membranes. The anode compartment is filled with 0.2 *M* sodium hydroxide, and the cathode compartment with 0.2 *M* sulphuric acid. The cation-exchange membrane lets through only the cations, with the exchange of H^+ ions, while on the anion-exchange membrane OH^- ions are exchanged for the anions of the solution to be made salt-free. In the low-volume middle cell, the solution from which the salt is to be removed is kept in motion by the passage of bubbles through it. In an apparatus with favourable geometry, the ion-exchange proceeds very rapidly at a potential of 6 V, as is readily seen from the fall in the current passing through the cell, as a consequence of the increase in resistance. The salts can be removed from the protein hydrolysate in 6–10 min. A certain degree of loss of basic amino-acids may occur; this may be avoided if the starting pH of the solution to be made salt-free is adjusted to 8.0 with 0.05 *M* potassium carbonate.

2.1.6 Centrifugation and centrifuges

One of the general laboratory methods used for separation of the dispersed particles of emulsions and suspensions is centrifugation. The development of preparative and analytical methods in protein chemistry is related in many respects to the development of laboratory centrifuges.

The essence of centrifugation is the faster sedimentation of the dispersed particles in a gravitational field that is increased as a consequence of rotary motion.

The Stokes sedimentation law applies to centrifugation:

$$V = \frac{2}{g}\,\frac{\rho^2(d-d_f)-g}{e}$$

where V is the rate of sedimentation, ρ is the density of the spherical particles, d is the density of the dispersed particles, d_f is the density of the dispersing medium, g is the gravitational acceleration, and e is the internal frictional coefficient of the medium.

A force $Mr\omega^2$ rad/s^2 (radial acceleration) acts on every particle of the dispersed system rotated in the centrifuge, where M is the mass of the particle (g), r is the rotation radius (cm), and ω is the angular velocity (rad/s).

In practice the relative centrifugal force is used, or, in another notation, the g value of the centrifugation. This may be calculated simply from the formula:

$$g = \frac{f^2 r}{89\,500}$$

where f is the rotation rate of the centrifuge in rpm, and r is the rotation radius measured in cm from the axis.

The time necessary for the sedimentation of some fraction with a definite degree of dispersity can be calculated approximately from the formula:

$$t = \frac{D-L}{D+L}\cdot\frac{N}{a^2(d-d_f)f^2}$$

where D is the length of the rotation radius from the axis to the meniscus of the suspension, N is the viscosity of the dispersing medium in poise ($g \cdot cm^{-1} \cdot s^{-1}$), d is the density of the material of the dispersed particles, d_f is the density of the dispersing medium, a is the average diameter of the dispersed particles, and f is the rotation rate of the centrifuge in rpm.

In preparative and analytical ultracentrifugation, attention must be paid to the following factors determining the sedimentation time.

Non-spherical particles sediment more slowly than spherical particles with the same mass and density, for which the Stokes formula holds.

The rate of sedimentation depends on the temperature. If the rotation rate and g value are increased, the sedimentation rate is decreased by the heat produced by friction. Accordingly, the cooling of the centrifuge must be adjusted to counterbalance the heating.

In description of laboratory methods, the conditions of centrifugation may be characterized unambiguously by giving the centrifugation time, the g value and the temperature.

For comparison of results obtained under different experimental conditions, the sedimentation rate of a definite fraction is reduced to a value of $g=1$ and a temperature of 20 °C. This value is known as the sedimentation constant; it is denoted by the symbol S_{20}, and has units of cm/s.

Centrifuges may be classified in terms of rotation rate and type of rotor. If the rotation rate is used as the criterion, we may distinguish:

low rotation-rate centrifuges:	0–5000 rpm
high rotation-rate centrifuges:	0–20000 rpm
ultracentrifuges:	0–60000 rpm

In terms of rotor construction, the classification is: angular centrifuges, swinging-head centrifuges, through-flow separators.

The rotor of an angular centrifuge is a truncated conical body made of a high-strength light metal or plastic; in this holes are drilled radially and symmetrically, at an angle of 15–60° to the rotation axis, for insertion of the centrifuge tubes. The g value varies along the length of the centrifuge tube, depending on the distance from the rotation axis. The maximum path-length for the dispersed substance to be sedimented is the horizontal diameter of the centrifuge tube, which depends on the angle of the drilling.

In swinging-head centrifuges, the centrifuge tubes are inserted into containers free to move in the forked supports from which they are suspended and which are arranged radially symmetrically with respect to the axle. As a consequence of the centrifugal force, the containers are horizontal during operation. The g value varies along the tube, depending on the distance from the rotation axis; it is customary to quote the value at the bottom of the centrifuge tube. The maximum sedimentation distance of the dispersed particles is the height of the suspension in the centrifuge tube.

The rotor of a through-flow centrifuge is a tube or drum, and the suspension is led into it from below through a hole in the axle. The separated fractions are led off at the upper rim of the cylinder, through ports situated at various distances from the rotation axis. The efficiency of sedimentation depends on the inflow rate of the suspension as well as on the technical parameters of the centrifuge.

Besides these preparative-type rotors, analytical type rotors are produced for ultracentrifuges; these permit continuous monitoring of the sedimentation by refractometric or interferometric means, on the basis of the principle elaborated by Svedberg and Pedersen [14] and subsequently further developed by numerous instrument companies. In recent years the instrument companies have constructed a very varied series of rotors for purposes of various biochemical preparative techniques, and particularly for the isolation of cell fractions. These permit centrifugation under conditions where the dispersing medium, e.g. a sugar solution, forms a density gradient in the drum rotor (preparative zonal centrifugation [15]) or in a centrifuge tube specially constructed for this purpose (Strohmaier fractionation tube [16]).

The technique of centrifugation may be combined with other separation procedures, e.g. filtration. For this purpose dismountable centrifuge tubes fitted with a special filter insert are produced.

Normal centrifuge tubes are made of glass or plastic (polyethylene, polypropylene or possibly cellulose acetate). It is not recommended that glass centrifuge tubes should be used at a rotation rate higher than 3000–4000 rpm.

Centrifuges with relatively low rotation rates are driven by a universal electric motor, with a variable resistance in series to ensure continuously variable rate regulation.

In centrifuges with high rotation rates, the effect of heat is diminished by using an oil-pump built into the centrifuge to keep the working area of the rotor under vacuum (the working area is enclosed by a strong protective sheath); constant low temperature is ensured by a cooling spiral surrounding the working area.

2.2 GENERAL CHARACTERIZATION OF MATERIALS FOR EXAMINATION

From the classification of the substances examined in protein analysis (Table 1.1) it may be seen that their detailed investigation differs to various extents, depending in part on their origin and in part on their use. However, because of their use, the complete analysis of certain protein-containing materials is pointless, e.g. in the routine quality-control of food and feed proteins it is quite sufficient to measure those properties which show the nutritional value, knowledge of which promotes their optimum utilization. These properties are the moisture, ash and protein contents. The methods for determination of the moisture and ash do not display great variety or involve any difficulty. The protein-determination methods are much more varied. For ease of survey, the methods and principles used are compiled in Table 2.7.

Lewis [17] has described a method for characterization of the variation in the sensitivity of some commonly used protein assay procedures.

Table 2.7
Survey of the more important protein-determination methods

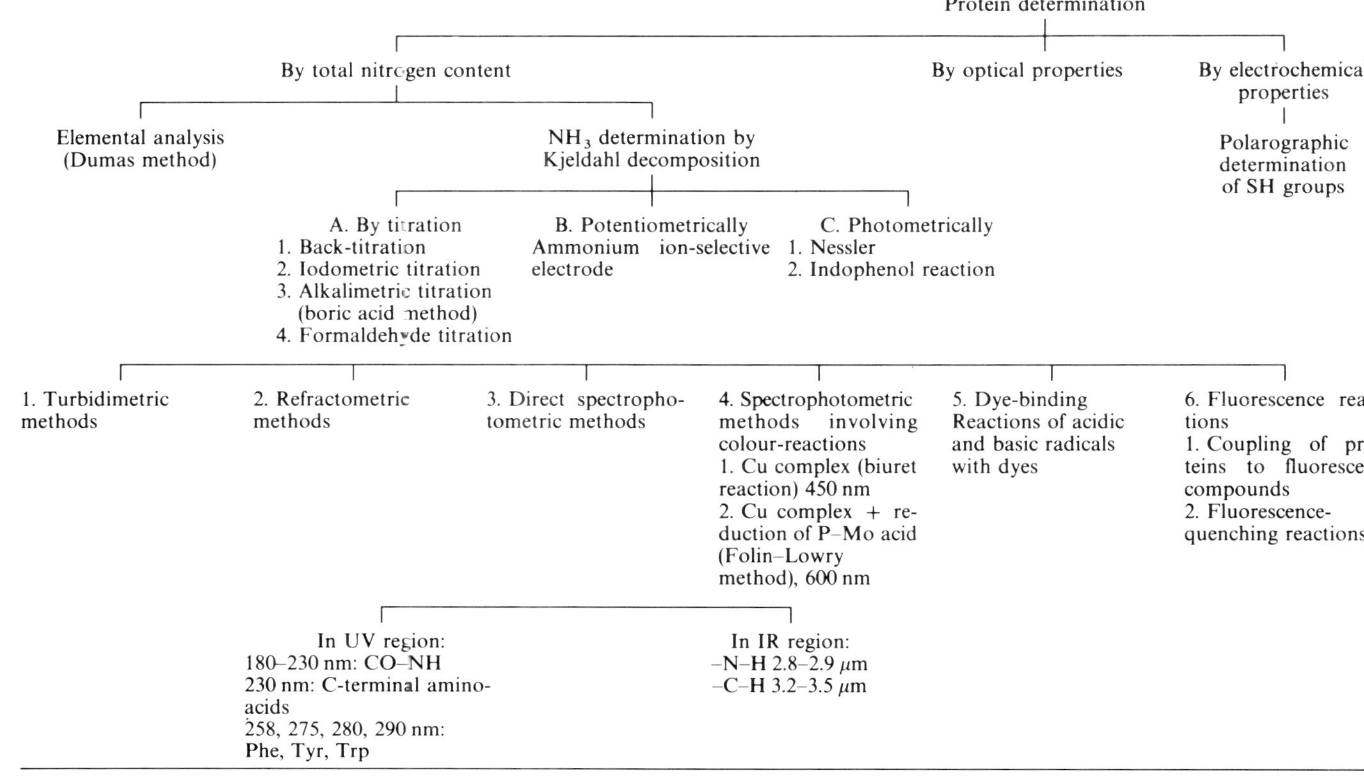

2.2.1 Dry matter

The analytical results are generally calculated on the dry weight. The basis of comparison of agricultural and industrial raw materials is not always the absolute dry weight, but that measured in accordance with conventions and standards; it is usually the moisture in the substance in equilibrium with the partial water vapour pressure of the atmosphere after a protracted period at room temperature in air with an average relative humidity of around 60%. For commercial goods, it is customary to use as the reference basis that dry matter or moisture content with which the goods can be stored without deterioration. In the majority of cases this value agrees with the 'air-dried' weight discussed above.

In naturally-occurring substances the moisture may be present as free water or intramolecular water. The most frequent method of determining the dry matter is drying at constant temperature. If the loss of water at constant temperature is examined as a function of time, a characteristic drying curve is obtained, in which there are three stages. In the first stage the water content decreases quickly as a result of the high evaporation rate from the surface and outer layers. In the second stage the rate of loss of water is determined by the increasing diffusional resistance of the drying substance. In the third stage the water-binding capacity of the third material is important, and the rate of loss decreases greatly. When we speak of drying at constant temperature to constant weight we are concerned with the attainment of this third stage. It must be stressed that, in terms of laboratory operations, determination of the dry matter by drying must be distinguished from drying of the substance. The drying employed for the determination, e.g. in the case of protein-containing materials, may lead to irreversible denaturing. In contrast, the drying of a substance in preparation for other examinations must be done in such a way that denaturing does not occur.

The main laboratory methods used for determination of the dry matter are classified into the following types:

(*a*) drying at atmospheric pressure and 100–105 °C or higher;

(*b*) drying in vacuum at below 100 °C;

(*c*) drying by distillation of the water;

(*d*) determination by chemical means (e.g. Karl Fischer titration).

A 2–5 g sample of air-dried finely-ground material may be dried practically to constant weight by drying for 3 h at 105 °C. With drying at 130 °C, weight constancy may be achieved within 1 h. A sample of a few g may be dried to weight constancy within 8–10 h in a vacuum drying oven at 70 °C and 6.67 kPa (50 mmHg). The moisture may be determined by distillation with toluene or xylene in a Dean and Stark apparatus. The sample taken should yield 2–5 ml of water. The solvent is added, and distillation conducted so that 2–4 drops condense per second. The distillation is continued for about 1 h. The volume of water collected below the solvent layer in the calibrated receiver tube is measured. The same method is used for determining water in oil.

The most popular chemical determination is the Karl Fischer titration [17] in one variant or another. The essence of the method is the reaction:

$$I_2 + SO_2 + 2\,H_2O = 2\,HI + H_2SO_4$$

When the concentration of sulphuric acid formed exceeds 0.05%, the reverse reaction, the oxidation of hydrogen iodide, begins:

$$H_2SO_4 + 2\,HI = I_2 + H_2SO_3 + H_2O$$

Naturally, the method works only in non-aqueous medium. The essence of the technique developed by Fischer is that the stabilities of the reagents are enhanced with the aid of methanol and pyridine, which form addition compounds with iodine and sulphur dioxide, and the reaction is driven to completion. The reaction is assumed to take place as indicated by the equation:

$$I_2 + 2\,H_2O + py_2.SO_2 + 2\,py \rightarrow py_2.H_2SO_4 + 2\,py.HI$$

The visual end-point is unsatisfactory and in practice electrometric methods are used, with potentiometric or dead-stop end-point detection.

2.2.2 Ash content

In the case of plant materials, a 2–3 g air-dried, finely-powdered sample is taken in a platinum, porcelain or quartz crucible and ashed initially over a low flame; as soon as the burning is over, ignition is continued over a strong flame until the ash is light in colour. In an ashing-oven, the maximum temperature used is 750 °C to avoid loss of alkali metals. Sometimes a temperature of 525–550 °C is recommended, to prevent loss of phosphorus. In plant raw materials the sand content is determined separately and the ash value corrected for any sand contaminating the outside of the plant material. In the case of foodstuffs, conserves, and especially meat products, the ash value must be corrected by subtraction of the sodium chloride content.

2.2.3 Fat content

The fraction extractable from materials of plant and animal origin with fat-dissolving agents is known as crude fat. Besides the triglycerides, this crude fat contains very many other substances: fatty acids, phosphatides, dyes, alkaloids, etc. The methods of determining these do not belong in the present topic.

The determination of the crude fat and the removal of fat from the sample are essentially the same process. In work on an analytical scale or for small-scale preparative purposes, it is done in a Soxhlet extractor. Peroxide-free diethyl ether or petroleum ether (40–60 °C) is used as the solvent. An air-dried, finely-ground sample is extracted for 5–6 hr at the rate of 8–10 syphon cycles per hour. An efficient condenser is needed, to give perfect reflux. The fat-containing solvent must be distilled off carefully, and the crude fat dried at 95 ±2 °C.

If the aim is merely the removal of fat from the sample, and no effort need be made to recover quantitatively either the fat-free sample or the crude fat, then the fat may be removed by homogenizing the air-dried material with a non-inflammable solvent (carbon tetrachloride, chloroform) in a blender, filtering off on a glass filter, and washing the residue several times with pure solvent.

2.2.4 Determination of nitrogen by the Kjeldahl method

Most indirect protein-determination procedures are based on determination of the nitrogen content, because the elemental compositions of the different proteins are very similar, independent of the nature and origin of the protein.

Naturally, the nitrogen content may be used only to determine pure proteins or the protein content of a material that includes no other nitrogen-containing matter than protein. In agricultural and food analysis, the total protein content of the sample is determined as 'crude protein', by multiplying the nitrogen content by 6.25, a factor corresponding to an average protein nitrogen content of 16%.

In practical analysis this yields accurate enough results, but if an exact result is required, the factor employed must correspond to the nitrogen content of the individual protein. The nitrogen contents and protein conversion factors of some proteins and protein-containing substances are given in Table 2.8.

The Dumas method for nitrogen determination was published in 1831 [18] and was the only reliable method until the Kjeldahl method was described in 1883 [19]. The latter, with its numerous modifications, has

Table 2.8
Nitrogen contents of proteins, and factors for calculation of the protein from the nitrogen content

Protein	Origin	Protein nitrogen content, %	Factor
	Proteins		
Plasma albumin	human	15.95	6.27
Plasma albumin	bovine	16.07	6.22
Ovalbumin	hen-egg	15.76	6.34
Globulin	human plasma	16.03	6.24
Lactoglobulin	cow's milk	15.60	6.41
Casein	cow's milk	15.63	6.39
Haemoglobin	equine blood	16.80	5.95
Fibrin	human plasma	16.90	5.91
Fibroin	pure silk	18.70	5.35
Keratin	feathers	15.00	6.66
Keratin	pig-bristle	16.60	6.02
Keratin	wool	16.30	6.13
Collagen	bovine skin	18.60	5.37
Gelatin	collagen	18.00	5.55
Histone	calf-liver	18.00	5.55
Histone	calf-thymus	18.20	5.49
Protamine	bird sperm	24.40	4.10
Protamine	salmon roe	30.70	3.25
Gliadin	prolamine from wheat	17.66	5.66
Hordein	prolamine from barley	17.21	5.81
Zein	prolamine from maize	16.20	6.17
Legumin	pea	16.04	5.54
Glycinin	soya-bean	17.30	5.78
Edestin	hemp-seed	18.55	5.39
	Naturally-occurring materials containing protein		
Cereal seeds			
barley		17.15	5.83
wheat		17.15	5.83
wheat-endosperm		17.54	5.70
wheat bran		15.84	6.31
wheat germ		17.23	5.80
maize		16.00	6.25
rye		17.15	5.83
oat		17.15	5.83
rice		16.80	5.95
Leguminous plant seeds			
bean, pea, lentil		16.00	6.25

Table 2.8 (cont'd)

Protein	Origin	Protein nitrogen content,%	Factor
Oleaginous plant seeds			
sunflower		18.86	5.30
soya-bean		17.51	5.71
coconut		18.86	5.30
cotton-seed		18.86	5.30
Materials of animal origin			
egg		16.00	6.25
meat		16.00	6.25
milk		15.67	6.38

remained the basis of biochemical nitrogen determination ever since; this has had a tremendous effect on the development of biochemical and applied biochemical analysis.

Kjeldahl decomposition

Much of the research on the Kjeldahl method has been devoted to perfection of the decomposition step. The chemistry of the decomposition with sulphuric acid is relatively simple. Concentrated sulphuric acid is an extremely effective dehydrating agent: at a given critical temperature it extracts the elements of water from oxygen-containing organic matter, and liberates carbon. At the decomposition temperature sulphuric acid also acts as an oxidant and oxidizes the liberated carbon to carbon dioxide.

In the case of saturated unbranched-chain primary amines, the reaction can be generalized as follows:

$$C_nH_{2n+1}NH_2 + mH_2SO_4 \rightarrow nCO_2 + (m-1)SO_2 + 4n\,H_2O + NH_4HSO_4$$

where n is the number of carbon atoms, and m is the sum of the numbers of carbon and hydrogen atoms in the alkyl radical.

For amides:

$$C_nH_{2n-1}CONH_2 + (m+1)\,H_2SO_4 \rightarrow (n+1)\,CO_2 + (m-1)\,SO_2 + (4n+1)\,H_2O + NH_4HSO_4$$

For amino-acids:

$$NH_2(CH_2)_nCOOH + (m+1)\,H_2SO_4 \rightarrow (n+1)\,CO_2 + m\,SO_2 + 4n\,H_2O + NH_4HSO_4$$

The work by Carpiaux [20] indicates that 1 g of protein consumes *ca.* 9.0 g of sulphuric acid.

The Kjeldahl decomposition is generally carried out in long-necked, pear-shaped Kjeldahl flasks. The long neck of the flask serves to condense the water vapour and to prevent too rapid evaporation of the decomposed liquid.

Several possibilities are available for acceleration and completion of the decomposition, such as raising the temperature, adding oxidants, adding reductants, using catalysts.

Table 2.9
Effect of K_2SO_4 in elevating the boiling point of sulphuric acid

K_2SO_4/H_2SO_4, g/ml	Boiling-point, °C
	0338
0.33	349
0.625	363
1.0	383
1.3	397
2.0	450

Elevation of the decomposition temperature. On heating, pure sulphuric acid boils and decomposes at 338 °C. By the addition of salts the boiling-point can be increased and the decomposition time for organic compounds shortened, and in this way the nitrogen content of certain substances that are otherwise difficult to break down can be converted quantitatively into ammonia. The salt generally used is ammonium sulphate. McKenzie and Wallace [21] have shown that the boiling point of concentrated sulphuric acid varies with the potassium sulphate concentration as illustrated in Table 2.9.

However, the addition of potassium sulphate not only raises the boiling point, but also consumes sulphuric acid by formation of potassium hydrogen sulphate:

$$K_2SO_4 + H_2SO_4 \rightarrow 2\,KHSO_4$$

On the basis of this equation, 1 g of potassium sulphate reacts with 0.56 g of sulphuric acid and this must be allowed for in the volume of sulphuric acid added.

Numerous other salts have been tried, including sodium sulphate, lithium sulphate and borax, but these do not display any particular advantage. The same may be said for phosphates, which consume even more sulphuric acid than potassium sulphate does, with the formation of acid salts.

The addition of salts is limited by their solubility in sulphuric acid. The decomposition temperature can also be raised by addition of a non-volatile acid, such as orthophosphoric acid, but there is often a decrease in oxidizing power and hence in the rate of oxidation. Experience has shown that it is not expedient for the proportion of phosphoric acid in the decomposition mixture to exceed 40%, and in terms of temperature this mixture has no advantages over a sulphuric acid–potassium sulphate solution. It is much better to add potassium sulphate to a sulphuric acid–phosphoric acid mixture, the best composition being 0.45 g of the salt per ml of 80% sulphuric acid–20% phosphoric acid mixture. The initial boiling point is 354 °C, and progressively rises during the decomposition. At the boiling point the rate of decomposition of the sulphuric acid is 0.081 ml/min, and during a 90 min decomposition the acid ratio changes from 80:20 to 76:24 and the boiling point to 364 °C.

Another possibility of raising the temperature is decomposition in a sealed glass tube [22], because of the increase in pressure. This technique is difficult and dangerous, so is only worthwhile if accompanied by a substantial advantage. It does have the advantage of needing no catalyst. Even a very small amount of ammonia can be determined by alkaline distillation from the decomposition solution and colorimetric determination, e.g. by the Nessler method, but catalysts may interfere in the Nessler reaction, which is why decomposition in a sealed tube without a catalyst is advantageous. With this technique, the decomposition temperature may reach 490 °C, even when potassium sulphate is not added, but care must be taken that it does exceed 500 °C, for in that case nitrogen might be formed, which would be lost when the tube was opened. Only dry samples may be decomposed in a sealed tube; if the material is in the dissolved state, it must be dried by vacuum-evaporation or lyophilization, most simply in the decomposition tube itself.

Addition of oxidants to accelerate the decomposition. In his original method, Kjeldahl used powdered potassium permanganate in the final stage of the decomposition to improve the oxidation of the organic matter. For a long time the necessity of adding permanganate was debated, on the basis of contradictory analytical findings, but finally it was considered to be unnecessary and was discarded. Numerous publications have appeared with regard to the application of other oxidants, mainly hydrogen peroxide, perchloric acid, persulphates and chromic acid. The question arises as to the value of the presence of an oxidant in the various stages of decomposition. It may be assumed that at the stage of dehydration and carbonization the nitrogen is still not completely reduced, and thus an oxidant may inhibit the reduction. Nevertheless, efforts to shorten the decomposition time make the use of oxidants necessary. With naturally-occurring materials, which do not contain a large quantity of oxidized nitrogen, it does not prove disadvantageous, particularly when hydrogen peroxide is used. The method of Kleeman [23], for example, gives very fast decomposition, and also diminishes fume formation. The procedure is as follows: about 1–5 g of fresh plant or animal material is weighed into a Kjeldahl flask and, with shaking, 25 ml of 30% hydrogen peroxide are added, followed by 40 ml of concentrated sulphuric acid. It is found that about 80% of the total nitrogen is reduced very rapidly to NH_3. If 15 g of potassium sulphate are also added, boiling for 10–15 min is sufficient for quantitative reduction of the nitrogen, but Kleeman recommends 45 min. In control experiments with known amounts of $(NH_4)_2SO_4$, there was no nitrogen loss under similarly intensive oxidizing conditions.

A number of modifications of the method are known. Heuss [24] added 15 ml of 30% hydrogen peroxide, 20 ml of concentrated sulphuric acid and 7–8 g of potassium sulphate to 1.75 g of dry material; complete decomposition can be attained within 30–45 min. Riehm [25] used a catalyst as well, adding 0.5 g of copper and 0.8 g of mercuric sulphate. Koch and McMeekin [26] use hydrogen peroxide in a micromethod. In the carbonization stage the heating is interrupted, 1–5 drops of hydrogen peroxide are added, and heating is resumed until white fumes again appear; this operation is repeated several times. Moore [27] adds hydrogen peroxide to the decomposed material after mild heating for 5 min.

Perchloric acid can be used in a fairly similar way. Mears and Hussey [28] add 1 g of copper sulphate and 2 ml of concentrated perchloric acid to 25 ml of concentrated sulphuric acid; the solution becomes clear within 3–7 min, and must then be boiled for at least a further 15 min.

A number of methods are available in which persulphates are used. Willard and Cake [29] employ persulphate in the determination of the nitrogen contents of flour, albumin, gelatin and peptone. The sample is heated with 15 ml of concentrated sulphuric acid until fuming ceases. The flask is cooled, a quantity of potassium persulphate weighing ten times as much as the sample is added, the solution is heated gently until it becomes clear, and then more strongly for 5 min to decompose the excess of persulphate. Too high an excess of persulphate may cause loss of nitrogen.

Swaminathan and Sud [30] proposed a procedure for rapid determination of total nitrogen which consists of weighing a 0.5–5 g sample (fruits and vegetables) into a 500 ml Erlenmeyer flask and adding to it 20 ml of 50% solution of chromic acid followed by 25 ml of concentrated sulphuric acid with swirling. After 30 minutes the contents of the flask are diluted with water, transferred to a 100 ml volumetric flask, cooled and made up to the mark. A 5 ml aliquot of the digest is pipetted into a beaker, a trace of potassium persulphate is added and boiled for a few seconds to oxidize the Cr^{3+} present to Cr^{6+}. About 20% of the nitrogen was found to be bound irretrievably as chromamine complex. Persulphate oxidation of the Cr^{3+} was found to be the only method whereby all the nitrogen could be recovered in the distillate quantitatively.

Application of reductants. In studies of nitrogen metabolism it may be necessary to determine the total nitrogen contents in plant materials containing nitrate. In this case the nitrate nitrogen can only be converted into ammonium ions in the Kjeldahl decomposition if a reductant is added. Reduction is also necessary in the determination of the nitrogen of azo, nitro, oxime, isoxazole, hydrazine and hydrazone groups. The reduction may be done either prior to or simultaneously with the sulphuric acid decomposition. Metallic zinc, metallic copper, titanous chloride, sodium dithionite or benzoic acid may be used for the preliminary reduction. According to Jodlbauer [31], the reduction of nitrate is more effective if phenol is used instead of benzoic acid. In German books on soil analysis, the Jodlbauer procedure is applied in general for determining the total nitrogen contents of substances containing nitrate. In the AOAC methods book [32], salicylic acid and metallic zinc are recommended for enhancement of the reduction.

Catalysts. During the long history of the Kjeldahl method, the catalytic effects of various metals, metal oxides and a whole series of salts have been investigated with a view to shortening the decomposition time. Osborn and Wilkie [33] found that the catalytic efficiency of the metals decreases in the sequence: Hg, Se, Te, Ti, Mo, Fe, Cu, V, W, Ag. However, numerous other examinations have proved Hg, Cu and Se to be the most effective [34]. Baker [35] has shown that catalyst mixtures do not have demonstrable advantages over the individual catalysts, but in practice the use of various catalyst mixtures is generally widespread.

The destructive decomposition passes through various stages: the dehydration, the carbonization and the clearing of the solution. The transformation of nitrogen into ammonia is typically exponential; the conversion is rapid at first, but becomes increasingly slower.

The stage at which the transformation curve is almost parallel to the time axis and the amount of nitrogen transformed approaches the absolute nitrogen value asymptotically, is termed after-boiling. According to Beet [36], it may be anything from 0 to 235 h. Kitto [37] found that the time required for the after-boiling of wheat-flour and milk is 1.75 times the time necessary for clarification of the solution. In the case of substances which are difficult to decompose (pyridine, nicotinic acid), Shirley and Becker [38] (using the AOAC method) considered a boiling period of 3–4 h after the clarification to be necessary in order to attain quantitative conservation. The potassium sulphate (added to raise the boiling point) and the catalysts are also of very great importance as regards the duration of the after-boiling period. The amount of acid used for the decomposition must be chosen so that, even after the after-boiling stage, sufficient remains to bind the ammonia formed. A loss will occur if, as a result of concentration of the potassium bisulphate and the catalyst, a solid phase is present when the decomposition is completed. Self [39] recommends that at the end of the decomposition at least 15 g of sulphuric acid should remain free in the solution from the 30 ml initially taken. Before the investigation of large series of samples of similar nature, even in the case of standard methods, it is advisable to carry out a few preliminary experiments to establish the optimum after-boiling time necessary; in this way, the analytical error may be decreased, and unnecessary time losses may be avoided.

Determination of the ammonia

The second stage in the Kjeldahl method is determination of the ammonia formed. This can be done titrimetrically (with or without distillation) or spectrophotometrically.

Two distillation methods are available: direct distillation and steam distillation. Separation of ammonia by diffusion may also be regarded as a distillation process.

In macrodeterminations, direct distillation is generally used (the Kjeldahl flask itself being used as the distillation flask), and the ammonia is liberated with concentrated alkali. An alternative is to dilute the decomposition solution to known volume and use an aliquot in a microdistillation. Mercury forms a complex with ammonia, so if a mercury catalyst is used, the mercury must be removed before addition of the alkali, or a low result may be obtained. Potassium sulphide or sodium thiosulphate is used to precipitate the mercury. About 0.2 g of thiosulphate is required for of 1 g of mercury. Salm and Prager [40] showed that, even in the presence of a mercury catalyst, loss of ammonia may be avoided by adding finely-divided zinc powder to the alkaline solution to be distilled. The zinc reduces the mercury salts to the metal. Another disadvantage of the mercury catalyst is that in micro and semimicro methods the precipitate of HgS may block the steam duct of the apparatus. By direct distillation with moderate heating, the ammonia can be distilled off quantitatively in 5–20 min, depending on the size of the apparatus. In the steam-distillation apparatus of Parnas-Wagner, Markham, Büchi or similar types for the micro and semimicro determination a maximum of 10 min is needed for the quantitative distillation of the ammonia.

In the classical Kjeldahl method, the ammonia is absorbed in sulphuric acid of known volume and normality, and the excess of acid is back-titrated iodometrically:

$$5KI + KIO_3 + 3H_2SO_4 = 3I_2 + 3K_2SO_4 + 3H_2O$$
$$I_2 + 2Na_2S_2O_3 = 2NaI + Na_2S_4O_6$$

A simpler method is absorption of the ammonia in acid of known volume and normality, and back-titration of the excess of acid with standard alkali.

Winkler [41] recommends absorption of the ammonia in 4% boric acid solution; this has a pH of 4.7, which changes to 8.6 on about 20% neutralization with ammonia. The ammonia thus absorbed in the boric acid solution can then be titrated with standard acid. Various indicators and indicator mixtures are used. Winkler suggested Methyl Orange, Methyl Red or Congo Red, and Scales and Harrison [42] Bromophenol Blue. The equivalence point at pH 4.7 is particularly well indicated by the Tashiro indicator: a 1:1 mixture of 0.2% Methyl Red and 0.1% Methylene Blue solutions. In pure boric acid this mixture is purple, but at pH above 4.7 it changes to apple-green. Reith and Klazinga [43] use a mixture of 1 volume of 1% Methyl Red solution, 3 volumes of 1% Bromocresol Green solution and 4 volumes of distilled water. The colour of this indicator mixture changes from bluish-grey to violet. The end-point may also be determined potentiometrically.

The thermal stability of ammonium borate has been studied. There is no ammonia loss from the ammonium borate solution during heating for 30 min at 50 °C; at 60 °C and 100 °C, however, there are losses of 1.3% and 62.3%, respectively.

The diffusion method is suitable for the determination of very small amounts of ammonia (0.1–100 μg). The procedure was developed in a form suitable for routine work by Conway and Byrne [44]. With the use of Conway dishes, this amount of ammonia may be absorbed in about 3 h at 20 °C. The method is rarely used in protein analysis, but may be employed very advantageously for ammonia determination in the investigation of microbial nitrogen metabolism.

A very rarely used method is formaldehyde titration:

$$2(NH_4)_2SO_4 + 6HCHO = (CH_2)_6N_4 + 2H_2SO_4 + 6H_2O$$

An aliquot of the diluted decomposition solution is neutralized in the presence of Methyl Red as indicator, and then, after the addition of formaldehyde, the titration is continued to a phenolphthalein end-point. The difference between the two titration volumes is equivalent to the ammonia.

In another method Willard and Cake [29] oxidized the ammonia with hypobromite, and determined the excess of hypobromite iodometrically.

The spectrophotometric methods are mainly used for direct determination of the ammonia in the decomposition solution, without preliminary distillation, but are also more advantageous than titration for the determination of ultramicro quantities in the distillation method. If the ammonia is absorbed in boric acid, for instance, and 0.01 *N* acid is used for the titration, the end-point can be determined with an error of *ca.* 0.02 ml with the Tashiro indicator under ideal conditions, which corresponds to 2.8 μg of nitrogen. In contrast, with the high-sensitivity ninhydrin reaction it is possible to measure 0.007 μg of nitrogen in a final volume of 1 ml. However, there are many interferences in the colour-reactions of ammonia, for example, by the catalysts used in the Kjeldahl method, and this must be taken into account in selection of the method.

The colour-reactions of ammonia with the Nessler reagent, indophenol and ninhydrin are generally used for the determination.

The Nessler reagent is an alkaline solution of potassium tetra-iodomercurate(II), from which a cyclic complex is formed by the action of ammonia:

$$2K_2HgI_4+3KOH+NH_3 \rightarrow O\begin{matrix} \diagup Hg \diagdown \\ \\ \diagdown Hg \diagup \end{matrix}N^+H_2\text{—}I^- +2H_2O+7KI$$

In the presence of a little ammonia, a yellow colour is obtained; with somewhat more ammonia, the colour is orange; and with much ammonia a brownish-red precipitate is formed. The sulphuric acid solution from the decomposition must be diluted and neutralized before addition of the reagents.

According to Roth [45], the Nessler method is advantageous if the quantity of nitrogen to be determined is in the order of 10 μg. One of the drawbacks of the methods is that the Nessler reagent cannot be stored for a prolonged period. Yuen and Pollard [46] made a study of the factors that govern the reaction. They found that the colour intensity produced depends on time, temperature, the amount of reagent, the nature and concentration of other ions, and pH. Hence identical experimental conditions must be used and a standard run simultaneously. Various wavelengths are recommended, but the measurement is usually made at 430–450 nm.

Even under optimum conditions, a turbidity may develop. This may be prevented with stabilizers such as tartrate, cyanide, and gum arabic [47].

It is advisable to choose decomposition conditions such that the concentration of the catalysts and potassium sulphate will not be too high in the final solution taken for the colour reaction.

The indophenol reaction is widely used and is also the basis of the automatic analyser methods. Phenol, salicylic acid and even thymol are used, with sodium or potassium hypochlorite or hypobromite as oxidant.

As reported by Berthelot as long ago as 1859 [48], the essence of the reaction is that in the presence of hypobromite *p*-nitrosophenol is formed from phenol and ammonia and in alkaline medium is then converted into a light-blue indophenol alkali metal salt. Riley [49] demonstrated that temperature and time are the critical factors. The reaction is catalysed by the Mn^{2+} ion. Exley [50] developed the phenol-hypochlorite method into an ultramicro procedure, and successfully utilized it for the analysis of insulin, casein, egg-white, bovine and human sera, etc.

Mann's method [51] is suitable for the determination of 1–15 μg of nitrogen. The final dilution must yield an ammonia concentration corresponding to 0.1–1.5 μg of nitrogen per ml. At this concentration, with measurement at 630 nm, the absorbance obeys the Beer–Lambert law, by an adaptation of the method of Ogg [52]. It is important that all experimental conditions be kept strictly constant. The colour intensity depends on pH, temperature and the reagent concentrations, but especially on the hypochlorite concentration, and this reagent should be checked by iodometric titration before every test series. Glebko *et al.* [47] consider the thymol-hypobromite reaction to have advantages over the phenol-hypochlorite reaction.

Holz and Kremers [53] developed an automatic analyser procedure. Sodium salicylate is used in place of phenol, and the oxidant is a triazine derivative, the sodium salt of dichloro-*s*-triazine-2,4,6(1H, 3H, 5H)-trione (sodium dichlorocyanurate), instead of a hypohalite, in the presence of sodium nitroprusside. The decomposition is done in test-tubes in an aluminium block. Beer's law is obeyed over the nitrogen concentration range 50–600 ppm. Experience obtained over 4 years by using an aluminium block digestor in automated total nitrogen analysis, using alkaline phenol-hypochlorite reagent has been described by Hambraeus *et al.* [54].

In the amino-acid ion-exchange chromatography procedure developed by Moore and Stein [55], the ammonium ion also gives the extremely sensitive ninhydrin reaction used for the determination of the α-amino group, and hence ammonia can be determined in the course of the separation of the basic amino-acids. The ninhydrin reaction is pH-sensitive but has the advantage that the effects of anions and cations may be neglected. At 570 nm the molar absorptivity of the ammonia–ninhydrin complex is $2.01 \times 10^4\ l \cdot mole^{-1} \cdot cm^{-1}$.

Whitaker *et al.* [56] have described a specific spectrophotometric assay for ammonia with trinitrobenzene sulphonic acid. 2,4,6-TNBS reacts stoichiometrically with ammonia in tetraborate buffer to give trinitrophenyl derivatives, with an absorbance maximum at 345 nm or 420 nm.

TNBS also reacts with primary amines, amino-acids and proteins. Therefore the ammonia must be distilled from the sample into an acid trap. This reaction occurs at room temperature and is more precise and more sensitive than the ninhydrin method.

2.2.5 Determination of nitrogen by combustion

For obvious reasons, the classical Dumas method and its micro and semimicro variants, the Pregl-Dumas nitrogen determination [57], have not become widely used in protein analysis. The original methods have the drawbacks that the determination is time-consuming and requires practice and careful technique. In recent years, however, automatic nitrogen analysers have been developed which operate on this principle.

The essence of the Dumas method is oxidation of the organic matter (in the present case protein) at 700 °C by heating with cupric oxide in a stream of carbon dioxide as carrier gas in a combustion tube. The nitrogen in the compound is released as the element; any oxides of nitrogen formed are reduced to the element on a hot copper spiral in the reducing zone of the combustion tube. The carbon dioxide is absorbed in 40% potassium hydroxide solution and the nitrogen is measured volumetrically in a nitrometer. The mass of nitrogen is calculated from the gas volume corrected to standard temperature and pressure and the amount of protein is obtained by multiplying the mass of nitrogen by 6.25, or by the precise conversion factor relating to a given protein. An analyst experienced in the method may achieve results within an absolute error of 0.2‰. Only rarely can the small amount of sample taken be regarded as an advantage of the method. In the Pregl-Dumas apparatus used in organic analytical laboratories, a sample of 2–5 mg is taken, and from this a volume of *ca.* 0.3–0.5 ml of nitrogen is formed from a protein. For substances with nitrogen contents less than 2‰, a sample of 8–10 mg is necessary if the error of the determination is to be within 0.2% absolute. The method is mainly of importance when only mg quantities of sample are available.

The Pregl-Dumas method has been automated in a number of instruments, which generally consist of two main units: the combustion unit and the nitrogen detector. The latter may be a gas chromatograph, a mass spectrometer, or a nitrometer. In modern instruments the results are usually presented as a digital display or a print-out, and the latest versions will be microprocessor-controlled with automatic calculation and correction facilities.

Some of the instruments are the CHNO elemental analysers generally used in organic analyses, whereas others are exclusively for the determination of nitrogen.

2.2.6 Determination of non-protein nitrogen (NPN)

A frequent operation in the protein-analysis laboratory is the determination of the NPN, or its separation from the 'true protein' nitrogen. By NPN we generally understand peptides with molecular weights less than 10 000, free amino-acids and amides. The methods of separation most often employed are dialysis (see p. 49), ultrafiltration (see p. 46) and precipitation.

A whole series of protein-fractionation methods has been developed from the partial precipitation techniques. The stabilities of protein solutions are governed by temperature, pH, the nature and concentration of ions in the solution, and the presence of stabilizing organic substances, e.g. lipids. If these factors are varied, the proteins are partially or totally precipitated.

The precipitation procedures employed most frequently are as follows.

By heat. The sample is held for 10 min in a water-bath at 100 °C, and then filtered.

With alcohol (Awapara [58]). Absolute alcohol is added to the solution to give a final concentration of 80%. The solution is left to stand for several hours, and then centrifuged.

With HCl/acetone (Boulanger and Biserte [59]). The NPH may be precipitated with acetone containing 1% of hydrochloric acid, from plasma dried over sulphuric acid in a desiccator.

With trichloroacetic acid (TCA). Two volumes of 10% TCA solution are added to 1 volume of protein solution (serum), the mixture is left to stand for 1 h at 0 °C, and then centrifuged.

With perchloric acid (Weber [60]). One volume of 10% v/v perchloric acid is added to 2 volumes of protein solution. After standing for 1 h, the mixture is centrifuged.

Sulphosalicylic acid (20% solution) is used similarly to TCA. Tungstic acid is employed in the method of Van Slyke and Hawkins [61]. A 1% tannin solution may be used, in accordance with the procedure of Mothes [62].

In feed examinations, the 'pure' ('true') protein is determined by the method of Barnstein [63].

Numerous publications have appeared on these various protein-precipitation methods, and on comparative investigations. Good surveys are provided by Hinsberg and Lang [64] and by Bell [65], who compared the effects of 20 protein-precipitating agents on 4 different materials with the results of dialysis.

2.2.7 Protein-determination methods

A number of methods are available for the determination of the protein contents of protein solutions. They may be classified on the basis of their measurement principles.

Ultraviolet spectrophotometry

Most proteins have an absorption maximum at 280 nm, which can be attributed to the tyrosine and tryptophan units. For different proteins with the same nitrogen concentration, the peak varies within comparatively narrow limits. Advantages of the method are the high sensitivity, and the fact that the determination may be performed without a reagent. These two properties are particularly convenient in connection with monitoring protein concentration in the eluate from chromatographic columns. Nucleic acids or their components interfere. Also, compounds containing purine and pyrimidine rings exhibit strong absorption at around 260 nm, and this broad absorption band interferes with the selective measurement of the proteins at 280 nm. Warburg and Christian [66] measured the absorbances at 280 and 260 nm of a series of nucleic acid–protein mixtures with various ratios, and constructed a correction table for the nucleic acid error. Kalckar [67] compared the Lowry protein results with the absorbances at 260 and 280 nm of protein solutions containing nucleic acids, and obtained the following formula for the calculation of the protein content:

$$\text{protein (mg/ml)} = 1.45\ A_{280} - 0.74\ A_{260}$$

The following equation can be fitted to the data of Warburg and Christian:

$$\text{protein (mg/ml)} = 1.55\ A_{280} - 0.76\ A_{260}$$

If the nucleic acid content is low, the concentration of non-turbid protein solutions can be determined in the range 10–1000 μg/ml from the absorbance at 280 nm.

For wavelength selection a prism or grating monochromator is used in conjunction with the continuous spectrum emitted by a deuterium lamp. In simpler detectors, low-pressure mercury vapour lamps are used; the 280 nm band is selected from the line spectrum with a filter combination. Bennett *et al.* [68] measure the absorbance of a protein solution at 285 nm, using a magnesium lamp, and a selective modulation technique. The visible light is absorbed by a Jena UG5 filter.

In the application of ultraviolet monitors for protein determination, it must be taken into consideration that many of the eluents used in chromatographic or gel-filtration methods themselves exhibit appreciable absorption at 280 nm. Background compensation by electrical means reduces the sensitivity of the measurement, and cannot be used at all in the case of gradient elution, since the absorption of the gradient varies continuously. In such a case, a double-beam photometer or monitor must be used.

Since photometers of appropriate sensitivity and stability have become available, methods have been developed in which the protein determination is based on the absorbance at wavelengths shorter than 280 nm, to avoid the effect of the nucleic acids. Groves *et al.* [69] use the absorbance at 224–240 nm. Although the proteins do not have an absorption maximum in this region, the nucleic acids display an absorption minimum. The method is more independent of the nature of the protein than the measurement at 280 nm, for the absorbance measured is due not only to the tyrosine and tryptophan units, but also the phenylalanine, methionine, cysteine and cystine, together with the contribution from the peptide bond at 185 nm. With this procedure, 5–180 μg of protein per ml may be determined. Ehresmann *et al.* [70] state that the interference of RNA can be eliminated even more advantageously at wavelengths of 228.5 and 234.5 nm. Kalb Bernlohr [71] have developed a sensitive UV-spectrophotometric protein assay which offers certain advantages over the method of Warburg and Christian. The protein concentration (μg/ml) $= 183\ A_{230} - 75.8\ A_{260}$, where A_{230} and A_{260} are the absorbances at 230 and 260 nm.

Proteins and peptides exhibit a well-defined absorption at 191–194 nm. For protein or peptide determination, this has the advantage of being independent of the quantity of aromatic amino-acids, and hence the absorption is almost independent of the nature of the protein or peptide. The method demands a spectrophotometer of high sensitivity and excellent optical quality, and also good technical conditions for the measurement (dust-free laboratory), because in the short ultraviolet region the measurements are subject to great variation due to scattered light. The sensitivity attainable with this procedure on a Beckman DBG spectrophotometer is 40–80 times better than that of the traditional absorption measurement at 280 nm.

For individual proteins, Kirschenbaum [72] published collected data of molar absorptivity and $A_{1\,cm}^{1\%}$ values at selected wavelengths in the ultraviolet and visible regions.

Visible region spectrophotometry

The colour-reactions of the proteins are in effect those of the amino-acids making up the proteins. Only a few of the detection reactions of value can be used quantitatively. The most important will be described below, in the order of frequency of use.

The biuret reaction. In alkaline medium the copper(II) ion is bound to four peptide nitrogen atoms: apart from the —$CONH_2$ group, the reaction is given by the —$CSNH_2$, —$C(NH)NH_2$, —CH_2NH_2, —$CRHNH_2$, —$CH(OH).CH_2NH_2$, —$CHNH_2.CH_2OH$ and —$CHNH_2.CH(OH)$-groups, and naturally by biuret itself, $OC(NH_2).NH.CONH_2$, after which the reaction is named. The violet copper complex has an absorption maximum at 540–560 nm, but can also be measured in the near ultraviolet at 310 nm. Gornall *et al.* [73] prolong the lifetime of the biuret reagent by preparing it as follows: 1.5 g of copper sulphate pentahydrate and 6 g of sodium potassium tartrate tetrahydrate are dissolved in 500 ml of water, 300 ml of 10% carbonate-free sodium hydroxide solution are added, and the volume is made up to 1000 ml with water. The reagent is stored in a polyethylene bottle, in the dark. The addition of 1 g of potassium iodide inhibits reduction of the copper, but has no effect on the biuret reaction. For the determination, 1 ml of protein solution, containing 10–100 mg of protein per ml, and 4.0 ml of biuret reagent are mixed and allowed to stand for 30 min at room temperature, and the absorbance is then measured at 540–550 nm against a mixture of 1 ml of water and 4 ml of biuret reagent. The calibration curve is prepared by use of bovine serum albumin solution. If greater accuracy is required, the protein content should be determined by the Kjeldahl method, and the determination referred to this. The heated biuret-Folin protein assay method of Dorsey *et al.* [74] has one important advantage over other biuret methods: freedom from protein character. Different proteins give almost identical absorbances per unit mass of nitrogen with this procedure. The NPN also reacts with the biuret reagent, and the determination must therefore be performed on an alkaline solution of the precipitate after precipitation with trichloroacetic acid. The presence of lipids causes a turbidity, and they must be removed by extraction.

Klungsöyr [75] passes the copper-protein complex formed in the biuret reaction through a Sephadex G-25 column, thereby separating it from lower molecular weight compounds giving the biuret reaction and from other interfering substances. The copper–protein complex is decomposed with a sulphuric acid-hydrogen peroxide mixture, the solution is made alkaline, and the copper is determined colorimetrically in the normal way with diethyldithiocarbamate. This procedure has the same sensitivity as the Lowry method, and is about 10 times more sensitive than the protein determination based on the absorption at 280 nm. Goldberg [76] likewise separates the copper-protein complex by Sephadex gel filtration, in this way eliminating the possible interference of various compounds. The copper(II) released from the decomposed complex is calculated from the absorption at 410 nm, arising from the copper-catalysed phenol–chloramine-T reaction. This copper determination is about 1000 times more sensitive than the atomic-absorption method. The sensitivity of the protein determination is 500 times higher than that of the Lowry procedure (0.01–0.2 μg of protein per ml). The absorbance in the biuret method can be measured to advantage in the near ultraviolet region. Koch and Putnam [77] recommend 330 nm: here the determination is not disturbed by the presence of nucleic acids or the ammonium ion. Ithaki and Gill [78] measure the absorbance of the copper complex at 310 nm. In the presence of DNA, they could determine 0.15–3.0 mg of protein per ml, and in a solution not containing nucleic acid less than 0.075 mg/ml could be measured. The biuret method of Ellman [79] takes as basis the absorption measured at 263 nm; the sensitivity of the procedure is 0.01 mg/ml, i.e. 10–15 times better than that of measurement in the visible region.

The protein determination is not always preceded by protein precipitation and dissolution in alkali. In many cases the determination must be carried out directly on the plant or animal tissue extract or homogenized material. The determination may also be disturbed by the detergent or buffer used for the extraction. In the presence of the detergents Triton X-100, Tween 80, digitonin or hexadecyltrimethylammonium bromide, the biuret reagent gives a precipitate with protein. Futterman and Rollins [80] dissolve the precipitate in propane-1,2-diol and measure the absorbance at 330 nm. A better result is obtained when the biuret reagent is prepared with potassium hydroxide instead of sodium hydroxide. The modified biuret method described by Watters [81] can be used advantageously for protein determinations in the presence of detergents. With the modified reagent different proteins produce colour of fairly uniform optical density. TRIS buffer is very frequently used in *in vitro* biochemical reactions, but its presence makes protein determination more difficult, as it disturbs both the biuret and the Lowry protein determinations. Because of the nitrogen content of the buffer, it is also difficult to determine the protein on the basis of the Kjeldahl nitrogen. Accordingly, the correct procedure is to record the biuret calibration curves at various TRIS concentrations, and apply an appropriate correction [82]. Goshev and Nedkov [83] recommend a biuret method for quantitative determination of insoluble proteins such as natural collagen and sheepskin creatine. The sample is oxidized with excess of H_2O_2, dried and subsequently dissolved in 0.05% sodium dodecyl sulphate–1 N NaOH at 100 °C. The clear solution is treated with biuret reagent.

Various macro variants of the biuret reaction can be used very advantageously for simple, practical protein determinations, e.g. for the study of the protein contents of cereal milling products. In these procedures the alkaline solution of copper(II) sulphate is employed not only as a reagent, but also as an extractant for the protein (Jennings [84]). Examination with wheat and barley milling products revealed a close correlation ($r = 0.97$–0.99) between the Kjeldahl nitrogen values and the absorbances. In the essentially similar method of Pinckney [85], glycerine is used instead of potassium sodium tartrate to stabilize the biuret reagent. The method of Johnson and Craney [86] likewise combines the extraction and the determination. The error arising from the tendency of the biuret reagent to undergo decomposition is eliminated by using finely-powdered copper carbonate in place of the prepared reagent. The alkaline aqueous solution is replaced by 0.1 N potassium hydroxide in 60% propan-2-ol. The colouring matter in the cereal milling products is less soluble in this solution, and thus the measurement of the absorbance will be more accurate in comparison with the traditional biuret reaction.

The most commonly used protein determination in biochemical analysis is the Folin–Lowry [87] procedure. The basis of the method is the biuret reaction of proteins with copper(II) in alkaline medium and the Folin–Ciocalteu phosphomolybdic-phosphotungstic acid reduction to heteropoly-molybdenum blue by the copper-catalysed oxidation of the aromatic amino-acids bound in the protein. A new modification of the Folin–Lowry method described by Ohnishi and Barr [88] appears superior to the original, having extremely stable colour development, good reproducibility and good storability of the reagents.

The Folin–Lowry protein determination depends strongly on the foreign substances present in the protein solution, the buffers, the detergents and the chelating agents. The course of the reaction is similarly changed by alcohols, sugars and polysaccharides. Further disturbances are caused by pigments, by higher concentrations of sulfhydryl groups and by sulfhydryl reagents.

Special procedures must also be employed for the determination of lipoproteins.

On the basis of publications that have appeared in the past few years, a survey is given below of the effects of the disturbing factors, and of the possibilities for eliminating them. Ji [89] showed that the effects of EDTA, Triton X-100, Bicin, Tricin and TRIS on the Lowry reaction are not linearly proportional to the concentrations of these compounds. The sensitivity of the determination is decreased by glycine, glycylglycine, citrates, succinates, carbamide, sodium dodecyl sulphate and sugars. If these compounds are present in definite concentrations, then their effects can be compensated for with a blank solution prepared under the same conditions.

The modified Lowry method of Peterson [90] is particularly applicable to soluble membrane and proteolipid proteins in dilute solution containing interfering substances. Polyalcohols, such as glycerine and sugars, protect the proteins from denaturing, and they are therefore employed in numerous preparative techniques. The investigations by Zishka and Nishimura [91] indicate that the absorbance increases linearly as a function of the glycerine concentration in both the biuret reaction and the Lowry protein determination, and hence the disturbing effect can be compensated for with an appropriate reference solution. Bonitati *et al.* [92] determined the protein equivalents of a whole series of reducing sugars and polysaccharides, i.e. the apparent quantities of protein that would be measured on the basis of the reaction of the sugars and the Folin–Lowry reagent. It was found that dextran and glycogen do not give colour-reactions, but the colour-reactions of monosaccharides, and particularly ketoses, may disturb the protein determination to a considerable extent. According to Chai-Ho Lo and Stelson [93], the synthetic sugar polymer "Ficoll" used to produce the gradient in the density gradient centrifugation fractionation of proteins, gives rise to a significant disturbance of the Folin–Lowry protein determination. A chloroform–methanol mixture is used for the extraction of lipoproteins and proteolipids, but the proteins prepared in this way dissolve in aqueous medium only with difficulty; in practice, dissolution is achieved with alkaline treatment overnight at 37 °C, or by treatment for 30 min at 100 °C. The pretreatment lengthens the time of the protein determination and decreases the accuracy. After the removal of the lipids, certain proteolipids form a precipitate with the protein reagent in the course of the Lowry protein determination. The decomposition products formed in the oxidation of lipids react with the Folin–Lowry reagent and thus increase the quantity of protein determined. In order to eliminate these difficulties, Lees and Paxman [94] developed a procedure for the determination of lipoproteins with the Lowry method. In this procedure, the protein is dissolved not in the customary 0.5 *M* sodium hydroxide, but in 0.5 *M* sodium hydroxide containing 5% sodium dodecyl sulphate.

Automated Folin–Lowry methods. A number of the Lowry determination procedures have been adapted for automatic measurement with the Technicon system. In the method of Huemer and Kyung-Dong Lee [95], the peristaltic pump, tube and spiral are selected so that the colour develops in about 10 min. In the automated micromethod of Gaunce and D'Iorio [96], the absorbance is measured at 660 nm after a reaction time of 5.5 min.

Osta *et al.* [97] reported the optimization of Folin–Ciocalteu reagent concentration in an automated Lowry protein assay, to obtain maximum sensitivity.

Infrared spectrophotometry

Infrared spectrophotometry is not only an important method in research into chemical structure: it can also be employed in quantitative analysis, including protein determination.

Some chemical bonds which occur with constant frequency in relation to the total nitrogen content in amino-acids, peptides and proteins, display infrared absorption bands in the near infrared at 0.8–2.5 μm. Such bonds include N=H and C=O.

The cause of the infrared absorption is the stretching vibration within the molecule. Since the absorption bands appear in the near infrared, measurements may be made even with glass or quartz optical equipment, as these still transmit in this region. The method can be used for the determination of dissolved proteins, on the basis of absorption photometry, for which the Beer–Lambert law holds; however, its application is particularly advantageous for insoluble or undissolved solid materials, on the basis of the diffuse reflection of a layer of theoretically infinite thickness, prepared from a finely-ground powder.

Since 1971 this method of measurement has been used for the determination of protein in agricultural products, with instruments developed on the basis of the principle of Norris [98]. According to Williams [99], the results agree with the results from the conventional method with a standard error of $\pm 0.22\%$.

Dye-binding

Under suitable experimental conditions, the acidic and basic groups of protein macromolecules interact with the dissociated groups of organic dyes, e.g. the acid radicals of sulphonic acids, and form coloured precipitates with them. The phenomenon of dye-binding can be utilized for quantitative analysis.

Amido Black 10-B binds well to protein in acid medium, and the protein can be determined on this basis. Führ and Hinz [100] developed procedures for the determination of the protein content of cerebrospinal fluid. The cerebrospinal fluid protein is stained and simultaneously precipitated with Amido Black in acetic acid–methanol mixture. The stained precipitate is dissolved in sodium hydroxide solution, and the colour intensity of the solution is measured.

Racusen [101] made an investigation of the binding of 9 different proteins by Amido Black under various experimental conditions: in solution (the dye excess being removed by dialysis), and with the protein (separated by electrophoresis) stained in the polyacrylamide gel. The dye was eluted from the gel with 1 *N* sodium hydroxide and the blue solution was measured photometrically. In the view of Racusen, the dye-binding procedure is less protein-specific than the Lowry method. Calculated on a nitrogen basis, the 9 different dye-bindings were almost the same. The adsorption of Amido Black on the protein in the polyacrylamide gel is linear up to a protein concentration of 1.5 mg/ml. The protein microdetermination method of Kihara and Kuno [102] is based on the fact that the protein binds quantitatively onto a nitrocellulose membrane filter, and the bound protein can be stained quantitatively with Amido Black. Kihara and Kuno report that a 0.1 *M* magnesium chloride solution promotes the binding of the protein on the nitrocellulose membrane. The NPN is also washed out with the magnesium chloride solution.

Bramhall *et al.* [103] dried a protein solution on a 2 × 2 cm Whatman 42 filter-paper, and then fixed it with 7.5% trichloroacetic acid solution. For 5–25 μg of protein, Xylene Brilliant Cyanine G (Michrome No. 1224) was used as dye, in a concentration of 10 mg/ml in 7% acetic acid solution, for 15 min at 50 °C. In a similar method, Naphthalene Blue Black (Michrome No. 1113) was used in a concentration of 10 mg/ml in acetic acid–methanol–water (1 : 4 : 5 v/v) solution, for 15 min at 50 °C.

A method for rapid quantitation of microgram amounts of protein was presented by Bradford [104]. The binding of the dye Coomassie Brilliant Blue G 250 to protein causes a shift in the absorption maximum of the dye from 465 nm to 595 nm. This assay is readily reproducible and gives rapid good colour stability. A simple protein determination method is described by Esen [105]. Protein solution is spotted on chromatography paper, stained with 0.1% Coomassie Blue solution. The dye–protein complex is eluted with

1% SDS, and the absorbance is measured at 600 nm. This assay is free from interference by non-protein substances.

In recent years, numerous rapid methods based on dye-binding have been developed for the study of the nature and content of proteins in cereal milling products. The dye-binding method was first employed in this field by Udy [106, 107] to determine the basic and acidic groups, and their ratios, in the proteins of wheat flour fractions. The dye-binding of the basic groups was examined with Orange G, in a 0.1% solution in McIlvaine buffer of pH 2.2, and that of the acidic groups with 0.2% Safranine O, dissolved in disodium hydrogen phosphate–sodium hydroxide buffer of pH 11.5. An indirect method was used for measurement of the dye. Not the bound dye, but the excess of dye was measured, from the decrease in the concentration of the dye solution taken. The colour of Orange G was measured at 472 nm, and that of Safranine O at 452 nm, in 50-fold dilution.

A new variant of the method was developed by Mossberg [108] who, from the dye-binding, sought a correlation relating to the proportion of basic groups, and indirectly the quantity of lysine.

Turbidimetry

Suspensions containing very small suspended particles, e.g. proteins precipitated in high dilution, scatter in all directions part of the light passing through them, and accordingly are turbid. There are two means whereby the turbidity may be measured. The intensity of light passing through the turbid solution is measured against that passing through the pure solution, in accordance with the rules of absorption photometry; this mode of measurement is termed turbidimetry. Alternatively the light scattered by the suspended particles is measured at a definite angle to the incident light-beam, e.g. 90°. This is called nephelometry. The turbidity depends on many factors: within certain limits, the intensity of the scattered light and the decrease in intensity of the light beam passing through the turbid solution are proportional to the number of suspended particles and, in the case of a given degree of dispersion, to the concentration of the suspended material. Turbidimetry is used as a simple protein-determination method. From what has been said, if all the experimental conditions are constant, the concentration can be measured only by mean of empirical calibration curves. Opal glass plates, or plates prepared from plastic polymerized together with finely suspended material, may be used as constant turbidity standards.

Turbidimetric methods suitable for protein determination, which are mainly applicable in clinical analysis, are reported by Layne [109].

The pepsin-decomposition of proteins, for instance, can be followed turbidimetrically [110]. Precipitation of protein with ammonium sulphate can be followed similarly. The concentrations of prolamines may be readily determined turbidimetrically. Craine *et al.* [111] determined the concentration of a zein solution by adding 6 ml of 1% sodium chloride solution to 2 ml of a 70% alcoholic solution of the zein; the turbidity was measured photometrically at 590 nm after 60 min. Feinstein and Hart [112] recommend a turbidimetric method for determination of the protein content of wheat flour.

Tappan [113] developed a turbidimetric procedure which uses a flow-cell spectrofluorometer to measure the light scattering at 478–485 nm. The experimental set-up can also be applied as a protein monitor in column-chromatographic or gel-filtration techniques. The protein-precipitating reagent is added to the eluted protein by a micropump, and the turbidity of the solution is measured continuously.

Fluorometry

There are two main possibilities for utilizing fluorescence in protein determinations: (a) the coupling of the proteins to fluorescent compounds; (b) the quenching of fluorescence by the addition of protein.

Sheth and Cohen [114] developed a sensitive fluorometric method for the determination of histones. With this procedure it is possible to determine 0.2–0.4 μg of histones rich in lysine or arginine.

The essence of the fluorescence quenching technique of Hiraoka and Glick [115] is that, corresponding to the method of dye–binding, a fluorescent dye is coupled to the protein. Various dyes have been tested: Eosin B, Erythrosin B, Floxin, Bengal Red, and Eosin Y; the last-named finally proved the most suitable. The fluorescence was excited at 518 nm, and measured at 540 nm.

Several sensitive fluorometric methods using orthophthaldialdehyde for unhydrolysed or hydrolysed samples have been published. Butcher and Lowry [116] have reported the measurement of nanogram quantities of protein by hydrolysis, followed by reaction with orthophthaldialdehyde in the presence of mercaptoethanol.

Udenfriend *et al.* [117] use fluorescamine to determine proteins, peptides, amino-acids and primary amines. At room temperature and in aqueous solution, fluorescamine reacts with the amino group. The fluorescence can be excited at 390 nm, and fluorescent light emitted at 475 nm can be measured. The method is suitable for the determination of 0.5 μg of protein. The fluorescence is stable in the pH range 4–10.

Stowell *et al.* [118] report an adaptation of an assay which is sensitive to submicrogram quantities of protein and not affected by the presence of several detergents, including Triton X-100.

Refractometry

Very old methods are known for the refractometric determination of certain proteins. Reiss [119] developed a procedure for the determination of blood serum with a Pulfrich immersion refractometer. The percentage of protein can be read off directly from a table of refractive indices.

Polarography

In his review, Müller [120] describes the Brdička reaction, the essence of which is that, when certain amino–acids, polypeptides and proteins are dissolved in a cobalt-containing buffer at a definite pH, they enter into a catalytic reaction with the dropping mercury electrode, and hence can be measured polarographically. The reaction is given essentially by the SH groups of the proteins. With the polarographic method of Paleček and Pechan [121], protein at a level of 10^{-9} *M* may be determined. These methods are not used very often in practice.

GENERAL REFERENCES

Schramm, W.: *Chemische und biologische Laboratorien, (Planung, Bau und Einrichtung)*, Verlag Chemie, Weinheim, 1957.

Edwards, J. A.: *Laboratory Management and Techniques*, Butterworths, London, 1960.

Buchholz-Meisenheimer, H.: *Anwendung physikalischer und physikalisch-chemischer Methoden im Laboratorium*, Ullmanns Encyklopädie der Technischen Chemie, Band 2 (1), Urban-Schwarzenberg, München, Berlin, 1961.

Keil, B., Sormová, Z.: *Laboratoriumstechnik für Biochemiker*, Akad. Verlag, Leipzig, 1965.

Rauen, H. M.: *Biochemisches Taschenbuch*, Springer-Verlag, Berlin, 1965.

Linskens, H. F., Tracey, M. V.: *Moderne Methoden der Pflanzenanalyse*, Vol. V. (Moor, H.: *Gefriertrocknung*, pp. 73–94), Springer-Verlag, Berlin, 1962.

Dechema Erfahrungsaustausch, Vakuumtechnik, Deutsche Gesellschaft für Chemische Apparatenwesen E. V., Frankfurt am Main.

Eckschlager, K.: *Fehler bei Chemischen Analysen*. Akad. Verlag, Leipzig, 1964.; *Errors, Measurements* and *Results in Chemical Analysis*, Van Nostrand, London, 1969.

Schendor, G. W.: *Statistical Methods*, 5th Ed., Iowa State College Press, Ames, Iowa, 1956.

Colowick, S. P., Kaplan, O. N.: *Methods in Enzymology*, Vol. III, 1957; Vol. VI, 1963, Academic Press, New York.

Peach, K., Tracey, M. V.: *Moderne Methoden der Pflanzenanalyse*, Vol. I, 1956; Vol. IV, 1955, Springer-Verlag, Berlin.

Alexander, P., Block, R. J.: *Analytical Methods of Protein Chemistry*, Vol. I, Pergamon Press, London, 1960.

Bradstreet, R. B.: *The Kjeldahl Method for Organic Nitrogen*, Academic Press, New York, 1965.

König, J.: *Die Untersuchung landwirtschaftlich wichtiger Stoffe*, Verlag Paul Parey, Berlin, 1923.

Horwitz, W.: *Official Methods of Analysis of the Association of Official Agricultural Chemists*, AOAC, 13th Ed., Washington, 1980.

REFERENCES

[1] Eckschlager, K.: *Collection. Czech. Chem. Commun.*, **25,** 987 (1960).
[2] Dean, R. B., Dixon, W. J.: *Anal. Chem.*, **23,** 636 (1951).
[3] Hübener, H. J.: *Biochem. Z.*, **331,** 410 (1959).
[4] Craig, L. C. Gregory, J. D., Hausmann, W.: *Anal. Chem.*, **22,** 1462 (1950).
[5] Osborne, T. B.: *The Vegetable Proteins*, 2nd Ed., Longmans-Green, London, 1924.
[6] Mertz, E. T., Bressani, R.: *Cereal Chem.*, **34,** 63 (1957).
[7] Kent-Jones, D. W., Amos, A. J.: *Modern Cereal Chemistry*, 4th Ed., Northern Publ., Liverpool, 1957.
[8] Helander, E.: *Acta Physiol. Scand.*, Suppl. 141, **41,** 1 (1957).
[9] Kuhn, A.: *Kolloidchemisches Taschenbuch*, Akad. Verlag, Leipzig, 1953, pp. 90–94.
[10] Buchholz-Meisenheimer, H.: *Anwendung physikalischer und physikalisch-chemischer Methoden im Laboratorium*, Ullmanns Encyclopädie der technischen Chemie, Band 2 (1), Urban-Schwarzenberg, München, Berlin, 1961, p. 183.
[11] Hospelhorn, V. D.: *Anal. Biochem.*, **2,** 180 (1961).
[12] von Hofsten, B., Falkbring, S. O.: *Anal. Biochem.*, **1,** 436 (1960).
[13] Blainey, J. D., Yardley, H. J.: *Nature*, **177,** 83 (1956).
[14] Svedberg, T., Pedersen, K. O.: *The Ultracentrifuge*, Oxford University Press, New York, 1940.

[15] Andersen, N. G.: *Preparative Zonal Centrifugation*, in *Methods of Biochemical Analysis*, Vol. XV, Interscience, New York, 1967, pp. 271–310.

[16] Strohmaier, K.: *A New System of Measuring the Sedimentation Constant of Active Macromolecular Dispersed Substances*, Achema, Frankfurt, 1964

[17] Lewis, N. A. H. R.: *Anal. Biochem.*, **99,** 136 (1979).

[18] Dumas, J. B. A.: *Ann. Chim. Phys.*, **47,** 198 (1831).

[19] Kjeldahl, J. Z.: *Z. Anal. Chem.*, **22,** 366 (1883).

[20] Carpiaux, E.: *Bull. Soc. Chim. Belg.*, **27,** 13 (1912).

[21] McKenzie, A., Wallace, H. S.: *Austral. J. Chem.*, **7,** 55 (1954).

[22] Jakobs, S.: *Determination of Nitrogen in Biological Materials, in* (Methods of Biochemical Analysis, Vol. XIII, Interscience, New York, 1965, pp. 241–263.

[23] Kleeman, Z.: *Z. Angew. Chem.*, **34,** 625 (1921).

[24] Heuss, R.: *Wochenschrift Brauerei*, **40,** 73 (1923).

[25] Riehm, H.: *Z. Zuckerind. Tschechoslov.*, **60,** 156 (1935).

[26] Koch, F. C., McMeekin, T. L.: *J. Am. Chem. Soc.*, **46,** 2066 (1924).

[27] Moore, R. H.: *Bot. Gaz. (Chicago)*, **100,** 250 (1938).

[28] Mears, B., Hussey, R. E.: *J. Ind. Eng. Chem.*, **13,** 1054 (1921).

[29] Willard, H. H., Cake, W. E.: *J. Am. Chem. Soc.*, **42,** 2646 (1920).

[30] Swaminathan, K. C., Sud, K. C.: *Anal. Biochem.*, **74,** 260 (1976).

[31] Jodlbauer, M.: *Landwirtsch. Versuchs Stationen*, **35,** 447 (1888).

[32] Horwitz, W.: *Official Methods of Analysis of the Association of Official Agricultural Chemists*, 9th Ed., AOAC, Washington, 1955.

[33] Osborn, R. A., Wilkie, J. B.: *J. Assoc. Offic. Agr. Chem.*, **18,** 604 (1935).

[34] Bradstreet, R. B.: *The Kjeldahl Method for Organic Nitrogen*, Academic Press, New York, 1965, pp. 80–83.

[35] Baker, P. R. W.: *Talanta*, **8,** 57 (1961).

[36] Beet, A. E.: *Fuel*, **11,** 406 (1932).

[37] Kitto, W. H.: *Analyst*, **59,** 733 (1934).

[38] Shirley, R. L., Becker, W. W.: *Ind. Eng. Chem., Anal. Ed.*, **17,** 437 (1945).

[39] Self, P. A. W.: *Pharm. J.*, **88,** 384 (1911).

[40] Salm, E., Prager, S.: *Chem. Z.*, **42,** 104 (1918).

[41] Winkler, L. W.: *Z. Angew. Chem.*, **26,** 231 (1913).

[42] Scales, F. M., Harrison, A. P.: *J. Ind. Eng. Chem.*, **12,** 350 (1920).

[43] Reith, J. F., Klazinga, W. M.: *Chem. Weekblad*, **38,** 122 (1941)

[44] Conway, E. J., Byrne, J.: *Biochem. J.*, **27,** 419 (1933)

[45] Roth, H.: *Mikrochim. Acta*, 663 (1960).

[46] Yuen, S. H., Pollard, A. G.: *J. Sci. Food Agr.*, **3,** 441 (1952).

[47] Glebko, L. I., Ulkina, J. L., Vaskovsky, V. E.: *Anal. Biochem.*, **20,** 16 (1967).

[48] Berthelot, M. P. E.: *Rept. Chim. Appl.*, **1,** 282 (1859).

[49] Riley, J. P.: *Anal. Chim. Acta*, **9,** 575 (1953).

[50] Exley, D.: *Biochem. J.*, **63,** 496 (1956).

[51] Mann, L. T.: *Anal. Chem.*, **35,** 2179 (1963).

[52] Ogg, C. L.: *J. Assoc. Offic. Agr. Chem.*, **43,** 689 (1960).

[53] Holz, F., Kremers, H.: *Stand und Leistung agrikultur-chemischer und agrarbiologischer Forschung, Vol. XX*, 26/1. Sonderheft zur *Z. Landw. Forschung*, 1971, pp. 177–191.

[54] Hambraeus, L., Forsum, E., Abrahamsson, L. and Lönnerdal, B.: *Anal. Biochem.*, **72,** 78 (1976).

[55] Moore, S., Stein, W. H.: *J. Biol. Chem.*, **176,** 337 (1948).

[56] Whitaker, R., Granum, P. E., Aasen, G.: *Anal. Biochem.*, **108,** 72 (1980).

[57] Pregl, F., Roth, H.: *Quantitative organische Mikroanalyse*, 6th Ed., Springer-Verlag, Vienna, 1949, pp. 85–105.

[58] Awapara, J.: *Arch. Biochem.*, **19,** 172 (1948).

[59] Boulanger, P., Biserte, G.: *Bull. Soc. Chim. Biol.*, **31,** 696 (1949).

[60] Weber, R.: *Helv. Chim. Acta*, **34,** 2031 (1951).

[61] Van Slyke, D. D., Hawkins, J. A.: *Biol. Chem.*, **79,** 739 (1928).

[62] Mothes, K.: *Planta*, **1,** 472 (1926).

[63] Barnstein, F.: *Landw. Versuchsstat.*, **54,** 327 (1900).

[64] Hinsberg, K., Lang, K.: *Medizinische Chemie*, Urban-Schwarzenberg, München, Berlin, 1957, pp. 978–982.

[65] Bell, P. M.: *Anal. Biochem.*, **5,** 443 (1963).

[66] Warburg, O., Christian, W.: *Biochem. Z.*, **310,** 384 (1941).

[67] Kalckar, H. M.: *J. Biol. Chem.*, **167,** 461 (1947).

[68] Bennett, P. A., Sullivan, J. V., Walsh, A.: *Anal. Biochem.*, **36,** 123 (1970).

[69] Groves, W. E., Francis, C. D., Sells, B. H.: *Anal. Biochem.*, **22,** 195 (1968).

[70] Ehresmann, B., Imbault, P., Weil, J. H.: *Anal. Biochem.*, **54,** 454 (1973).

[71] Kalb, Jr. V. F. and Bernlohr, R. W.: *Anal. Biochem.*, **82,** 362 (1977).

[72] Kirschenbaum, D. M.: *Anal. Biochem.*, **90,** 309 (1978).

[73] Gornall, A. J., Bardawill, C. S., David, M. M.: *J. Biol. Chem.*, **177,** 751 (1949).

[74] Dorsey, T., McDonald, P. W. Roels, O. A.: *Anal. Biochem.*, **78,** 156 (1977).

[75] Klungsöyr, L.: *Anal. Biochem.*, **27,** 91 (1969).

[76] Goldberg, M. L.: *Anal. Biochem.*, **51,** 240 (1973).

[77] Koch, A. L., Putnam, S. L.: *Anal. Biochem.*, **44,** 239 (1971).

[78] Itzhaki, R. F., Gill, D. M.: *Anal. Biochem.*, **9,** 401 (1964).

[79] Ellman, G. L.: *Anal. Biochem.*, **3,** 40 (1962).
[80] Futterman, S., Rollins, M. H.: *Anal. Biochem.*, **51,** 443 (1973).
[81] Watters, C.: *Anal. Biochem.*, **88,** 695 (1978).
[82] Robson, R. M., Goll, D. E., Temple, M. J.: *Anal. Biochem.*, **24,** 339 (1968).
[83] Goshev, I., Nedkov, P.: *Anal. Biochem.*, **95,** 340 (1979).
[84] Jennings, A. C.: *Cereal Chem.*, **38,** 467 (1961).
[85] Pinckney, A. J.: *Cereal Chem.*, **38,** 501 (1961).
[86] Johnson, R. M., Craney, C. E.: *Cereal Chem.*, **48,** 276 (1971).
[87] Lowry, O. H., Rosenbrough, N. J., Farr, A. L., Randall, R. J.: *J. Biol. Chem.*, **193,** 265 (1951).
[88] Ohnishi, T. S., Barr, J. K.: *Anal. Biochem.*, **86,** 193 (1978).
[89] Ji, T. H.: *Anal. Biochem.*, **52,** 517 (1973).
[90] Peterson, G. L.: *Anal. Biochem.*, **83,** 346 (1977).
[91] Zishka, M. K., Nishimura, J. S.: *Anal. Biochem.*, **34,** 291 (1970).
[92] Bonitati, J., Elliott, W. B., Miles, P. G.: *Anal. Biochem.*, **31,** 399 (1969).
[93] Chai-Ho Lo, Stelson, H.: *Anal. Biochem.*, **45,** 331 (1972).
[94] Lees, M. B., Paxman, S.: *Anal. Biochem.*, **47,** 184 (1972).
[95] Huemer, R. P., Kyung-Dong Lee: *Anal. Biochem.*, **37,** 149 (1970).
[96] Gaunce, A. P., D'Iorio, A.: *Anal. Biochem.*, **37,** 204 (1970).
[97] Osta, G. M., Mathewson, N. S., Catravas, G. N.: *Anal. Biochem.*, **89,** 31 (1978).
[98] Norris, K. H., Hart, J. R.: *Proc.* 1963 *Intern. Symp. Humid., Moist.*, Vol. 4, Reinhold, New York, 1965, pp. 19–25.
[99] Williams, P. C.: *Cereal Chem.*, **52,** 561 (1975).
[100] Führ, J., Hinz, O. S.: *Klin. Wochenschr.*, **31,** 153 (1953).
[101] Racusen, D.: *Anal. Biochem.*, **52,** 96 (1973).
[102] Kihara, H. K., Kuno, H.: *Anal. Biochem.*, **24,** 96 (1968).
[103] Bramhall, S., Noack, N., Wu, M., Loewenberg, J. R.: *Anal. Biochem.*, **31,** 146 (1969).
[104] Bradford, M. M.: *Anal. Biochem.*, **72,** 248 (1976).
[105] Esen, A.: *Anal. Biochem.*, **89,** 264 (1978).
[106] Udy, D. C.: *Cereal Chem.*, **31,** 389 (1954).
[107] Udy, D. C.: *Cereal Chem.*, **33,** 190 (1956).
[108] Mossberg, R.: *Evaluation of Protein Quality and Quantity by Dye-binding Capacity: a Tool in Plant Breeding; New Approaches to Breeding for Improved Plant Protein.* International Atomic Energy Agency, Vienna, 1969, pp. 151–160.
[109] Layne, E.: *Methods in Enzymology*, Vol. III, Academic Press, New York, 1957, p. 447.

[110] Rona, P., Kleinmann, H.: *Biochem. Z.*, **140,** 478 (1923); **150,** 444 (1924).
[111] Craine, E. M., Jones, C. A., Boundy, J. A.: *Cereal Chem.*, **34,** 456 (1957).
[112] Feinstein, L., Hart, J. R.: *Cereal Chem.*, **36,** 191 (1959).
[113] Tappan, D. V.: *Anal. Biochem.*, **14,** 171 (1966).
[114] Sheth, K., Cohen, L. H.: *Anal. Biochem.*, **37,** 142 (1970).
[115] Hiraoka, T., Glick, D.: *Anal. Biochem.*, **5,** 497 (1963).
[116] Butcher, E. C., Lowry, O.: *Anal. Biochem.*, **76,** 502 (1976).
[117] Udenfriend, S., Stein, S., Böhlen, P., Dairman, W.: *Science*, **178,** 871 (1972).
[118] Stowell, C. P., Kuhlenschmidt, T. B., Hoppe, C. A.: *Anal. Biochem.*, **85,** 572 (1978).
[119] Reiss, E.: *Ergebn. Inn. Med. Kinderheilkd., Berlin*, **10,** 531 (1913).
[120] Müller, O. H.: *Methods of Biochemical Analysis*, **XI,** 329 (1963).
[121] Palaček, E., Pechan, Z.: *Anal. Biochem.*, **42,** 59 (1971).

CHAPTER 3

Electrophoretic methods

Ö. Takács, I. Kerese

3.1 THE BASIS AND TECHNIQUE OF ELECTROPHORESIS

In electrophoresis the charged components to be separated migrate in an electric field at different rates which are governed primarily by the charges and by the strength of the electric field. The charge on a species may be intrinsic or produced by an ion bound to the surface of the species.

The motion of particles in a medium is impeded by the frictional resistance, which increases in proportion to the diameter of the particles and to the viscosity of the solution. Besides friction, other factors may inhibit the motion, such as ions of opposite charge to that of the particles. The resulting ionic sheath, or 'ion cloud' is itself associated with a solvation sheath, which likewise impedes the motion. This braking force is the larger, the higher the ionic strength of the buffer, and thus the rate of migration decreases with increase in the ionic strength in the electric field.

Electrophoresis may be used for analytical and preparative separation of mixtures of molecular or colloidal particles. The direction and distance of migration may be important data, and by the new methods developed, especially in recent years, it may be possible to determine the molecular weights and isoelectric points of the components of a mixture.

The technical variants of electrophoresis can be classified in various ways, but for protein work the most useful is according to the field of application: analysis or preparation. There are three main types of each.

Analytical electrophoresis

(a) Free, frontal electrophoresis in a liquid, with a density gradient.

(b) Electrophoresis on porous-structured supports such as paper, cellulose acetate foil, powders.

(c) Electrophoresis in gels, e.g. starch, agar or polyacrylamide gels.

Preparative electrophoresis

(a) Continuous free electrophoresis without a support.

(b) Continuous or zonal electrophoresis on paper or on a gel support.

(c) Continuous column electrophoresis on a convection-inhibiting, non-adsorbing, granulated or powder support.

In terms of the technique itself, however, the medium used for the separation is a more convenient basis, and in this respect the resolving power of an individual system is an important criterion. Depending on the medium and method used, for example, serum proteins may be separated into different numbers of fractions, as follows:

Type of electrophoresis	Number of fractions
Paper	5–7
Agar gel	8–10
Cellulose acetate foil	9–12
Starch gel	18–20
Polyacrylamide gel	20–50

3.2 FREE ELECTROPHORESIS

In free electrophoresis the separation is performed in solution, without a support. The Tiselius apparatus [1] is a U-tube containing a homogeneous solution of the sample (Fig. 3.1). If an alkaline buffer is used, the protein

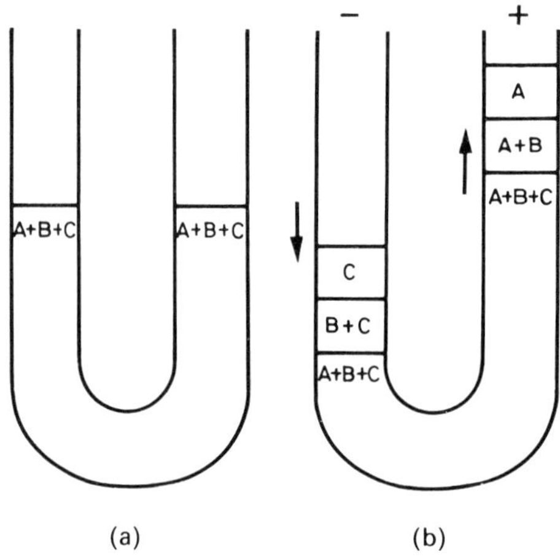

Fig. 3.1. Outline of the Tiselius free electrophoresis process. (a) Before separation. (b) During separation

molecules migrate towards the anode in the order of their isoelectric points. Detection is based on change in refractive index.

The Tiselius apparatus is now outmoded by instrumental development, but is still useful for determination of the mobilities of ions in free solution.

A significant advance in support-free electrophoresis was the introduction of Hannig's continuous, support-free apparatus [2,3]. Free-flow electrophoresis without a support has the obvious advantage that there is no support to cause loss of certain macromolecules by irreversible adsorption; even more importantly, however, the constituents of fibrous proteins (e.g. collagen molecules in solution), viruses, plant and animal cell suspensions, and haemocytes can be separated into various fractions.

The principle of the apparatus (Fig. 3.2) is similar to that of the continuous-flow paper electrophoresis apparatus. The separation compartment consists of two vertical cooled glass plates, 0.6 mm apart, the space between them being filled with buffer solution. A solution of the substances to be separated is added continuously to the compartment. The separated fractions are collected in 48 vessels situated at the bottom of the compartment. The nature of the fractions, which may be obtained in preparative amounts, is of particularly great importance in cell research. Hannig *et al.* carried out separations by free-flow electrophoresis (without support) on liver homogenates [4], blood [4], tumour cells [5] and plant (spinach) cell homogenates [6].

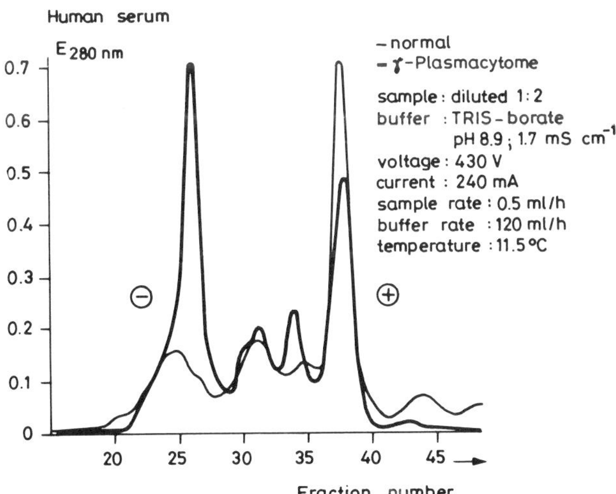

Fig. 3.2. Preparative electrophoresis of human serum proteins with a free-flow electrophoresis apparatus (Desaga FF 48), according to Hannig. (By kind permission of Prof. Hannig)

3.3 PAPER ELECTROPHORESIS

The spread in use of electrophoresis is largely due to the paper electrophoresis methods developed by Grassmann and Hannig [7]. Although developments in this field in the past 25 years have outstripped this simple method and provided indispensable tools for protein analysis, this method is still useful for medical diagnostics and rapid separation of peptides and oligonucleotides.

Because of the pore-size of the support, the separation effect in paper electrophoresis is close to that in free liquid electrophoresis. The spaces between the paper fibres are of such a size that they exert a filtering effect on the particles migrating in the electric field.

During the electrophoresis, the filter paper adsorbs neither the globular particles to be separated nor the ions of the buffer used, and the fibres do not dissolve in the buffer.

The horizontal compartment necessary for paper electrophoresis is cheap and simple, and may be made in the laboratory. The electrodes are immersed in buffer solutions in two labyrinth-system compartments, and between them is stretched the support, a filter-paper strip (Whatman No. 1; Schleicher & Schüll No. 2043 a and b; Macherey & Nagel No. 819) saturated with buffer. The sample is applied in a narrow zone on the filter paper, and in the electric field the component proteins migrate towards the electrode of opposite charge in accordance with their charges. This feature determines the position at which the sample is placed on the strip.

The technique can be conveniently exemplified by the electrophoresis of serum proteins [7]. A 4 × 40 cm paper strip is saturated with buffer solution, and 3–6 μl of serum (0.2–0.4 mg of protein) are placed on it, 4 cm from the cathode end. It is best to use a Michaelis veronal buffer of pH 8.6 and ionic strength 0.1 [8]. (For other proteins, the composition, ionic strength and pH of the buffer are selected in accordance with the properties of the proteins.) The electrophoresis is performed for 3–14 h at a potential of 110 V with cooling. The proteins can be detected on the strip by thermal coagulation and staining with a suitable dye (Amido Black 10 B, Ponceau S, Lissamine Green) (Fig. 3.3).

Paper electrophoresis can yield reliable values for five basic types of serum protein, and indicate whether their amounts correspond to a normal or pathological state.

Grassmann and Hannig [9] further employed a much higher potential for the electrophoresis, which necessitated cooling the paper. With the paper vertical, two-dimensional separation became possible; with continuous-flow electrophoresis on a preparative scale, involving continuous addition of the substances to be separated and continuous flow of the buffer, 10 g of protein mixture could be separated into its basic components daily. In the electric field between electrodes attached to the vertical edges of a 42 × 40 cm filter paper, the individual protein fractions are selectively moved sideways in accordance with their charges, at the same time as they are carried downwards by the continuous flow of buffer under gravitation. The separated

Fig. 3.3. Electrophoresis of human serum proteins on paper. Support: MN 214 paper, 36 × 350 mm. Applied: 0.01 ml of serum. Buffer: veronal–sodium acetate [8], pH 8.6, $I = 0.1$. Electrophoresis: 200 V, 8 mA, for 70 min. Fixing: with 10% acetic acid for 15 min. Staining: with 0.3% Acid Fuchsin solution for 30 min. Destaining: with acetic acid–methanol–water (1 : 4 : 5) solution for 15 min, then three times with 10% acetic acid. (By kind permission of Cs. Ambrus, First Department of Medicine, University Medical School, Budapest.)

protein fractions are collected in vessels situated at intervals along the bottom of the paper (Fig. 3.4).

The possibilities of paper electrophoresis are limited by the physical properties of the fibrous support. A more modern form of electrophoresis on a cellulose-based support is cellulose thin-layer electrophoresis. Its resolution is better because the support material, cellulose powder, is more homogeneous than paper. However, neither two-dimensional analytical electrophoresis on filter paper or a cellulose thin layer, nor the combination of electrophoresis with chromatography in the second dimension, has proved able to compete with the newer methods.

3.4 CELLULOSE ACETATE FOIL ELECTROPHORESIS

The use of cellulose acetate foil for zone electrophoresis has two advantages over paper electrophoresis: (*i*) greater separating power; (*ii*) greater sensitivity of detection because the foil is transparent.

In principle, the methods used are the same as for paper electrophoresis. As little as 20–50 μg of protein may be applied to the cellulose acetate strip. The limits of detection are 5–1000 μg of separated protein. The electrophoresis takes 4–5 h at 200 V, and 1.5–2 h at 400 V.

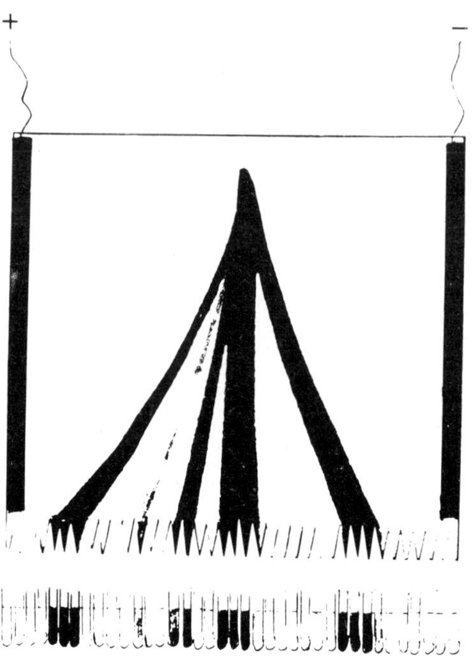

Fig. 3.4. Outline of Hannig preparative continuous-flow paper electrophoresis

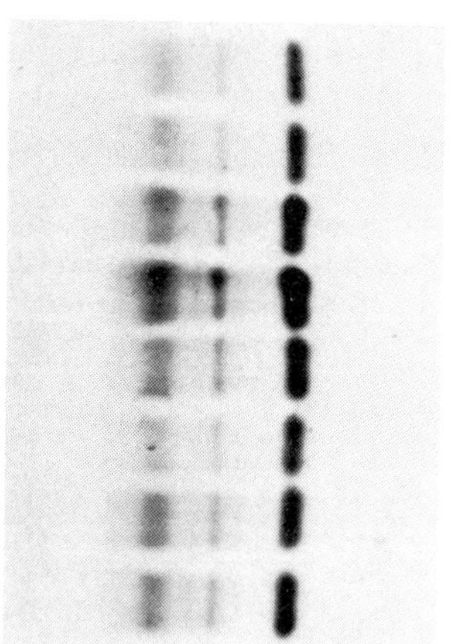

Fig. 3.5. Electrophoresis of human serum proteins on cellulose acetate foil. Support: Sartorius cellulose acetate foil. Applied: 1.5 μl of serum. Buffer: veronal–sodium acetate [8]. Electrophoresis: 250 V, 4 mA, for 20 min. Fixing: with 3% trichloroacetic acid. Staining: with 0.2% Ponceau S for 5 min. Destaining: with 5% acetic acid, then with acetone–tetrahydrofuran. (By kind permission of F. Péterfy, "HUMAN" National Institute for Serobacteriological Production and Research, Gödöllő, Hungary.)

With cellulose acetate foil, good separations can be achieved in the electrophoresis of serum (9–12 components) (Fig. 3.5), urine, cerebrospinal fluid, haemoglobin, lipoproteins, glycoproteins, esterases, casein, whey, various plant proteins, and the proteins of animal meat extracts.

3.5 STARCH GEL ELECTROPHORESIS

The intermicellar spaces of colloid-structured gels prepared from hydrolysed starch are small enough to inhibit the electrophoretic migration of large and asymmetric protein molecules to a greater extent than that of smaller, globular protein molecules. These opposing effects, the electrophoretic motion as a consequence of the charge, and the sieving effect of the pores of the starch gel, give rise to enhanced separation and resolution.

The high degree of resolution characteristic of starch gel electrophoresis (18–20 fractions may be obtained from a serum protein) depends on the extent of hydrolysis of the starch. The method was developed by Smithies [10], who has presented [11] a detailed account of the conditions for optimum hydrolysis of the starch, and for the electrophoresis.

In starch gel electrophoresis, uniformity of the hydrolysed starch is a prerequisite for reproducible separation. Suitable starch can be bought, or made in the laboratory. The latter possibility is an attractive feature, for a cheap basic material displaying constancy over a long period can be obtained by careful processing of a large enough quantity of potato starch. From starch hydrolysed with hydrochloric acid dissolved in acetone according to Smithies, it is convenient to prepare a 13–14% gel with the appropriate buffer solution.

The electrophoresis may be performed horizontally, and sedimentation of the sample during electrophoresis may then be prevented by using a thinner gel (1–2 mm thick). A thicker gel block (6 mm) may be used for vertical electrophoresis, and even preparative amounts may then be separated. For example, a vertical gel with a cross-section of 6.5×20 mm is suitable for the separation of 50 μl of serum. The sample can also be soaked up on filter paper, which is then placed in a groove in the gel; the quantity of sample to be used depends on the thickness of the gel. An example is shown in Fig. 3.6.

The field strength of 70 mA necessary for the electrophoresis requires a potential gradient of 6 V/cm, under which conditions an electrophoresis is completed within 6 h. The Amido Black 10 B method is generally applicable for staining the gel. Special dyes may be used for specific separated proteins.

Because of its advantages, starch gel electrophoresis has been further developed by several researchers; the work of Scopes [12] is worth particular attention, because of its wide scope. The vertical cooled starch block is prepared with TRIS-borate and TRIS-EDTA buffers of pH 8.4, containing 6% urea.

Starch gel electrophoresis is excellently suited for study of the polymorphism of proteins, and for genetic examination: haemoglobin, haptoglobin

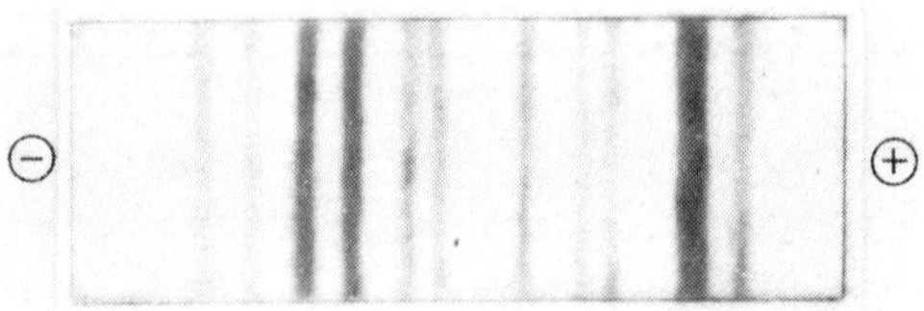

Fig. 3.6. Electrophoresis of human serum proteins in starch gel. Support: 35 × 100 × 6 mm 13% starch gel. Applied: 0.04 ml of serum. Buffer: TRIS–EDTA–borate, pH 8.6. Electrophoresis: three parallel starch gels, 320 V, 18 mA, for 17 h. Staining: with a 0.2% solution of Amido Black in acetic acid–methanol–water (1:4:5) for 30 min. Destaining: with acetic acid–methanol–water (1:4:5) for 24 h. (By kind permission of Mrs. M. Horváth, National Institute of Oncology, Budapest.)

and various isoenzymes have been separated [13, 14]. The methods of detection of enzymes are listed in the work of Scopes [12]. The basis of the detection is that NAD(P) reduced by the enzyme is capable of reducing Nitrotetrazolium Blue. In the *positive method*, the NAD(P) reduced by the enzyme or the auxiliary enzyme reduces the Nitrotetrazolium Blue in the reaction mixture to yield a purple colour. In the *negative method*, those sites where the enzymes or auxiliary enzymes oxidize NADH remain light, while the colour of the Tetrazolium Blue appears where there is no enzyme.

3.6 AGAR GEL ELECTROPHORESIS

A gel prepared from agar-agar (composed of galactan sulphate and agarose) may be utilized for the successful electrophoresis of virtually all proteins. The possibilities and technical details of the method are given in the monograph by Wieme [15].

Galactan sulphate is electrically charged, and by transferring charge to certain proteins it promotes their movement in the direction of the cathode. Accordingly, to obtain clear-cut effects it is best to prepare an 8.5% gel from the agar used for bacteriological purposes. The gel is cut into small cubes after cooling, and kept for 3 days in tap-water and then for 3 days in distilled water, the water being changed twice daily. The washed cubes are melted, then stored in a refrigerator until use. Gel prepared from such washed agar has a lower electroendosmosis than a gel prepared from the unwashed agar.

The gel layer is generally formed by pouring the melted gel onto a microscope slide, and 50 μl of sample may be added onto the layer by a sample applicator. At a potential gradient of 10 V/cm the electrophoresis

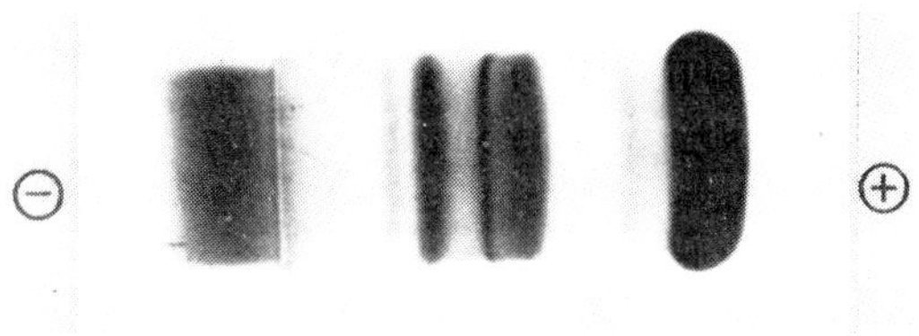

Fig. 3.7. Electrophoresis of human serum proteins in agarose gel. Support: 0.8% agarose gel, 26 × 76 × 1 mm. Applied: 2 μl of serum. Buffer: veronal–veronal sodium $I = 0.05$, pH = 8.6. Electrophoresis: 5 V/cm, 4 mA, for 50 min. Fixing: picric acid–acetic acid. Staining: 0.2% Amido Black 10B. Destaining: 7% acetic acid. (By kind permission of Mrs. I. Szamosvölgyi, "PHYLAXIA" Veterinary Biological and Feedstuffs Co., Budapest.)

takes 60 min. Before staining, the separated protein fractions must be fixed by soaking the slide for 4 h in a solution containing 5% trichloroacetic acid and 2.5% formaldehyde. An example is shown in Fig. 3.7. Wieme [15] provides complete information on preparation of the gel, and the techniques and applications. Agar gel is particularly important because it is the basic material for immunoelectrophoresis by various methods.

3.7 IMMUNOELECTROPHORESIS

The method developed by Grabar and Williams [16, 17] is a combination of gel electrophoresis and immunodiffusion. In the micromethod introduced by Scheidegger [18], the sample is subjected to electrophoresis in a purified agar gel on a microscope slide then a thin strip is cut out of the gel parallel to the direction of the separation, and antiserum is placed in the groove. The gel layer is next incubated in a moist chamber for 24–48 h, during which immunoprecipitation occurs between the antibody and antigen molecules. An example is shown in Fig. 3.8.

Numerous variants of the immunoelectrophoretic method are given in the literature, and are used primarily for the study of plasma proteins; because of the immuno-effects, these form a separate topic of discussion, and are not treated in detail here. Comprehensive accounts of electrophoretic methods

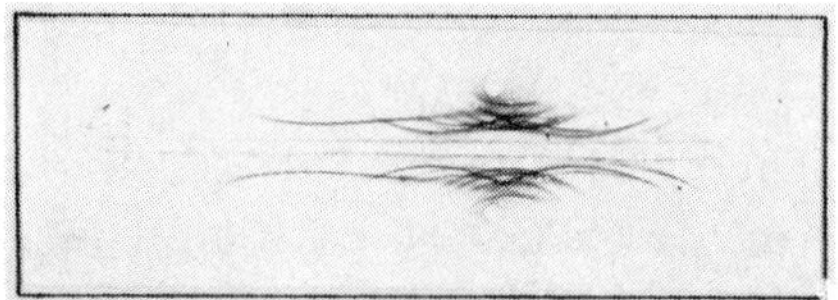

Fig. 3.8. Immunoelectrophoresis of human serum proteins. Support: 26 × 76 × 1.5 mm (3 ml) 1.5% agar gel. Applied: 2 μl of normal serum. Buffer: veronal–sodium acetate [8], pH 8.6, $I = 0.1$. Electrophoresis: 150 V, 50 mA, for 180 min. Incubation at room temperature for 24 h, then washing-out with 0.9% NaCl solution. Staining: with 0.3% Acid Fuchsin for 30 min. Destaining: with a solution of acetic acid–methanol–water (1 : 4 : 5). (By kind permission of Cs. Ambrus, First Department of Medicine, University Medical School, Budapest.)

based on immuno-reactions, and description of the most widely recommended procedures, may be found in the books by Backhaus [19] and by Allen, Hill and Stokes [20].

3.8 POLYACRYLAMIDE GEL ELECTROPHORESIS (PAGE)

The unique advantages of polyacrylamide gel electrophoresis (PAGE), namely the simple technique and the extremely high resolution, clearly give this method priority over all previous electrophoretic procedures. PAGE is generally suitable for the separation and analysis of all proteins over the range from polypeptides to ribosomes and viruses. If the buffer composition is appropriately chosen, proteins of neutral, basic or acidic character can be separated with equally good results. As a support, polyacrylamide has the outstanding property that it is completely inert, neither changing nor adsorbing the protein molecules. The possibility of concentration also gives rise to special fields of application, such as the direct electrophoresis of very dilute protein solutions.

Most striking is the high resolving power, exemplified by biochemical and clinical laboratory results that could be achieved only by this method, such as the exact 'mapping' of the basic units of proteins of high molecular weight (e.g. myosin, various enzymes, virus proteins) and the detection of polymeric proteins, which otherwise appear homogeneous, and their mono-, di-, tri- and tetrameric forms. With a differential method, Wright and Mallmann [21] were able to detect 44–50 serum proteins in blood plasma, while with a two-dimensional slab method they were able to find 90–95 serum protein components.

An important field of application of PAGE is investigation of the purity of protein preparations. If enzyme preparations are analysed by means of

PAGE, the effectiveness of the purification method may be monitored during purification, and the protein composition of the pure preparation can be analysed.

Electrophoretic separations are governed fundamentally by the physico-chemical characteristics of the proteins. Factors influencing the separation are the concentration and steric network of the gel, the composition of the gel and electrode buffers, and the magnitude of the electric field applied.

3.8.1 The basis of PAGE

Polyacrylamide gel is a polymerization product with a steric network structure, formed from acrylamide and *N,N'*-methylene-bis(acrylamide) (BIS). The polyacrylamide chains can connect through their free functional groups with the neighbouring chains, and thus a three-dimensional gel structure results. The steric network produced is irregular in form, corresponding to maximum entropy.

The network structure, which is determined by the concentrations of acrylamide and BIS and also by the degree of polymerization, gives a sieving effect which fundamentally influences the electrophoretic separation, as shown in Fig. 3.9.

Polyacrylamide gel has a number of advantages over other gel materials:

(a) chemical stability and inertness;

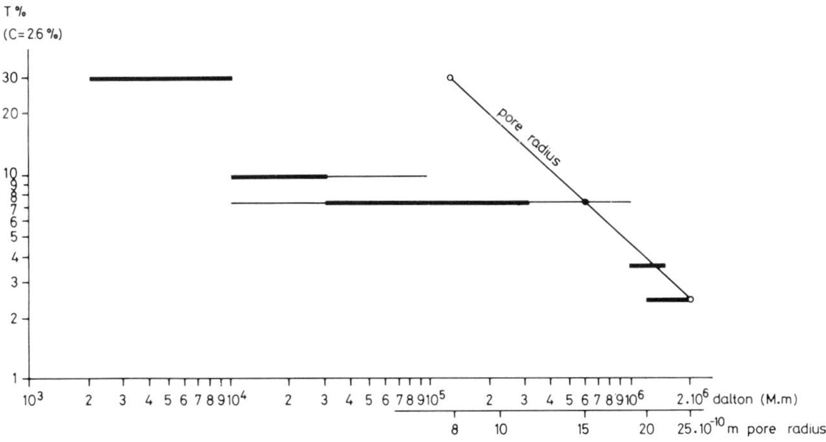

Fig. 3.9. Correlations between the acrylamide concentration of the gel and the pore radius, and the various concentrations and molecular weights of the proteins passing through the pores (indicated by horizontal lines; thick line: optimum range). (Compilation by I. Kerese Jr.)

(b) high degree of transparency;
(c) pore-size may be selected within wide limits;
(d) no adsorption or electro-osmosis;
(e) insolubility in most solvents.

The rod form of disc electrophoresis developed by Ornstein and by Davis [22] uses three different gel compositions. The proteins to be separated are contained in the sample gel, which is large-pored and has a low acrylamide concentration. Below this is the collecting or spacer gel, which is likewise large-pored and has a low acrylamide concentration, in which the material to be separated is concentrated in the first phase of the electrophoresis. Below this in turn is the smaller-pored, more concentrated separating gel, in which the material is separated into its individual fractions (Fig. 3.10).

In the electrophoresis, the ions of the buffer in the gel are the 'leading' ions, while the ions originating from the buffer in the electrode vessel are the 'following' ions. At the beginning of the separation, there are leading ions in

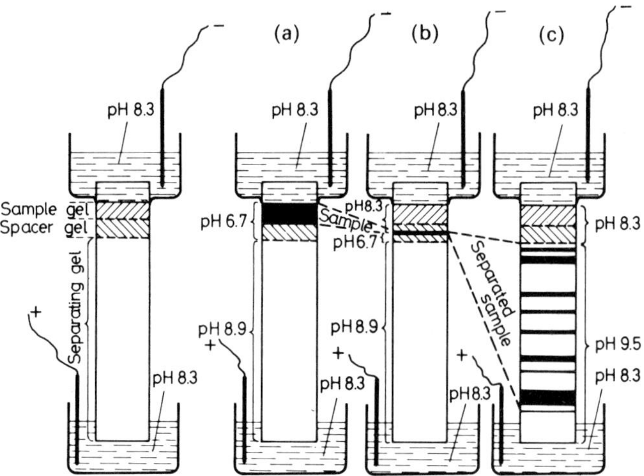

Fig. 3.10. Process of rod electrophoresis according to Ornstein. (a) Start. (b) Protein concentration. (c) Separation of protein

the sample and spacer gels; the following ions are only in the electrode buffer. When the current is switched on, the higher mobility of the leading ions causes them to move in advance of the proteins and the following ions in the spacer gel, leaving behind them a zone of lower conductivity and higher field strength. This accelerates the proteins and the following ions, so that they migrate at the same rate behind the leading ions and hence accumulate in the spacer gel. When the migrating protein zone reaches the dividing line of the spacer and separating gels, it comes into a buffer of such pH that the rate of the following ions almost attains that of the leading ions and exceeds that of any of the protein molecules; here, therefore, the following ions migrate behind the leading ions, but in front of the proteins.

3.8.2 The chemical materials for PAGE

Acrylamide and bis-acrylamide (BIS)

The matrix-forming substances of the polyacrylamide gels are acrylamide ($CH_2{=}CH{-}CO{-}NH_2$) and *N,N'*-methylene-bis(acrylamide) ($CH_2{=}CH{-}CO{-}NH{-}CH_2{-}NH{-}CO{-}CH{=}CH_2$). Numerous companies market preparations of very high purity, which may be used without purification. If necessary, acrylamide can be purified by precipitation with benzene from an acetone solution [23], and BIS by crystallization from acetone solution [24]. After the acrylamide and BIS solutions have been mixed, any free acrylic acid may be bound by adding Elgalite (1 g/100 ml of solution) or Amberlite MB 1 ion-exchange resin [25] and storing the mixture in a refrigerator.

Substances catalysing and regulating polymerization

Catalysts are used to initiate and accelerate polymerization of the acrylamide solution. It is essential that they should not change either the pH or the composition of the buffer solution. The quantity of catalyst is chosen so that gel formation is complete within 15–45 min.

N,N,N',N'-Tetramethylethylenediamine (TEMED) in very low concentration accelerates the polymerization. 3-Dimethylaminopropionitrile (DMAPN) displays a weaker effect than that of TEMED. Ammonium persulphate has the advantages that it can be prepared in high purity, is relatively stable at around 0 °C, and there is no tendency for it to form molecular oxygen. As a consequence of its strong oxidative effect, it has the drawback that it forms artefacts with certain of the proteins to be separated, and inactivates some enzymes. This can be avoided in part by removal of the catalyst from the gel by electrophoresis for one hour before application of the sample [25]; alternatively, an antioxidant [dithiothreitol (DTT) or 2-mercaptoethanol] can be added along with the persulphate before the polymerization [26].

It is also customary to use the various catalysts in combination: persulphate + DMAPN, persulphate + TEMED, riboflavin + TEMED.

In acidic medium the polymerization proceeds very slowly, e.g. in 8–10 h at pH 2.9, and Jordan and Raymond [27] therefore developed a special catalyst for the preparation of an acidic gel: 0.1% ascorbic acid, 0.0025% iron(II) sulphate and 0.03% hydrogen peroxide. The peroxide must be added to the acrylamide solution immediately before it is poured out, and the gel solution must then be cooled to 5–10 °C. The polymerization takes place in 3–5 min.

A substance that can regulate the polymerization is potassium ferrocyanide, which is employed on occasions to ensure uniform effect of the catalyst.

Surfactants

Surfactants such as urea and sodium dodecyl sulphate (SDS) further extend the possibilities of application of PAGE.

Urea considerably decreases the adsorptive interactions of the protein monomers bound to one another; in a urea solution of high concentration, therefore, these polymers dissociate into their sub-units. Urea treatment denatures protein molecules, however, destroying their original properties. Reductants must be added to the gel solution simultaneously with the urea, to prevent formation of disulphide bonds, i.e. the reaggregation of the sub-units, and hence the production of artefacts. 2-Mercaptoethanol or dithioerythritol (DTE) in a concentration of 1–5% is generally used as reducing agent.

The effect of SDS arises from its enhancement of the hydrophilic character of proteins. The size of a protein molecule does not change significantly when SDS is bound to it. During electrophoresis in the presence of SDS, the effect of the charge of the protein molecules on the migration rate is not manifested; only the molecular-sieve effect governed by the pore size of the gel is displayed.

Buffer solutions

A basic requirement of the buffer solutions used is that their composition and pH should not alter the chemical and biological properties of the substances to be separated, and the interaction between protein molecules should be minimal. During migration of the sample in the gel, the charges of the leading and following ions must have the same sign. It is not permissible for the mixture to be separated to change the pH of the gel appreciably in the course of the separation; it is therefore recommended that the pH of the sample solution should be adjusted to within ± 0.5 pH unit of the pH of the electrode buffer, before the sample is applied to the gel.

Richards *et al.* [28] give equations for calculation of the concentrations of the leading and following ions, and also the counter-ions, together with the pH of the gel and the electrode buffer.

The necessity for buffers of higher ionic strength, and the lack of any interaction between the protein molecules and the buffer, are two opposing requirements; attempts have been made to solve this problem by the addition of EDTA or other special reagents to the buffer [29].

3.8.3 Analytical application of PAGE

Apparatus

There are two methods for performing the electrophoresis: the rod and the slab procedures. The various types of apparatus can therefore be divided into two groups. The apparatus for the simple rod method consists of an upper and a lower buffer compartment, connected by the gel rods: gel-filled glass tubes dipping into the lower compartment are fitted into equidistant holes in the bottom of the upper buffer compartment. The platinum electrodes are introduced into the buffer compartments.

Many authors have striven to replace these cylindrical gel tubes with square or rectangular cross-section tubes made from glass or 'Plexiglas', because the gels can then be more accurately evaluated densitometrically. Such types of apparatus represent a transition towards the slab types, as gels prepared in this way can be regarded in essence as parts of a cut-up slab.

The slab apparatus types consist of two parts, a cell for the pouring of the gel slab, and the electrophoresis apparatus, which is similar to that for horizontal paper electrophoresis.

Preparation of the gel

The most important aspect in selection of the gel is the character of the protein or protein mixture to be examined. Attention must be paid to the size of the molecules and their charge, pH-sensitivity and tendency to undergo association. Thus, the buffer and gel used for the separation of proteins which are predominantly of low molecular weight are of completely different composition from those used for proteins of high molecular weight (see Table 3.3). Deciding features are the pore-size of the gel, and the pH and ionic strength of the buffer.

Stock solutions are generally used for preparation of the polyacrylamide gel and the electrode buffers. The compositions of the basic solutions necessary for the gels most frequently employed are listed in Tables 3.1 and 3.2.

With the exception of the persulphate solution, which is best prepared fresh at least weekly, the solutions may be stored for even 2–3 months at 4 °C,

Table 3.1
Base solutions for polyacrylamide gel
(Quantities given to be dissolved and made up to 100 ml)

No. of stock. soln.	Acrylamide, g	BIS, g	Urea, g	Persulphate, g	Riboflavin, g	DMAPN, ml	TRIS, g	Gly, g	$K_3Fe(CN)_6$, g	Sucrose g
1	30.0	0.8								
2	60.0	1.4								
3	60.0	1.2								
4	10.0	2.5								
5	28.0	0.735	48.0							
6	10.0	2.5	48.0							
7				0.14						
8				0.18						
9				0.28						
10				0.48						
11				1.00						
12				2.8						
13				10.0						
14					0.004					
15						1.6	0.3	1.44		
16									0.03	
17			48.0							
18										40.0

but the auxiliary sucrose solution readily undergoes mould formation, even if kept in the cold.

Blattler *et al.* [30] have reported their detailed studies on the gel composition. Hjertén [31] uses the following formulae to express the proportion of BIS in the gel:

$$T = \text{g of acrylamide} + \text{g of BIS in 100 ml of solution}$$

$$C = \frac{\text{g of BIS in 100 ml of solution} \times 100}{T}$$

As an example, gel 3 in Table 3.4 (7.5 g of acrylamide and 0.2 g of BIS per 100 ml) would be described as $T = 7.7\%$, $C = 2.6\%$.

From the limiting values for the optimum amount of BIS, Blattler *et al.* compiled the data presented in Table 3.3.

The compositions of the gels most frequently employed are given in Table 3.4.

Preparation of gel rods

Gel rods may vary in thickness from 0.1 to 10 mm, with a length of 5–20 cm. The most frequent dimensions are a diameter of 5–6 mm and a length of 8–12 cm. The appropriate quantities of the stock solutions are measured into a small suction flask and carefully mixed. Room temperature is generally suitable for this operation, but in certain cases temperatures of 4–10 °C are prescribed. Before addition of the catalysts, the monomer solution is freed from dissolved gases by application of a vacuum. In the preparation of the layered gel by the method of Ornstein [22], a solution of the separating gel is first poured into the tubes, up to the appropriate height, distilled water is layered to a height of 0.5–1 cm on the still liquid monomer solution, care being taken that no mixing occurs, and the monomer solution is left to polymerize in natural, fluorescent or ultraviolet light. In the formation of neutral or alkaline gels, the polymerization is generally complete within 10–20 min. In acidic solution the process is very slow with the customary catalysts; to shorten the polymerization time, it is necessary to use either a larger quantity of catalyst or the catalyst of Jordan and Raymond [27] (see p. 99). The completion of the polymerization is indicated by a sharp demarcation line between the water and the gel. The water is then sucked off, the residual traces being absorbed with a strip of filter paper. The spacer gel is then layered onto the separating gel in a similar way, and finally the sample gel is poured onto the spacer gel. In the last stage, however, care must be taken that the solution of sample gel added should contain an exactly known

Table 3.2
Buffer solutions for PAGE

No.	pH	TRIS, g	Gly, g	β-Ala, g	DL-Val, g	TEMED, ml	12 *M* HCl, ml	1 *M* HCl, ml	1 *N* H_3PO_4, ml	H_3BO_3, g
A. GEL BUFFERS Quantities given are to be diluted with distilled water to 100 ml.										
19	8.9	36.6				0.23		48		
20	6.9	5.7							25.6	
21	7.5	6.85				0.46		48		
22	4.3					4.0				
23	2.9	1.15								
24	5.5	4.95				0.46			30	
25	5.9									
26	8.9	12	3							
27	6.7	1.2	0.3						*	
28	6.7	0.75					0.4			
29	8.9	4.6					0.4			
30	2.9					0.4				
31	2.9	1.45								
32	9.2	2.23								0.37
33	8.9	36.6					4.8			
34	6.7	5.98					4.8			
35	6.7					0.46				
B. ELECTRODE BUFFERS Quantities given are to be diluted with distilled water to 1000 ml										
36	8.3	6	28.8							
37	8.3	215.5								110
38	7.0	1								
39	4.5			31.2						
40	4.0		28.1							
41a	4.0				35.13					
41b	4.0		22.53							
42	4.0		27.78							
43	9.0	22.31								0.73

* = adjust pH with necessary amount

CH_3COOH, ml	Diethyl-barbituric acid, g	Citric acid, g	1 *M* KOH, ml	Na_2 EDTA, g	Urea, g	Note
						for separating gel
						for spacer gel
						for separating gel
17.3			48			for separating gel
53.25			12			for separating gel
						for spacer gel
2.95			48			for spacer gel
						for separating gel
						for sample gel
						for sample gel
						for separating gel
*			4.2			for separating gel
		*				for separating gel
				0.19		for separating gel
					48	for separating gel
					48	for separating gel
2.87			48			for separating gel
						to be diluted 10x
				37		to be diluted 20x
	5.52					
8.0						to be diluted 10x
3.06						to be diluted 10x
*						anode buffer (upper)
*						cathode buffer (bottom)
		*				electrode buffer for No. 32 buffer
				1.92		to be diluted 2x

amount of protein. Depending on the diameter of the tube, this may be 50–200 μg. The experiments of Hjertén [32] indicate that the sample and spacer gels may be dispensed with if the protein solution to be separated is diluted with buffer, or if 0.01–0.02 ml of distilled water is layered onto the sample applied to the separating gel. Hence, since the potential gradient increases as

Table 3.3
Quantities of BIS to be used for the preparation of polyacrylamide gels of various concentrations

Acrylamide, g/100 ml	BIS, mg/100 ml	C%
3.75	80–330	2.1–8.8
5.0	65–290	1.3–5.8
7.5	46–210	0.6–2.8
15.0	30– 88	0.2–0.6

the sample proceeds downwards in the gel rod, the various protein zones are sharply differentiated. Michl and Pastuszyn [33] state that a 1–2% agarose solution is also suitable for the preparation of a spacer layer. It is likewise possible to employ only a separating layer if the sample, mixed in 1 : 1 ratio with 20% sucrose solution, is layered under the electrode buffer onto the surface of the gel by means of a micropipette or a microsyringe. It should be ensured that the sample solution does not mix with the electrode buffer. Narayan *et al.* [34] report that the omission of the sample and spacer gels presents no drawbacks if a sucrose solution is used; indeed, the preparation of the sample gel may even be a source of inaccuracies.

A concentration-gradient gel will give a better separation (on the basis of the sizes of the protein molecules) than will a separating gel with homogeneous concentration [35]. A continuous concentration-gradient can be prepared with a gradient mixer fitted with two reservoirs containing solutions with different acrylamide concentrations (see p. 120).

Preparation of gel slabs

The preparation of gels suitable for electrophoresis by the slab technique varies to a certain extent, depending on whether the electrophoresis will be performed vertically or horizontally.

During vertical electrophoresis, the gel slabs remain between the two glass plates used for pouring of the gels; otherwise, the same general rules hold

Table 3.4
Compositions of more frequently used polyacrylamide gels

Compositions of gels						Electrode buffers			
No. of gel	Acrylamide content of gel, %	pH of gel	Stock solutions	No of. stock soln.	Proportions by vol. for gel	Buffers	No. of buffer	pH of buffer	Notes
1	3.75	8.9	Acrylamide soln.	1	1	TRIS–Gly	36	8.3	For separation of proteins with M.W. = 10^6–2×10^6
			Gel buffer	19	1				
			Riboflavin soln.	14	1				
			H_2O		5				
2	7.5	8.9	Acrylamide soln.	1	2	TRIS–Gly	36	8.3	For separation of proteins with M.W. = 10^4–10^6
			Gel buffer	19	1				
			H_2O		1				
			$(NH_4)_2S_2O_8$ soln.	7	4				
3	15	8.9	Acrylamide soln.	2	2	TRIS–Gly	36	8.3	For separation of proteins with M.W. lower than 3×10^4
			Gel buffer	19	1				
			H_2O		1				
			$(NH_4)_2S_2O_8$ soln.	7	4				
4	20	8.9	Acrylamide soln.	2	4	TRIS–Gly	36	8.3	For separation of proteins with M.W. = 2×10^3–10^4
			Gel buffer	19	1				
			$(NH_4)_2S_2O_8$ soln.	8	3				
5	3.12	6.9	Acrylamide soln.	4	2				Spacer gel for separating gels Nos. 1, 2, 3 and 4
			Gel buffer	20	1				
			Riboflavin soln.	14	1				
			Sucrose soln.	18	4				
6	7.5	8.5	Acrylamide soln.	1	2	TRIS–Gly	36	8.3	Similar to separating gel No. 2
			DMAPN soln.	15	2				
			$K_3Fe(CN)_6$ soln.	16	2				
			$(NH_4)_2S_2O_8$ soln.	10	2				

Table 3.4 (cont'd)

	Compositions of gels					Electrode buffers			
No. of gel	Acrylamide content of gel, %	pH of gel	Stock solutions	No of. stock soln.	Proportions by vol. for gel	Buffers	No. of buffer	pH of buffer	Notes
7	7.5	7.5	Acrylamide soln.	1	2	TRIS–diethyl-barbituric acid	38	7.0	For separation of enzymes under pH 8.0
			Gel buffer	21	1				
			H_2O		1				
			$(NH_4)_2S_2O_8$ soln.	7	4				
8	3.12	5.5	Acrylamide soln.	4	2				Spacer gel for separating gel No. 7
			Gel buffer	24	1				
			Riboflavin soln.	14	1				
			Sucrose soln.	18	4				
9	7.5	4.3	Acrylamide soln.	1	2	β-Ala–AcOH	39	4.5	For separation of basic proteins with M.W. higher than 2×10^4
			Gel buffer	22	1				
			H_2O		1				
			$(NH_4)_2S_2O_8$ soln.	9	4				
10	15	4.3	Acrylamide soln.	2	2	β-Ala–AcOH	39	4.5	For separation of proteins with M.W. *ca.* 2×10^4 (e.g. histones)
			Gel buffer	22	1				
			H_2O		1				
			$(NH_4)_2S_2O_8$ soln.	9	4				
11	3.12	6.7	Acrylamide soln.	4	2				Spacer gel for separating gels Nos. 9 and 10
			Gel buffer	35	1				
			Riboflavin soln.	14	1				
			H_2O		4				
12	7.5	2.9	Acrylamide soln.	1	2	Gly–AcOH	40	4.0	For separation of basic proteins with M.W. higher than 2×10^4
			Gel buffer	23	4				
			$(NH_4)_2S_2O_8$ soln.	12	2				
13	3.12	5.9	Acrylamide soln.	4	2				Spacer gel for separating gel No. 12
			Gel buffer	25	1				
			Riboflavin soln.	14	1				
			Sucrose soln.	18	4				

14	7.5	8.9	Acrylamide soln.	1	80	TRIS–Gly	36	8.3	Similar to separating gel No. 2
			Gel buffer	26	175				
			$K_3Fe(CN)_6$	16	40				
			DMAPN (conc.)		1				
			$(NH_4)_2S_2O_8$ soln.	13	4				
15	5	8.9	Acrylamide soln.	5	48	TRIS–Gly	36	8.3	For separation of macromolecules
			Gel buffer	33	40				
			Urea soln.	17	232				
			DMAPN (conc.)		1.5				
			$(NH_4)_2S_2O_8$ soln.	13	1.5				
16	7.5	8.9	Acrylamide soln.	5	80	TRIS–Gly	36	8.3	For separation of proteins with M.W. $= 3 \times 10^4$–3×10^5
			Gel buffer	33	40				
			Urea soln.	17	200				
			DMAPN (conc.)		1				
			$(NH_4)_2S_2O_8$ soln.	13	1				
17	10	8.9	Acrylamide soln.	5	136	TRIS–Gly	36	8.3	For separation of proteins with M.W. $= 10^4$–10^5
			Gel buffer	33	40				
			Urea soln.	17	164				
			DMAPN (conc.)		1				
			$(NH_4)_2S_2O_8$ soln.	13	1				
18	6.5	6.7	Acrylamide soln.	6	160				Sample gel for separating gels Nos. 15, 16 and 17
			Urea soln.	17	106				
			Gel buffer	34	40				
			DMAPN (conc.)		1				
			$(NH_4)_2S_2O_8$ soln.	13	1				
19	18	2.9	Acrylamide soln.	3	5	Anode buffer	41a	4.0	For separation of basic proteins (e.g. histones)
			Gel buffer	30	5				
			H_2O		5	Cathode buffer	41b	4.0	
			$(NH_4)_2S_2O_8$ soln.	13	2				

for their preparation as in the case of gel rods. Grooves may be made by various means in the upper part of the gel slab, for application of the sample.

In a horizontal gel slab, the sample groove is either produced by the mould used in preparation of the slab, or cut out of the polymerized gel with a special tool. Only the separating gels are usually prepared in this way.

Depending on the thickness of the gel, 10–40 μg of protein in 5–20 μl of solution is a reasonable amount to apply to one sample site; it is mixed in 1 : 1 ratio with 20% sucrose or 8 *M* urea solution. In the event of a protein consisting of very many components, even 200 μg may be applied to one site.

The slab method has certain advantages over the rod gel technique:

(*i*) many samples may be run in parallel;

(*ii*) the Joule heat is more easily dissipated;

(*iii*) a two-dimensional variant is possible;

(*iv*) densitometric evaluation is more accurate, and photographic documentation is more successful;

(*v*) the slabs can be employed more easily for an autoradiographic method.

Preparation of the sample and its application to the gel

The pH and electrolyte concentration of the sample solution should if possible be the same as those of the buffer of the spacer gel. Before the sample is applied to the gel, the concentration of the protein solution to be examined must be determined. The sample solution should not mix with the electrode buffer, and the protein should enter the gel quantitatively. Depending on the presumed number of components in the sample, 50–200 μg of protein are applied to one site. A single protein (e.g. albumin) can easily be detected even in a quantity of only 0.1 μg. In the case of a sample containing 20–30 or even more components, the examination is begun with gel rods with comparatively large diameters, and 200–300 μg of protein are applied. Biological fluids (serum, cerebrospinal fluid, etc.) can be applied directly to the gel.

Serum is best prepared by the following method: 0.1 ml of electrode buffer and 0.2 ml of 40% sucrose solution are added to 0.1 ml of serum. A little 0.1% Bromophenol Blue solution (in a 9 : 1 mixture of methanol and acetic acid) is previously added to the buffer. As a marker, this indicates the advance of the front during the electrophoresis. (For an acidic gel the indicator dye may be basic fuchsin.) Then 10–20 μl of the serum mixture is layered onto each sample site. Solutions with high ionic strengths (proteins extracted with salt solution of high molarity) cannot be applied directly to the gel; before electrophoresis, it is essential to dialyse them against the diluted buffer.

Instead of dialysis, which requires a long time, gel-filtration on a 9.1 ml Sephadex G-25 gel column (Prepacked Disposable Column PD-10, Pharmacia) is suitable for the deionization of solutions of low volume.

The electrophoresis

After application of the sample, the electrophoresis may begin. In alkaline buffers the protein migrates towards the anode; with vertical gels, therefore, the cathode is located at the top. The d.c. supply source should be such that the potential and the current strength can be constantly controlled. When gel rods are to be used analytically, it is advisable to concentrate the proteins with a current strength of 1 mA per gel for 10–15 min, and then to increase the current strength to 2–5 mA per gel. The gel slabs used horizontally generally have larger surface areas than the vertical ones, and a higher potential is therefore necessary for the electrophoresis. The horizontal gel slab can be placed on a cooling surface, to conduct away the Joule heat produced during the electrophoresis. A different potential and current strength are necessary for every buffer; for the Ornstein [22] TRIS-glycine buffer, a potential of 250–300 V is appropriate. If the potential is less than the optimum, the separation of the zones will not be sharp, while if it is higher than the optimum, the Joule heat produced will impair the separation. The problem of cooling can also be solved by circulating buffer at 4 °C through the electrode compartments. It is recommended that an apparatus not fitted with a circulatory cooling system should be used in a cold room or a refrigerator. The electrophoresis must be stopped when the indicator dye has progressed through the gel and has approached to within 0.5–1 cm of its edge. After the voltage supply has been switched off, the gels must be transferred to the fixing or staining solution without delay, to avoid diffusion of the separated substances. The electrode buffers may be used several times, but only with the same electrodes.

The quality of a separation depends on the amount of protein applied, and also on the distance of electrophoretic migration. A good separation can be achieved within an upper limit of 30 μg for each component; the lower limit is governed by the intensity of the staining dye employed. A typical separation is shown in Fig. 3.11.

The gel rods may be removed rapidly from the tubes with the aid of an injection syringe needle [13]. The method of removing gel slabs depends on the construction of the apparatus employed.

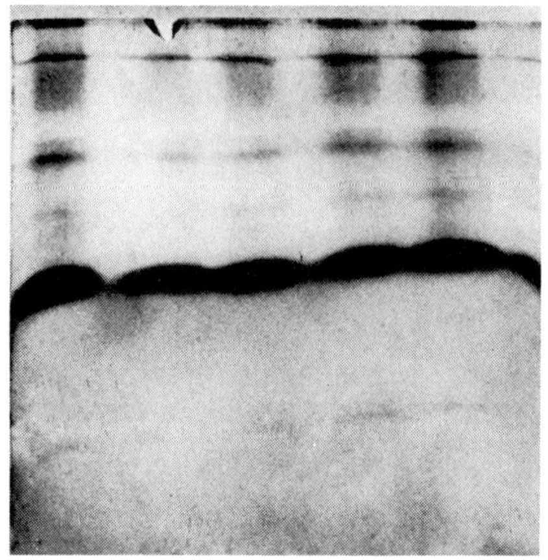

Fig. 3.11. Electrophoresis of human serum proteins in homogeneous polyacrylamide gel slab. Support: 78 × 78 × 2.7 mm 7.5% ($C = 2.6\%$) polyacrylamide gel slab (see Table 3.5, gel 3). Applied: 5, 10 or 25 μl of serum. Electrode buffer: TRIS–glycine, pH 8.3 (see Table 3.4, No. 36). Electrophoresis: 100 V, 80 mA, for 40 min; 200 V, 120 mA, for 180 min. Fixing: in 7% acetic acid for 60 min. Staining: in 0.1% Amido Black solution in acetic acid (7% v/v) solution. Destaining: with 7% acetic acid. (By kind permission of Mrs. I. Péter, Central Laboratory, Agricultural College, Kaposvár, Hungary.)

Fixing, staining and washing of the gel

Before staining, it is generally necessary to fix the separated proteins, not only to prevent diffusion of the proteins, but also to prepare it for the dye solution. The composition of the fixing solution may vary, depending on the staining method. After fixation, the protein zones are made visible by staining, and can then be evaluated from the intensity of staining.

The dyes most frequently used are Amido Black and Coomassie Blue. In a study of the binding of dyes to proteins, Fischbein [36] demonstrated that the most commonly used dye, Coomassie BB, is bound to different extents to the individual proteins. If Coomassie BB R-250 is used, the lower detection limit for the staining corresponds to less than 1 μg of protein. The quality of staining is affected by the distance covered by the protein during electrophoresis, and by other factors. The most frequently employed dyes are listed in Table 3.5.

The fixing, staining and destaining of the gels can be carried out simply in test-tubes in the case of gel rods, and in a glass or plastic bath in the case of slabs. Destaining of the dye may be accelerated by electrophoresis or by diffusion. One of the oldest methods for electric destaining of the dye is that of

Table 3.5
Staining methods for proteins in polyacrylamide gels

No.	Fixing solution	Staining solution	Destaining	Ref.
1	7% AcOH, 60 min	1% Amido Black 10 B in 7% AcOH at 95 °C. Allowed to cool in stain for 12 h.	7% AcOH; by diffusion or electrophoresis	[37]
2	None	0.01% Amido Black in 0.06 *M* AcOK (cathode buffer). Electrophoresis	None	[38]
3	MeOH:AcOH: H_2O, 5:5:1,	0.25% Coomassie BB R-250 in MeOH:AcOH:H_2O, 5:5:1	3.75% AcOH	[39]
4	12.5% TCA, 60 min	1% Coomassie BB R-250 diluted 20-fold with 12.5% TCA	10% TCA followed by 7% AcOH	[40]
5	454 ml 50% MeOH + 46 ml AcOH	0.25% Coomassie BB R-250 in fixing solution. 60 min. (For SDS gels)	75 ml AcOH + 50 ml MeOH + 875 ml H_2O	[41]
6	12.5% TCA, 10 min	0.25% Coomassie BB G-250 in 10% TCA	None	[42]
7	None	1% Coomassie BB R-250 in abs. EtOH diluted 20-fold with 12% TCA, gently shaken overnight	Dist. H_2O and MeOH alternately	[36]
8	None	Remazol BB prestaining in 1 *M* phosphate buffer and 10% SDS, 85 °C, 5 min	None	[43]
9	None	Remazol BB prestaining in NP-30 detergent, 48 °C, 3 h.	None	[44]
10	7% AcOH	1% Fast Green in 7% AcOH, 20 °C, 120 min	7% AcOH	[45]
11	20% sulphosalicylic acid	1% Procion BB RS in 10% AcOH–50% MeOH, 120 min	10% AcOH–50% MeOH, followed by 7% AcOH	[46]
12	None	0.04% Coomassie BB G-250 in 3.5% $HClO_4$, 120 min	None, or 7% AcOH	[47]
13	None	0.1% Coomassie BB R-250 in 25% TCA, 120 min	8% AcOH for 2–3 days	[48]
14	1st staining, 60 min	0.25% Coomassie BB R-250 in 25% propan-2-ol, 10% AcOH	10% AcOH	[49]
	2nd staining, 60 min	0.25% Coomassie BB R-250 in 10% propan-2-ol, 10% AcOH		
15	1st staining, 60 min	1% Amido Black in EtOH:H_2O:AcOH, 30:60:10	5% MeOH, 10% AcOH	[50]
	2nd staining, 60 min	0.25% Coomassie BB R-250 in MeOH:AcOH:H_2O, 50:50:10,		

Table 3.6
Specific staining

Protein	Dye	Ref.
Haemoglobin, myoglobin	Benzidine, H_2O_2	[22]
Haemoglobin, myoglobin	Benzidine, H_2O_2	[53]
Haptoglobin	Dimethoxybenzidine, H_2O_2	[54]
Coeruloplasmin	*o*-Dianisidine	[54]
Lipoprotein	Sudan Black	[55]
Lipoprotein	Sudan Black	[56]
Lipoprotein	I_2 in glacial acetic acid	[57]
Glycoproteins	Fuchsin–sulphite	[58]
Glycoproteins	Periodic acid	[59]
Glycoproteins	Alcian Blue	[60]
Glycoproteins	PAS reagent	[61]
Proteoglycans, glycosaminoglycans	Toluidine Blue	[62]
Phosphoproteins	Ammonium molybdate	[63]
Ribonucleic acid	Lanthanum acetate, Acridine Orange	[28]
Ribonucleic acid	Lanthanum acetate, Cyanine, chrome alum	[64]
Ribonucleic acid	Methylene Blue	[65]
Desoxyribonucleic acid (native)	Methyl Green, Crystal Violet	[66]
Desoxyribonucleic acid (denatured)	Pyronine-B	[66]

Schwabe [51], in whose apparatus the dye not bound to protein can be washed out from the gel rods within 6–8 min. Maurer's apparatus [52] is based on a similar principle.

Specific staining

Complex proteins may be detected with staining procedures in which the dyes bind to the non-protein component. A compilation of these procedures is given in Table 3.6.

3.8.4 Evaluation of electropherograms

Direct evaluation in ultraviolet light

If the electrophoresis was done in quartz tubes or cells, then it can be evaluated immediately after the completion of the separation, by examination with ultraviolet light, without staining. Radola and Delincée [67] claim that quantitative measurement is more reliable for slabs than rods. The individual proteins have different aromatic amino-acid contents, so each needs its own calibration curve.

Fluorometric evaluation

This method is based on staining of the protein with a fluorescent reagent before or after separation; its sensitivity is similar to that of direct ultraviolet evaluation. The fluorometric reagents that have proved of value are listed in Table 3.7.

Table 3.7
Fluorometric reagents

Reagent	Ref.
N-2-Propenyl-1-dimethylamino-5-sulphonamide	[68]
1-Anilino-8-naphthalenesulphonic acid, Mg salt	[69–71]

Evaluation of stained electropherograms

Quantitative evaluation of stained gels is very difficult, and in general only relative differences can be measured. There are two reasons for this: (*i*) the same amounts of different proteins give different absorbances because they bind to different extents with a given dye; (*ii*) because of the structural properties of the protein molecules, the different dyes are bound in different ways to a given protein molecule.

The stained fractions can be evaluated quantitatively only on the basis of a calibration curve prepared for the same protein. Amido Black gives unstable products and the intensities of the protein fractions must be measured within a short time. Coomassie BB is more sensitive, but the background staining is more intense.

The most common errors in the quantitative densitometry are:

(*i*) the separation of the peaks is not perfect;

(*ii*) the detector response is not a linear function of the light-absorption;

(*iii*) most integrators are inaccurate, particularly for small areas.

Only fast electronic integration meets requirements; mechanical integration is not satisfactory.

Instead of densitometry, elution and photometric measurement of the dye can be used. Fenner *et al.* [72] cut out the stained strips accurately, extracted them by shaking at room temperature overnight with a standard amount (4–20 ml) of 25% aqueous pyridine solution, and measured the absorbance of the extract at 650 nm. In the photometric method of Kahn and Rubin [48], 0.8 mm thick gels, the protein content of which is in the range 0.05–2.75 μg,

Table 3.8
Correlation of values obtained by densitometry and by elution [73]

Protein	Regression line	95% confidence band	Correlation coeff.
Actin	$y=0.05x+0.01$	0.0132	0.9958
Albumin	$y=0.08x+0.02$	0.0328	0.9845
CPK	$y=0.016x+0.14$	0.0384	0.9972
F_2-Histone	$y=0.06x+0.03$	0.0259	0.9380
Myosin	$y=0.05x+0.03$	0.0252	0.9246
Lysozyme	$y=0.04x+0.04$	0.0282	0.9712
Pyruvate kinase	$y=0.04x+0.05$	0.0456	0.9476
Troponin	$y=0.07x+0.05$	0.0427	0.8914

are measured at 590 nm: the cut-out pieces of gel are smeared onto the walls of the photometer cell, which is then filled with 8% acetic acid.

In our own work the integrated densitometric values and the results obtained by pyridine elution have been compared for 1–40 μg of purified preparations applied on a gel slab 3 mm thick and subjected to electrophoresis in the presence of SDS. The integration values varied from protein to protein, but were dependent on concentration. Correlation of the densitometric and elution results for individual proteins gave the regression equations in Table 3.8.

Rodbard and Chrambach [74] developed a procedure for testing the homogeneity and identity of the spots obtained. The method of Hsieh and Anderson [75] may be used for quantitative measurement. A known amount of lysozyme (as internal standard) is mixed with the protein to be determined. After the electrophoresis, staining is performed (by method 5, 10, 12 or 14 of Table 3.6), the gel is subjected to densitometry, and the areas given by the protein are determined by planimetry, triangulation or electronic integration. Within certain limits, the integrated areas are directly proportional to the amount of protein. Since the proteins are stained to different extents, only an 'empirical weight' of the unknown protein can be determined (using Coomassie staining). The Fast Green staining method of Gorovsky [45] can give the 'true' quantities of protein.

Fractionation of the gel

After separation, it is frequently necessary to identify or recover the fractions, to measure various activities, etc. This may entail cutting-up of the gel. Most equipment for this [76–82] operates on the principle of the egg-slicer or the guillotine. There are separate procedures for the fractionation of gel rods and slabs. A histological microtome can be adapted for high-accuracy gel fractionation [83]. With the sharp knife of the microtome, soft and frozen gels alike can readily be cut into 0.5–1 mm segments.

Preservation and storage of gels

It is customary to store electropherograms for a longish period. Gel rods may be preserved in test-tubes and slabs in polyethylene bags, with 7% acetic acid, without any change in the intensity of the staining over a very long time [84]. Larger or thicker gel slabs may be preserved by dehydration. For this, it is recommended that the slabs, cut into thin strips, be dried first in 90% ethanol, 5% 2-propanediol, then placed between two cellulose sheets and dehydrated further with silica gel.

Interpretation of one-dimensional electropherograms

It is well known that the number of fractions obtained by means of a single separation is not always a faithful reflection of the actual number of different protein components in the sample. For various reasons, a given protein or enzyme may even appear in the form of several bands. In some cases there is a real difference in the sizes of isomeric molecules, but the cause of the isomerism may be a difference in charge, or a lower degree of ligand binding, and the conformations of the fractions may also differ. It may happen that there is not a real heterogeneity of the molecule or fraction, the apparent difference resulting from the means of separation. There are a number of possible reasons for this:

(*i*) interaction between buffer and protein;

(*ii*) protein–protein interactions;

(*iii*) partial dissociation of the given protein polymer during the separation;

(*iv*) partial aggregation of the protein;

(*v*) linkage or interaction between the conformational isomers.

To give a correct picture, therefore, a one-dimensional separation is never enough; a number of different methods must be employed to ensure that the heterogeneity of the preparation is possible and real. Through assessment of the various factors (isomerism, conformational interactions, ligand binding, molecular weight, sub-unit composition), the combined methods give overall and supplementary information for the evaluation of electropherograms.

3.8.5 Preparative methods of recovering protein fractions

Separated fractions may be obtained either by dissolution from the cut-up gel, or by continuous elution during electrophoresis. Cutting up the gel is not a satisfactory method for recovery of larger amounts of protein, and continuous elution is better. Numerous instruments suitable for this are described in the literature and commercial prospectuses. The compositions of special buffers and gels necessary for preparative separation are described by Jovin *et al.* [85], Gordon and Louis [86], and Jacobson and Lodish [87]. Very many advantages are displayed by the elution method of Hjertén *et al.* [88]; further development of this has permitted the separation of 1 g of a protein sample containing several components, the resolving power approaching that of analytical methods.

Methods involving elution with a constant flow of buffer have in common a high degree of dilution of the low-mobility components, because of the slowness of the diffusion. This effect can be diminished by reducing the length of the separating gel. Another possibility is manually or automatically controlled discontinuous elution. Dilution of the eluate was successfully avoided by van Jaarsweld *et al.* [89] with a simple instrumental modification. The apparatus of Boyde and Remtulla [90] deserves special mention. This can be operated continuously for 10 days, and is capable of separating 1 g of protein within 2 h.

3.8.6 Special applications of PAGE

Electrophoresis in divided gel

Clarke [53] introduced a two-sample method of electrophoresis in gel rods for the identification of proteins. A strip cut from thin waterproof paper or plastic is pressed onto the surface of the gel, dividing it into two parts. The two protein samples to be compared are placed one on each half of the divided surface, and then subjected to electrophoresis side by side. In the 'split gel' method of Leboy *et al.* [91], the gel rod is divided into two parts by a small separating wall. It is also possible to apply the protein sample by first soaking the protein solution onto a little filter-paper and drying this (if the protein can withstand drying).

Gradient PAGE

With the aim of attaining increased resolution, Slater [92] and subsequently Margolis and Kenrick [93] prepared gels with increasing acrylamide concentrations. The concentration gradient may be stepwise or continuous.

Preparation of a stepwise gradient is the simpler, and does not require special apparatus. Allen's procedure [94] can be employed for both rods and slabs, on this basis. We have separated serum proteins on such a gel, the result being presented in Fig. 3.12.

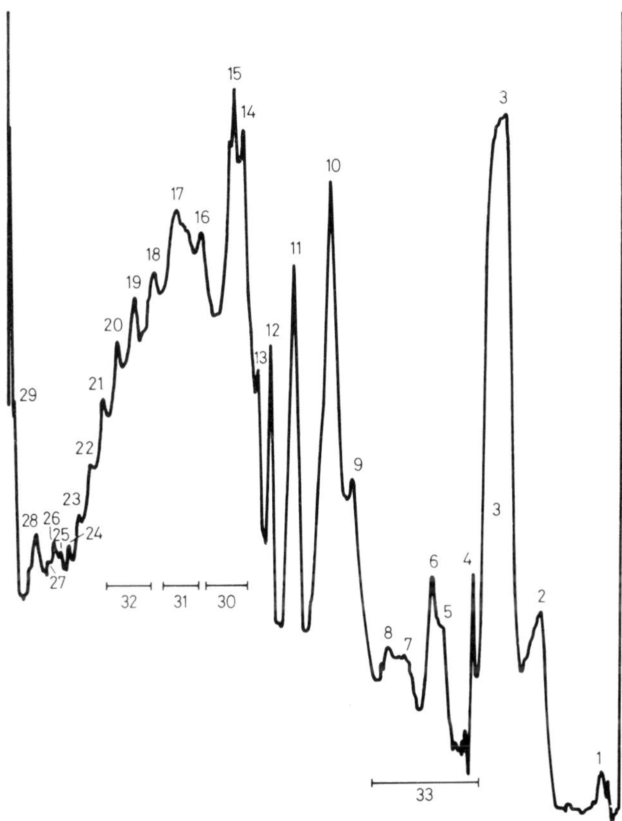

Fig. 3.12. Densitogram of fractions of human serum proteins separated in a stepped gradient polyacrylamide gel slab. Support: 260 × 150 × 3 mm 6–10% polyacrylamide gel ($C = 3.5\%$) slab. Applied: 15 μl of serum. Electrode buffer: 0.1875 *M* TRIS–citrate, pH = 9. Electrophoresis: 150 V, 10 mA/cm^2, for 4 h. Fixing: with a solution of methanol–water–acetic acid (5:5:1). Staining: 0.25% Coomassie BB R-250 in methanolic acetic acid solution. Destaining: with aqueous solution of 7.5% acetic acid + 5% methanol. Densitometry: with a Kipp and Zonen BD 1 integrator. Numbering of the proteins: 1, prealbumin; 2, acid α_1-glycoprotein; 3, albumin; 4, α_1-antitrypsin; 5–7, Gc-globulins; 8, coeruloplasmin; 9, haematopexin; 10, transferrin; 11–13, haptoglobin polymers of 2–1 genetic type; 14, β-glycoprotein; 15, α_2-macroglobulin; 16–18, haptoglobin polymers of 2–1 type; 19, β-lipoprotein; 20–29, unknowns; 30, γA-globulin; 31, γG-globulin; 32, γM-globulin; 33, α_1-lipoprotein. (By Ö. Takács, Department of Biochemistry, University Medical School, Szeged, Hungary.)

The preparation of a continuous gradient necessitates special apparatus. In the case of gel rods, the methods of Kopperschläger *et al.* [95, 96] may be used. A pore gradient may also be prepared, in capillary tubes, by the methods of Rüchel [97] or Dames and Maurer [98]. A typical continuous-gradient separation is shown in Fig. 3.13.

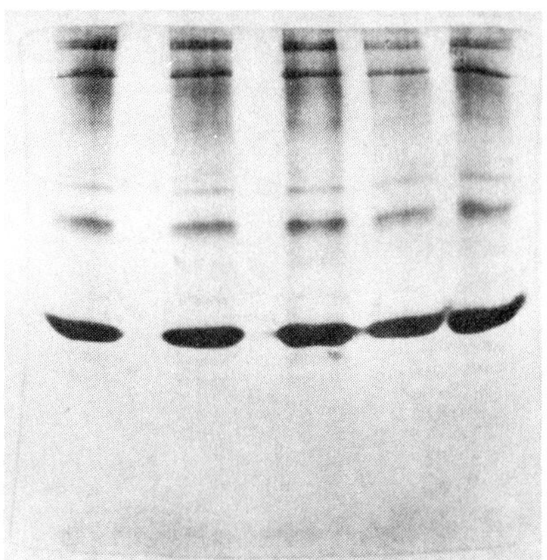

Fig. 3.13. Fractions of human serum proteins separated by homogeneous gradient PAGE. Support: Pharmacia Polyacrylamide Gradient Gel PAA 4/30. Applied: 5, 10 or 25 μl of serum. Electrode buffer: TRIS–glycine, pH 8.3 (see Table 3.4, No. 36). Electrophoresis: 500 V, 140 mA, for 17 h. Staining: with 0.2% Amido Black in 7% acetic acid solution, for 60 min. Destaining: with 7% acetic acid. (By kind permission of Mrs. I. Péter, Central Laboratory, Agricultural College, Kaposvár, Hungary.)

The gradient mixing apparatus of Foissy [99] is suitable for the slab technique. A pore-gradient gel was successfully employed for the isolation of proteins by Wright *et al.* [100], who also isolated a soluble tumour antigen from renal carcinoma cells [101].

Numerous compositions are in use for the preparation of an acrylamide concentration gradient gel. Those used by Margolis and Kenrick [35] are reported in Table 3.9.

The pore-gradient gel of Lorentz [102] contains SDS, which permits an accurate molecular weight determination.

Application of surfactants in PAGE for molecular weight determination

The applications of PAGE are considerably extended by the fact that the molecular weights of proteins treated with SDS can be determined rapidly, with a simple technique, and with an accuracy satisfying practical needs, in polyacrylamide gel prepared with an SDS-containing buffer. Treatment with

Table 3.9
Solutions necessary for preparation of 4–20% polyacrylamide gradient [35]

	Solution A $T=20\%$	Solution B $T=2\%$
Acrylamide	42.75 g	4.3 g
BIS	2.25 g	0.2 g
Sucrose	25.0 g	–
DMAPN	1.0 ml	1.0 ml
$K_3Fe(CN)_6$, 1% solution	4.5 ml	3.0 ml
Buffer solution, pH 9.2*	112.5 ml	112.5 ml
H_2O	to 213.7 ml	to 213.7 ml
$(NH_4)_2S_2O_8$, 10% solution	11.3 ml	11.3 ml
	225.0 ml	225.0 ml

* TRIS 22.31 g, Na_2EDTA 1.92 g, H_3BO_3 0.73 g, H_2O to 1000 ml. Electrode buffer: gel buffer: $H_2O = 1:1$.
Note: After dissolution of the first seven components (at $\leq 40\,°C$), the solution is cooled and subjected to vacuum, and then the $(NH_4)_2S_2O_8$ solution is added before the pouring.

SDS converts proteins into 'random' chains of identical charge, as a consequence of which the electrophoretic migration of the SDS–protein complex is proportional to the molecular weight. However, the migration distance is proportional to the molecular weight only if variation in the hydrodynamic forms and charge/mass ratios of the macromolecules is correlated. These conditions are met if the different proteins are bound to the same extent to SDS. In practice, a great many proteins satisfy this requirement, but it must always be ensured that the protein in question is not an exception [103].

In the standard methods, the proteins are incubated with SDS, mercaptoethanol being added to protect the SH group. Certain authors use dithiothreitol in place of mercaptoethanol [104].

The general conditions for the gel preparation are as follows:

minimum SDS concentration	0.4%
minimum ionic strength	0.005
time necessary for SDS–protein binding at 100 °C	1–5 min
T	5–15%
C	2–5%

The combination of SDS and PAGE for molecular weight determination was first used by Shapiro *et al.* [105] and by Weber and Osborn [41]. The most frequently used procedures are given in Table 3.10. For the determination of molecular weights with PAGE, standards of known molecular weight are necessary. These must be dissolved in a concentration of 1 mg/ml in the SDS-containing buffer.

About 5 μg each of the standard proteins and the material under examination are applied to the gel. For complex protein solutions, the optimum quantity of total protein and of the individual protein in question must be established empirically. After electrophoresis, the length of the gel and the migration distance of the marker dye are measured. After staining (Table 3.6), the migration distances of the standards and the unknown

Table 3.10
SDS–polyacrylamide gel compositions

No.	Gel pH	Composition of gel buffer		Dilution of gel buffer in gel	Composition of electrode buffer		Ref.
1	7.1	$NaH_2PO_4 \cdot H_2O$	7.8 g	1:1	gel buffer : water, 1 : 1		[105] [41]
		$Na_2HPO_4 \cdot 2H_2O$	38.6 g				
		SDS	2.0 g				
		H_2O	to 1000 ml				
2	8.4	TRIS	12.0 g	1:4	gel buffer : water, 1 : 4		[106]
		Glycine	58.0 g				
		SDS	1.0 g				
		H_2O	to 500 ml				
3	8.2	TRIS	24.22 g	1:2	gel buffer : water, 1 : 2		[107]
		SDS	2.0 g				
		AcOH	to pH 8.2				
		H_2O	to 100 ml				
4	8.8	TRIS	32.6 g	1:8	TRIS 3.0		[108]
		SDS	0.8 g		Glycine 14.4 g		
		6*N* HCl	to pH 8.8		SDS 1.0 g		
		H_2O	to 100 ml				
5	8.96	1*N* HCl	18.72 ml	1:4	Upper buffer:	Lower buffer:	[109]
		Ammendiol	4.63 g		Ammendiol 4.25 g	Ammendiol 6.57 g	
		H_2O	to 100 ml		Glycine 3.0 g	1 *M* HCl 50.0 ml	
		(Separating gel)			H_2O to 1000 ml	H_2O to 1000 ml	

proteins are measured, together with the length of the gel after staining. The relative mobilities, R_f, of the proteins can be established on the basis of the formula [41]:

$$R_f = \frac{\text{protein migration distance}}{\text{gel length after staining}} \times \frac{\text{gel length before staining}}{\text{marker migration distance}}$$

If the logarithms of the molecular weights of the standards are plotted as a function of the R_f values, a linear correlation is obtained. After the R_f values of the proteins of unknown molecular weight have been determined, their molecular weights can be read off the plot. An example is shown in Fig. 3.14.

Some workers have found a linear correlation between the relative migration distance and the logarithm of the molecular weight [106, 110]. According to Bryan [111], however, the correlation between the retardation coefficient, K_R, and the molecular weight is not linear in the case of polymeric

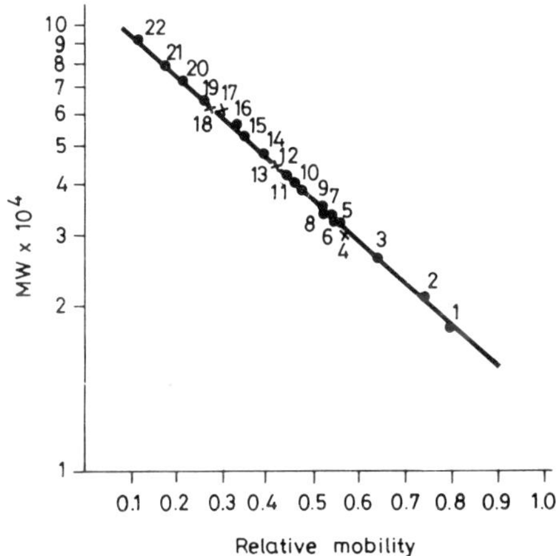

Fig. 3.14. Standards suitable for determination of the molecular weights of proteins separated by SDS–PAGE. Support: 260 × 150 × 3 mm polyacrylamide gel slab (T = 7.5%, C = 5%) containing 1% SDS. Applied: 10 μg per sample. Gel buffer: 0.2 *M* TRIS–acetate, pH = 8.2. Electrode buffer: 0.1 *M* TRIS–acetate, pH = 8.1. Electrophoresis: 125 V, 10 mA/cm^2, for 4 h. Fixing: with a solution of methanol–water–acetic acid (9 : 9 : 1). Densitometry: with a Kipp and Zonen BD 1 integrator; calculation according to Weber and Osborn [41] (see Table 3.6). 1, myoglobin; 2, myokinase; 3, triose phosphate isomerase; 4, experimental sample; 5, phosphoglycerate mutase; 6, α-glycerol phosphate dehydrogenase; 7, lactate dehydrogenase; 8, experimental sample; 9, glyceraldehyde-3-phosphate dehydrogenase; 10, aldolase; 11, creatine kinase; 12, enolase; 13, experimental sample; 14, phosphoglycerate kinase; 15, phosphoglucose isomerase; 16, pyruvate kinase; 17, 18, experimental samples; 19, adenosine deaminase; 20, phosphofructose kinase; 21, phosphorylase b kinase; 22, phosphorylase b [180]. (By Ö. Takács, Department of Biochemistry, University Medical School, Szeged, Hungary.)

proteins. The deviation can be eliminated by the use of the Ferguson plot [112] (log R_f vs T).

In the absence of a standard protein, an internal standard suitable for the determination of molecular weight can be prepared from the dansyl derivative of insulin or lysozyme, according to Inouye [113]; the derivative is mixed with the solution of proteins of unknown molecular weight. After electrophoresis, the dansylated proteins are detected in ultraviolet light and marked on the gel, and the whole gel is then stained conventionally. The unknown proteins can also be dansylated, and can then be detected and photographed in ultraviolet light after electrophoresis. Determination of the molecular weight on the dansylated sample becomes simpler with the use of the 'logarithmic sample groove preparer' of Catsimpoolas and Kenney [114].

Chiga *et al.* [115] compared the utility of the ultracentrifugation and SDS–PAGE techniques for the characterization of human serum proteins in accordance with their molecular weights. They concluded that PAGE gave the better resolution.

It was found by Williams and Gratzer [116] that SDS–PAGE is reliable for determination of molecular weights above 15 000, but below this value it is no longer accurate. Peptides with molecular weights less than 6000 migrate together in the front irrespective of their molecular weights.

In a study of the electrophoretic mobilities of membrane proteins in SDS–polyacrylamide gel, Fessenden-Raden [117] observed that the fatty acids in the preparation alter the migration and staining properties of the protein molecules; this circumstance must therefore be taken into consideration in the separation of samples containing fatty acids.

Lambin *et al.* [118] report that SDS–polyacrylamide gel can be employed in either stepwise or continuous gradient form.

SDS–PAGE is also very suitable for isolation purposes. The SDS may be rapidly removed from the isolated preparation by electrodialysis, by the method of Tuszynski and Warren [119].

Triton X-100 may be used in place of SDS for use with PAGE. A method of molecular weight determination has been described by Hearing *et al.* [120], which preserves the structure and function of the protein; this makes use of Rodbard and Chrambach's upper buffer A [121], to which 0.1% Triton X-100 is added. The sample is treated with 1% Triton X-100 solution for several hours, then an equal volume of 20% sucrose solution diluted twofold with the upper buffer containing 1% Triton X-100 is added; the current for the electrophoresis is 2 mA per gel rod. The method is of great advantage for the investigation of intact membrane proteins. There is a linear correlation between the effective molecular radius and the square root of K_R [122].

Nuclear processing of gels containing radioisotope-labelled proteins

Proteins and nucleic acids can in general be labelled with the isotopes ^{131}I, ^{32}P, ^{64}Cu and ^{14}C. The γ-activity can be measured with any γ-detector after fractionation of the gel. A gel containing soft β-emitting isotopes must be homogenized and then transferred into solution. Procedures for the dissolution of the gels and measurement of the radioactivity have been published by Alpers and Glickman [123], Leon and Bohrer [124], Paus [125] and Tykva and Votruba [126].

The positions on gel slabs of protein molecules labelled with the radioisotopes ^{131}I, ^{59}Fe and ^{32}P can be located by autoradiography. If this requires a long time, the gel and the the film must be protected against drying-out, by being wrapped in polyethylene foil. The sharpness of the spot can be enhanced by performing the autoradiography at 0–2 °C.

For the autoradiography of proteins labelled with soft β-emitters, it is advisable first to dry the gel. The technical details of the preparation and the autoradiography are given by Fairbanks *et al.* [127] and Uriel and Lavaille [128]. An effective and very economical method for measurement of ^{14}C and ^{32}P activities has been described by Leinen and Witliff [129]. After electrophoresis, the gel is cut up and dried on Whatman No. 1 paper strips, and the samples are then measured in a scintillation cocktail. The extraction method of Bouisson *et al.* [130] may be employed after two-dimensional separation.

Separation of enzymes and isoenzymes by PAGE

The separation of enzymes by PAGE does not generally differ from other PAGE methods. An essential aspect is to keep the electrode buffer and the gels at low temperature (2–4 °C) during the electrophoresis, so that the activities of the heat-sensitive enzymes are not decreased.

For the detection of enzymes and the measurement of their activities, the gel must be homogenized after it has been cut up, so that the proteins can be eluted with buffer from the individual fractions. The eluates may be concentrated by ultrafiltration and then lyophilized, and the enzymatic activities may be measured in the normal way.

The following possibilities are available for the detection of isoenzymes:

(*i*) the acrylamide gel slab is covered with filter paper or cellulose acetate membrane, and the reaction mixture is poured onto this;

(*ii*) the substrate is dissolved in agar or agarose gel, which is then poured onto filter paper; the agar gel slab obtained after congelation is placed on the polyacrylamide slab;

(*iii*) the substrate is dissolved in agarose gel and poured onto the acrylamide gel slab that has been subjected to electrophoresis, and the agarose gel congeals on it;

(*iv*) the gel is placed in the incubation solution.

Incubation is generally carried out at 37 °C, for a period determined by the rates of all the processes catalysed by the isoenzymes to be separated; that is, the incubation must be continued until all of the isoenzymes have reached maximum activity.

The correlation between the amount of the product and the activity of the enzyme is generally linear. In that case densitometry and peak-area evaluation can be used. However, non-linearity may be caused by diffusion of the reacting components, the temperature and duration of the incubation, and the pH of the buffer employed.

For specific detection of enzymes the corresponding histochemical methods are generally used. It is very important that the diffusion of the reagents should not be impeded, in order for the reaction products to accumulate in the outer gel layers. For the detection of enzymes or isoenzymes, therefore, a gel thicker than 3 mm should not be used if possible. Some of the more important specific methods of enzyme detection are collected in Table 3.11.

Table 3.11

Specific enzyme detection

Enzyme	Ref.
Cholinesterase	[131]
Lactate dehydrogenase	[132]
Peroxidase	[133]
Cytochrome oxidase	[134]
Aldolase	[135]
Alkaline phosphatase	[136]
Acid phosphatase	[137]
Arginase	[138]
Alcohol dehydrogenase	[139]
Amylase	[140]
Hexokinase, glucose phosphate isomerase, phosphoglucomutase	[141]
Proteinases	[142]
Glycosidases	[143]
6-Phosphofructokinase	[144]
Superoxide dismutase	[145]
Nuclease	[146]

Immunological applications of PAGE

Acrylamide gel is also suitable for immuno-precipitation investigations. Some of its advantages are that preparation of the gel is simple, there is no endo-osmotic effect, and the electrophoretic separation can be effected rapidly. It has the drawback that the immuno-reaction requires a comparatively long time.

Of the numerous immuno-electrophoretic methods, the disc-electrophoresis procedure of Louis-Ferdinand and Blatt [147] is particularly noteworthy; in this the antiserum is polymerized into the spacer gel, and the antigen into the sample gel. After electrophoresis, the tubes are immediately placed in ice-cold water, to inhibit diffusion of the protein; 10–15 min later, the gel is taken out of the tubes and left to stand overnight in an acetic acid solution of Amido Black. The result may be read off after destaining. From comparison of control and experimental samples by the method of Makonkawkeyoon and Haque [148], it may clearly be seen which are the immunologically related proteins; these may also be measured quantitatively too. Detection of the antigens is also possible after SDS–PAGE, by the method of Olden and Yamada [149].

Two-dimensional PAGE

By one-dimensional electrophoresis in a homogeneous gel, the proteins are separated fundamentally in accordance with their charges, whereas in gradient and SDS gels they are separated in accordance with their sizes and molecular weights. However, these methods are insufficient when the sample contains a number of different molecules of the same charge and size. The reason for this is that, in a one-dimensional electropherogram, the proteins with similar physico-chemical properties migrate together, and the bands overlap. In their examination of the methods suitable for the separation of complex protein mixtures, Wright *et al.* [150] obtained most data with two-dimensional PAGE, first in a homogeneous gel, and then a continuous 2–30% concave pore-gradient.

The resolution attainable in practice with one- and two-dimensional PAGE methods is governed by the peak capacity and by the sensitivity of detection. In the one-dimensional method, the various limiting factors strongly reduce the maximum resolving power theoretically possible. If a correctly selected two-dimensional method is employed, the ideal theoretical plate number of 2000 may be approached.

Consecutive use of the different electrophoretic methods ensures good possibilities for the isolation of the components of complex protein mixtures which are otherwise difficult to separate. Some examples of the combinations that may be employed are presented in Table 3.12.

Microscale PAGE

In principle, any 'macrosystem' PAGE apparatus may be adapted for a microprocedure. One of the oldest microcell PAGE instruments is that of Pun and Lombrozo [151]. The extremely simple method developed by Hyden *et al.* [152] permits the analysis of proteins even at a cell level. The gel tube is a capillary 55 mm long and 0.215 mm in diameter, with a volume of 2 μl, into which the gel solution is sucked. Making use of the concentrating effect of PAGE, Burr *et al.* [153] developed a micromethod which is also suitable for the electrophoresis of very dilute protein solutions. The sample is placed in the sample collector, constructed from the funnel-like end part of an

Table 3.12
Combined electrophoretic methods

First dimension	Second dimension
PAGE	PAGE, different pore size
	PAGE, different pH and pore size
	pore gradient electrophoresis
	transverse pore gradient electrophoresis
	isoelectric focusing in polyacrylamide gel
	immunoelectrophoresis
	isotachophoresis in polyacrylamide gel
pore gradient electrophoresis	PAGE
	pore gradient electrophoresis
	transverse pore gradient electrophoresis
	isoelectric focusing in polyacrylamide gel
	immunoelectrophoresis
	isotachophoresis in polyacrylamide gel
cellulose acetate electrophoresis	PAGE
	SDS–PAGE
	pore gradient electrophoresis
	immunoelectrophoresis
isoelectric focusing in polyacrylamide gel	PAGE
	SDS–PAGE
	pore gradient electrophoresis
	immunoelectrophoresis
isotachophoresis in polyacrylamide gel	PAGE
	SDS–PAGE
	pore gradient electrophoresis
	immunoelectrophoresis

Eppendorf pipette; its volume may be 0.5–0.6 ml. As a result of the concentrating effect, the protein in the dilute sample solution is concentrated into a very small area of the capillary.

Micro-rod electrophoresis was employed by Rüchel [97], Neuhoff [154] and then by Dames and Maurer [155], with and without a pore-gradient gel.

For purposes of slab-microelectrophoresis, Been and Rasch [156] modified the vertical apparatus of Reid and Bielski [157], in which the cell contains a thin, film-like gel slab. In the micro-gel apparatus of Maurer and Dati [158], the dimensions of the gel slabs are $75 \times 18 \times 0.75$ mm, the volume of sample that can be applied is 1–5 μl, and the lower limit to the amount of protein detectable is 0.1 μg. The slab-microelectrophoresis method may also be used for the preparation of a stepwise gradient, and it may be utilized for enzymatic microinvestigations, and immuno-precipitation and radioactive techniques.

3.8.7 The applications of PAGE in biochemical research

The PAGE methods have yielded significant results in enzymology, primarily in the field concerning the sub-unit structures of enzymes and the study of changes in them.

Detailed investigations have been made with PAGE of lactate dehydrogenase and malate dehydrogenase in blood [159], sperm [160] and tissues [161], of glyceraldehyde-3-phosphate dehydrogenase [162], of phosphorylase *a* and *b* [163], of DNA polymerase [164], of ribonuclease [165], of carbonic acid hydrase [166], of hyaluronidase [167], and also of various esterases, phosphatases, amylase, proteolytic enzymes and aldolase. In protein biochemistry, the results of studies of histones with PAGE are extremely important [168–175].

The peptide isolation technique of Wada and Shell [176] is of significance as a biochemical method, as is the fingerprint method of Bray and Brownlee [177] for the identification of peptides. In Houston's procedure [178], the protein zones cut out of stained gels can be utilized for amino-acid analysis. *N*-Terminal sequence analysis on nanomole quantities can be carried out by the method of Weiner *et al.* [179].

The results obtained with PAGE in muscle biochemistry have led to an understanding of the sub-unit structures of structural proteins. From our own work [180, 181], mention may be made of the qualitative and quantitative SDS–PAGE analysis of the muscle proteins, in which the compositions of the myofibrillar, sarcoplasmatic and mitochondrial fractions of rabbit and porcine muscles were clarified. An electropherogram of the mitochondrial proteins of the soleus, psoas and gastrocnemius muscles of rabbit is illustrated in Fig. 3.15.

3.8.8 Clinical chemical applications of PAGE

In a detailed treatment of the clinical chemical applications of PAGE, Maurer and Allen [182] described the standardization of sample collection and preparation. Hoffmeister [183] and Felgenhauer [184] have discussed the possibilities of utilizing PAGE for clinical diagnosis.

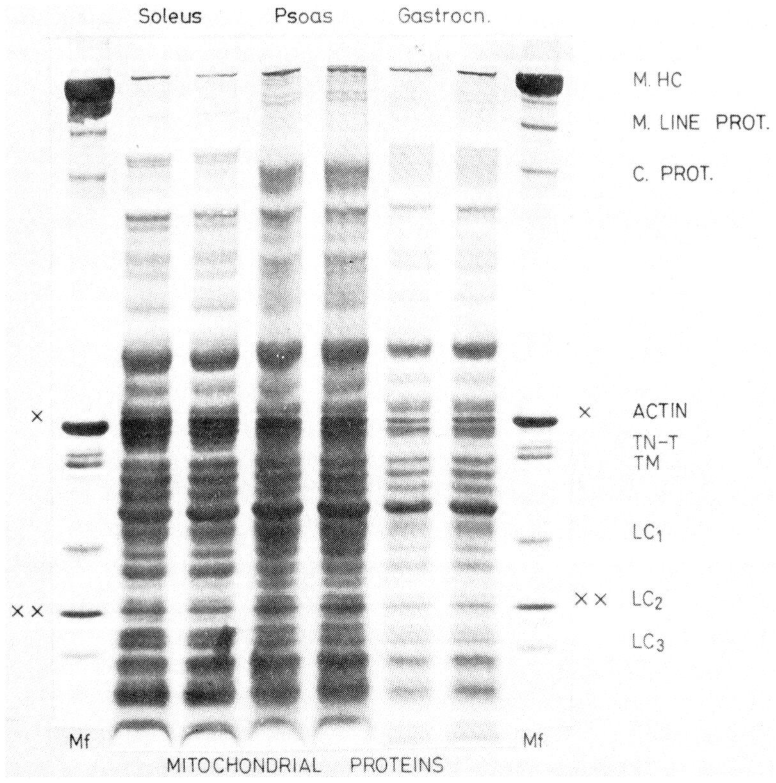

Fig. 3.15. Electropherograms of mitochondrial proteins in rabbit soleus, psoas and gastrocnemius muscles. Support: 260 × 150 × 3 mm stepped gradient polyacrylamide gel slab. Monomer concentration: from top to "x", $T = 6\%$, $C = 5\%$; from "x" to "xx", $T = 10\%$, $C = 5\%$; from "xx" to front, $T = 15\%$, $C = 5\%$. Applied: 200 μg. Gel buffer: 0.2 *M* TRIS–acetate, pH = 8.2. Electrode buffer: 0.1 *M* TRIS–acetate, pH = 8.1. Electrophoresis: 125 V, 10 mA/cm^2, for 5 h. Fixing: with a solution of methanol–water–acetic acid (9 : 9 : 1). Staining: 0.25% Coomassie BB R-250. Destaining: with 7.5% acetic acid solution containing 5% methanol. (By Ö. Takács, Department of Biochemistry, University Medical School, Szeged, Hungary.)

In clinical applications, the collection and storage of the samples to be examined do not differ from those customary in protein analysis. In selection of the type of apparatus, the following aspects must be borne in mind: the slab method is more advantageous for the simultaneous analysis of a large number of samples, and the gel-slabs can easily be combined with other separation techniques, e.g. various types of immunoelectrophoresis, and isoelectric focusing. In the qualitative and quantitative evaluation of serum proteins, knowledge of the haptoglobin type of the individual is essential, for without it only a few peaks can be identified. The electropherograms of the serum of individuals of the same age and sex, typified for haptoglobin, may be used as reference. In our work we have determined the quantitative distribution of the individual haptoglobin types, and the relative amounts of the individual components. The results are given in Table 3.13.

The haptoglobin pattern is imprinted on the protein spectrum, as it were, and conceals certain components if a general protein-staining method is employed. Accordingly, it is more correct if, after parallel samples have been

Table 3.13
Distribution of components of haptoglobin types

Type	Components						
	I	II	III	IV	V	VI	VII
Hp 2–2							
R_f	0.02	0.08	0.13	0.18	0.28	0.39	
%	24.2	7.35	11.91	25.98	32.39	8.54	
Hp 2–1							
R_f	0.05	0.09	0.13	0.19	0.29	0.45	0.73
%	1.12	2.58	6.67	13.03	25.16	33.85	22.83
Hp 1–1							
R_f	0.75						

run, some strips are stained for total protein, and the others specifically for haptoglobin.

On the basis of a large number of serum analyses, Pastewka *et al.* [185] gave the relative proportions of the individual fractions. For the qualitative analysis of serum proteins, the method of Allen [94] may be made use of with discontinuous gel-pore and buffer-ion concentrations but constant pH.

Isolated lipoproteins were first investigated by Narayan and Kummarow [186]. The PAGE of serum lipoproteins was dealt with by Muniz [187].

3.8.9 Applications of PAGE in food and feed protein research

The various forms of PAGE have proved extremely suitable for the study of food and feed proteins, both for solution of certain basic questions of research, and for quality control. In Table 3.14 we give a brief selection of the possible uses of PAGE in this field.

In our own work it has been found that for the analysis of the myofibrillar proteins of striated muscle, PAGE in combination with isoelectric focusing (see Table 3.15) provides more data than any of the previous methods (see Figs. 3.16 and 3.17).

3.9 ELECTROPHORESIS WITH POROUS SUPPORTS

Besides the fibrous supports and those with steric networks, porous sorbents find special uses in electrophoretic separation, owing to their physical characteristics and also to their technical applications. Sorbents suitable for this purpose may be either hydrophobic or hydrophilic, but a requirement for

Table 3.14
Food and feed proteins examined by PAGE

Plant proteins	Ref.
Lemon, fruit	[188]
Potato, tuber	[189]
Wheat, seed	[190]
Carrot, rhizome	[191]
Soya, seed	[192]
Bean, soluble proteins of seed	[193]
Rice, seed, γ_3 globulin	[194]
Aspergillus oryzae, enzymatically lysed	[195]
Vicia faba, globulins extracted from seed	[196]
Bean, globulins of seed	[197]
Potato, proteins of tuber	[198]
Soya, 7S and 11S globulins of seed	[199]
Proteins of various cereal seeds	[200]
Soya, seed proteins (in meat products)	[201]
Animal proteins	**Ref.**
Tryptic cystine peptides of ovalbumin	[202]
Extracted proteins of muscle tissues of various domestic animals	[203]
Proteins of meat of various fish	[204]
Proteins of hen-egg, chromatographic fractions	[205]
Emu egg yolk, separation of apovitellenin I	[206]
Porcine muscle structure proteins	[180]

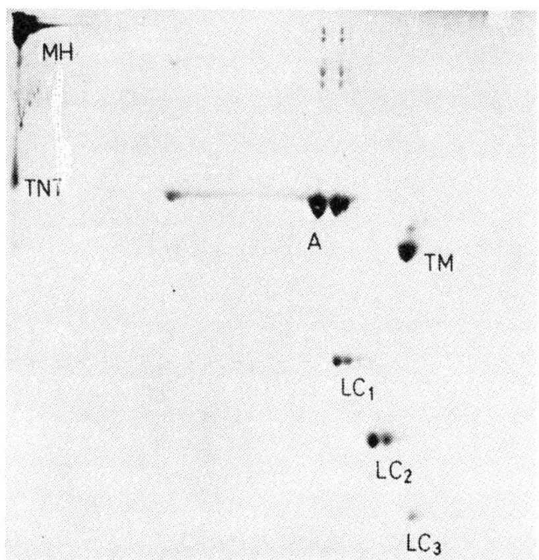

Fig. 3.16. Two-dimensional electropherogram of myofibrillar proteins of soleus muscle. First direction: IEF: 120 × 2.5 mm 1% SDS-containing polyacrylamide gel rod ($T = 5\%$, $C = 5\%$), with a urea concentration of 9.2 *M* and an Ampholin content of 2%. Applied: 200 μg. Cathode electrolyte: 0.2 *M* NaOH. Anode electrolyte: 0.2 *M* H_3PO_4. Electrofocusing: 400 V, 20–0 mA (for 6 rods), for 12 h. Second direction: in 260 × 150 × 3 mm polyacrylamide gel slab ($T = 5\%$, $C = 5\%$). Electrode buffer: 0.1 *M* TRIS–acetate, pH = 8.1. Electrophoresis: 125 V, 10 mA/cm^2, for 4 h. Fixing: with a solution of methanol–water–acetic acid (9:9:1). Staining: 0.25% Coomassie BB R-250. Destaining: with an aqueous solution of 7.5% acetic acid + 5% methanol. (By Ö. Takács, Department of Biochemistry, University Medical School, Szeged, Hungary.)

both types is that they should be able to stabilize a large volume of buffer by virtue of their granular structure or their hydration.

Of the hydrophobic substances suitable for the electrophoresis of proteins and peptides, kieselgel [207], kieselguhr [207], aluminium oxide [207], silica

Table 3.15
Electrophoretic separation methods combined with IEF

Methods employed	Materials examined	Ref.
IEF + PAGE	protein solutions of heterogeneous composition	[233]
IEF + SDS–PAGE	non-histone proteins	[175]
IEF + SDS–PAGE	soluble proteins	[234]
IEF + SDS–PAGE	peptides of membrane proteins	[235]
IEF + SDS–PAGE	normal and virus-infected cells	[236]
IEF + PAGE	plasma proteins	[237]
IEF + SDS–PAGE	peptides of myofibrillar muscle proteins, non-histone proteins of muscle cell nuclei	[180]

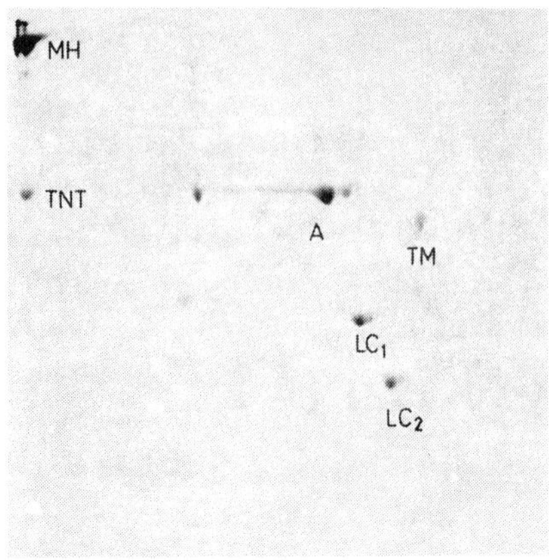

Fig. 3.17. Two-dimensional electropherogram of myofibrillar proteins of gastrocnemius muscle. Separation method as in Fig. 3.17. (By Ö. Takács, Department of Biochemistry, University Medical School, Szeged, Hungary.)

gel [208], controlled-pore glass [209] and Pevikon C-870 [210, 211] have proved to be good supports, as have cellulose [211, 212], Sephadex and polymerized acrylamide powder amongst the strongly hydrophilic substances. It is a further basic requirement for hydrophobic supports that they should have a very small but uniform grain-size, for these properties determine the water-binding capacities of the support layers prepared from them.

3.10 ISOELECTRIC FOCUSING (IEF)

In contrast with electrophoresis, for which a buffer of constant pH is used, in isoelectric focusing (IEF) a uniformly increasing pH gradient from the anode to the cathode is created by means of carrier ampholyte molecules with different charges. In the pH gradient the individual proteins migrate (under the action of the electric current) to the point where the pH of the medium corresponds to the isoelectric points of the individual protein molecules, and since at the isoelectric points the net charges of the proteins are equalized, and the proteins are outwardly neutral, they are focused there. Figure 3.18 shows an example.

The direct aim of IEF may be the isolation of certain component proteins, or the establishment of their isoelectric points; in addition, however, the IEF results may give a guide to the most expedient means of further preparative and analytical processing of the material under examination (see. p. 24).

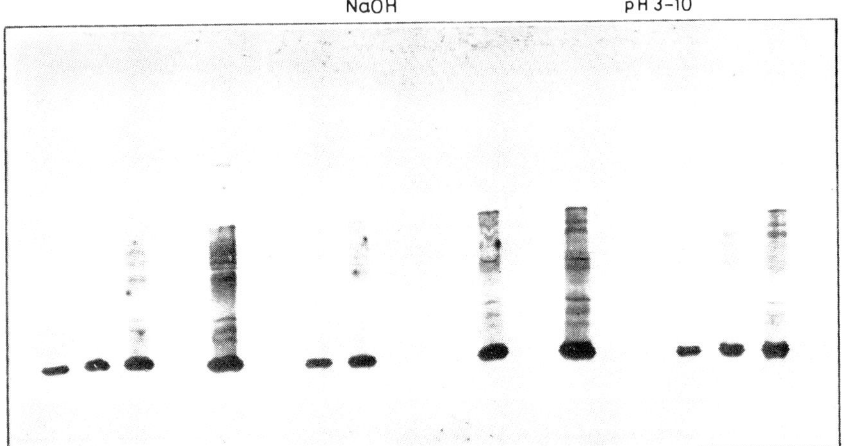

Fig. 3.18. Isoelectric focusing of a protein sample applied at different sites of a gel slab. Composition of gel slab: 18 ml of 10% acrylamide solution ($C = 5\%$), 4 ml of glycerine, 1.9 ml of Pharmalyte pH 3–10, H_2O to 30.0 ml, 0.2 ml of $(NH_4)_2S_2O_8$ solution (228 mg/10 ml). Dimensions of gel slab: 230 × 115 × 1 mm. Sample applied: 5, 10 or 25 μl of 0.6% ovalbumin solution. Apparatus: Flat Bed Apparatus FBE 3000, Pharmacia Fine Chemicals. Cathode solution: 0.1 *M* NaOH. Anode solution: 0.1 *M* H_2SO_4. Electrofocusing: 900–2000 V, 33 mA, for 90 min. Fixing: with a solution containing 10% trichloroacetic acid and 5% sulphosalicylic acid, for 1 h. Staining: 0.1% Coomassie BB R-250 in 40% ethanol–water solution, overnight. Destaining: with a solution of methanol–water–acetic acid (9 : 9 : 1). (By kind permission of T. Låås and I. Olsson, Pharmacia Fine Chemicals, Uppsala, Sweden.)

The principle of the method was established by Kolin [213] and by Vesterberg and Svensson [214, 215], who produced the carrier ampholyte preparations from mixtures of various polyamino- and polycarboxylic acids. The aliphatic ampholytes synthesized by Vesterberg and Svensson dissolve very well in water, the pH varying in the range 3–10. The various commercially-available carrier ampholytes (Ampholine, Bio-Lyte, Pharmalyte, Servalit) have narrower pH ranges and allow selection of the product with the most appropriate pH interval for the protein concerned (pH 2.5–5, 4–6.5, 6.5–9, 8–10.5). The ampholytes should have high buffer capacities, low conductivity throughout the entire gradient, low and uniform absorption in the ultraviolet, no toxicity, zero or minimal interaction with proteins, and no chelation properties. The desirable physical properties are exemplified in Fig. 3.19.

Like electrophoresis, IEF can be carried out by a column method in a sucrose gradient without a support, or in an acrylamide gel column, by a polyacrylamide gel rod method, and also by a layer method in an acrylamide gel slab, or in a granulated gel layer.

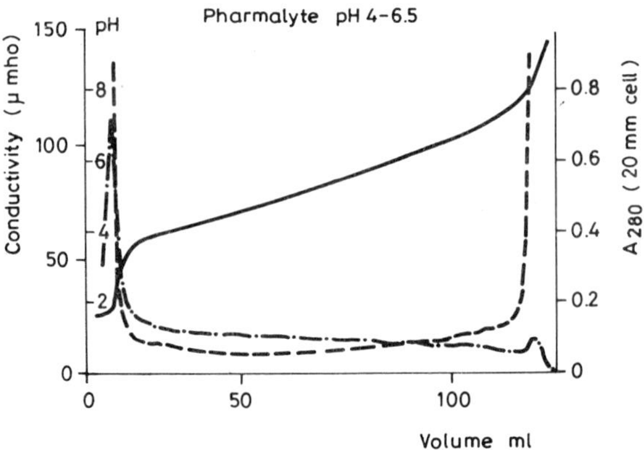

Fig. 3.19. The physical properties of a typical ampholyte. (By kind permission of Pharmacia Fine Chemicals, Uppsala, Sweden.)

3.10.1 IEF by a column method

IEF without a support, in a sucrose gradient

Commercial industrial focusing apparatus consists of cylindrical tubes of volume 110 or 440 ml, that can be temperature-controlled both inside and outside [216]. In the original method of Haglund [217], the sample mixed with ampholyte of appropriate pH interval is layered into the separating space with a sucrose gradient preparer. At the beginninig of the focusing, a potential of 500 V is generally employed; as the current falls, the potential is increased to 600–1000 V. The separation requires 24–72 h. After completion of the electrofocusing, the sucrose solution containing the focused proteins is led through an ultraviolet detection-unit into a fraction collector for further processing and analysis. Certain ampholyte preparations strongly absorb light of wavelength 280 nm; if these are used, low protein concentrations cannot be detected by ultraviolet absorption. Before protein determination by the Følin reaction, the fractions must be dialysed for at least 72 h, with repeated exchange of the water. By gel filtration on a Sephadex G-50 column, the carrier ampholytes can be removed more rapidly than by dialysis.

It is essential that the pH should be measured at the temperature at which the focusing was performed. The isoelectric points (pI) of the proteins detected in the fractions may be established on the basis of the pH of the solution. If a suitable ultraviolet monitor and the pH-meter are connected in series, the protein zones can be detected and their pH values recorded automatically [218].

The experience acquired with high-volume preparative apparatus has been utilized to develop various analytical-scale, sucrose-gradient column IEF instruments; of these, that of Osterman [219], with a volume of 12 ml, is noteworthy. Godson's apparatus [220], consisting of ten J-shaped 0.9×25 cm glass tubes, allows the simultaneous analytical processing of 10 samples.

IEF on a preparative gel column

Finlayson and Chrambach [221] carried out preparative IEF on a polyacrylamide gel column, the capacity of which was 2 mg per protein band. O'Brien *et al.* [222] used a 2.5×20 cm water-cooled Sephadex G-15 column for preparative IEF. The separation began with a current of 15 mA (at 800 V), and continued until it diminished to 2.5 mA.

3.10.2 Analytical IEF in a polyacrylamide gel rod

The method of analytical IEF in polyacrylamide gel was described by Wrigley [223]; the rod technique was modified by Fawcett [224] and numerous other authors [225–229], and employed for the separation of the most varied substances.

The gels used for rod IEF may be prepared by either chemical or photo-polymerization. Of the large number of gel recipes, the compositions of those most widely used are given below.

Gels prepared by chemical polymerization:

7.5% gel:	
acrylamide solution ($T = 30\%$, $C = 3.3\%$)	7.6 ml
ampholyte	0.9 ml
urea	14.4 g
3-dimethylaminopropionitrile (DMAPN)	0.1 ml
$(NH_4)_2S_2O_8$, 10% solution	0.1 ml
distilled water	to 30.0 ml
5% gel:	
acrylamide solution ($T = 30\%$, $C = 3.3\%$)	4.5 ml
ampholyte	0.9 ml
urea	14.4 g
3-dimethylaminopropionitrile (DMAPN)	0.1 ml
$(NH_4)_2S_2O_8$, 10% solution	0.1 ml
distilled water	to 30.0 ml

The oxidizing effect of the persulphate on proteins [26] can be eliminated by storing the chemically polymerized gels overnight before use, or by subjecting them to prefocusing for about 1 h before addition of the sample [25].

In photo-polymerization, 25% v/v of 0.1% riboflavin solution may be added to the acrylamide solution instead of *N*, *N*, *N'*, *N'*-tetramethylethylenediamine (TEMED) and $(NH_4)_2S_2O_8$ catalysts. Photo-polymerization takes place on application of a fluorescent light-source.

The sample must be transferred into the polymerized gel rods under a protective solution containing 2% ampholyte and 5% sucrose, layered 2–5 mm thick on top of the rod. However, very dilute protein samples may be added to the prepared acrylamide solution even before polymerization. The amount of protein may be 10–20 μg per band.

The anode solution is usually 0.2% sulphuric acid, and the cathode solution 0.4% ethanolamine (or triethanolamine). In rod IEF, the initial potential of 200–300 V must gradually be raised to such an extent (generally to 400–500 V) that during the process of separation in 10 gel rods the current should remain at 40 mA. The IEF is complete when the current no longer decreases; this occurs after about 90–120 min.

The pH gradients of the gel rods may either be determined at 2–5 mm intervals on the surface of the reference gel with a microelectrode, or measured after elution of the cut-up sections of the gel. If the resulting pH values are plotted as a function of distance, the pI values of the individual protein fractions may be read off the curve. Before staining, the protein fractions in the focused gels must be fixed with trichloroacetic acid or trichloroacetic acid + sulphosalicylic acid solution; the ampholyte must then be washed out of the gel before the use of the traditional staining methods. Even without washing-out of the ampholyte, staining may be performed with a solution of Coomassie BB R-250, either by the method of Reisner *et al.* [47] (Table 3.5, No. 12), or with the staining solution of Wellner and Hayes [226], which contains not only the dye but also the trichloroacetic acid + sulphosalicylic acid used for fixing.

Microscale IEF methods have been developed by Bispink and Neuhoff [230] and by Gainer [227], on the basis of the pattern of microrod electrophoresis. The detection of even 10^{-10} g of protein is possible in gel polymerized in 2, 5 or 10-μl capillaries.

3.10.3 Analytical IEF in a polyacrylamide gel slab

IEF in a polyacrylamide gel slab possesses similar advantages to those observed in PAGE, with respect to the gel rod method. The gel slab is generally 1 mm in thickness, while it may vary in length from 9 to 20 cm between the two electrodes. If good types of gel pourer and IEF instrument

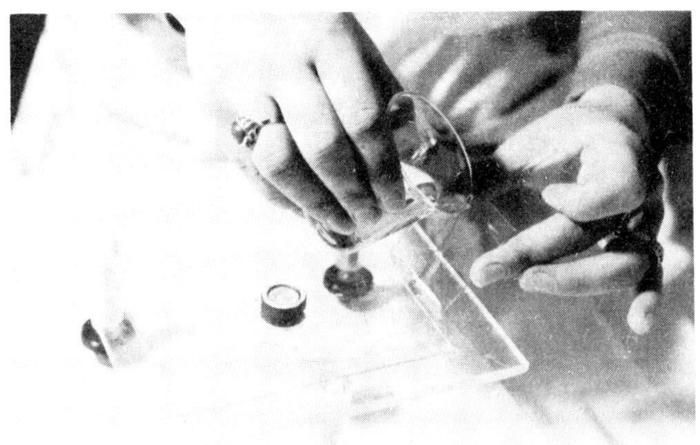

Fig. 3.20. Pouring apparatus for the preparation of gel slabs [231]

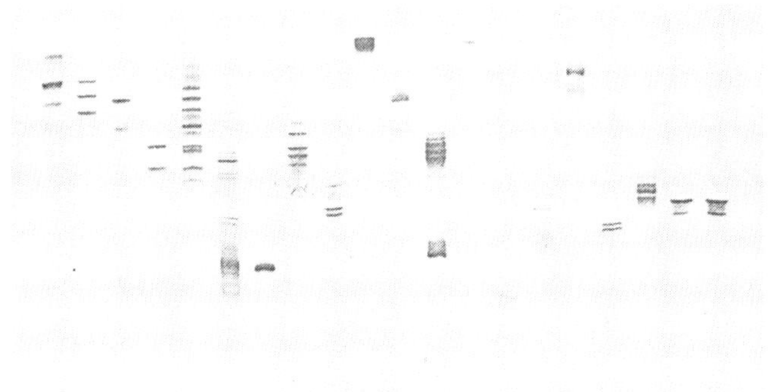

Fig. 3.21. Analytical isoelectric focusing on polyacrylamide thin layer with Pharmalyte pH 3–10, 30 W, 90 min. Samples (left to right): ADH (horse liver), lentil lectin, sperm whale myoglobin, horse myoglobin, *Helix pomatia* lectin, *Helix pomatia* blood, ovalbumin, plasminogen (ovine), conalbumin, chymotrypsinogen A, subtilisin, lysed human blood, dextranase, fetuin, chymotrypsin, lactoglobulin, soya-bean lectin, carbonic anhydrase (two lanes). (By kind permission of Pharmacia Fine Chemicals, Uppsala, Sweden.)

are not available, the gel may be prepared very easily with the apparatus described by Uriel and Berges [231] (Fig. 3.20).

A cooling plate must definitely be inserted below the glass plate supporting the gel, in order to cool the gel slab; 2–20 μl of a solution containing 1–2% protein is applied to the surface of the gel.

With unknown samples it is advisable to begin the IEF with an ampholyte displaying a wide pH interval (3–10). The result gives information regarding the correct choice of ampholyte with narrower pH interval, which yields better resolution. A very good picture of this is provided by Fig. 3.21, which illustrates the results of the IEF of 18 proteins in polyacrylamide gel containing Pharmalyte with a pH interval of 3–10.

3.10.4 Preparative IEF on a polyacrylamide gel flat bed

With a gel slab of preparative dimensions, the electrofocusing of 10–20 mg of protein can be achieved within 6 h, e.g. in a 2–4 mm thick 7.5% gel slab by the method of Graesslin *et al.* [232]. The protein zones are located by staining the strips cut off the two edges of the slab, and on this basis the desired zones can be cut out of the slab. The proteins to be further studied can be obtained in preparative amounts from these by elution, and then by concentration of the eluates.

3.10.5 IEF on a granulated gel bed

The most appropriate substances for IEF in granulated gels are Sephadex IEF and Sephadex G-200 SF; these may be used to prepare thin layers of analytical dimensions, and also gel beds 5 mm in thickness, suitable for preparative separations. IEF carried out on the gel thin layer has no advantages over IEF on the polyacrylamide layer, either with regard to operation (preparation of the thin layer and detection are the same) or resolution. However, preparative IEF in a 5-mm thick Sephadex G-200 SF gel bed yields a fractionation that cannot be approached either with a sucrose gradient in a column system, or separation in a preparative polyacrylamide gel column. The method has the very great advantage that after IEF in a broad pH interval in a suitable apparatus, the desired fraction or fractions may be selected and removed from the gel bed for further purification with an ampholyte displaying a narrower pH interval.

3.10.6 Combined IEF methods

IEF is a procedure which may be coupled with other electrophoretic methods, e.g. conventional gel electrophoresis; in certain cases, such a combination may result in a better resolution.

The examples given in Table 3.15 and Fig. 3.22 demonstrate the possibilities of combined IEF methods.

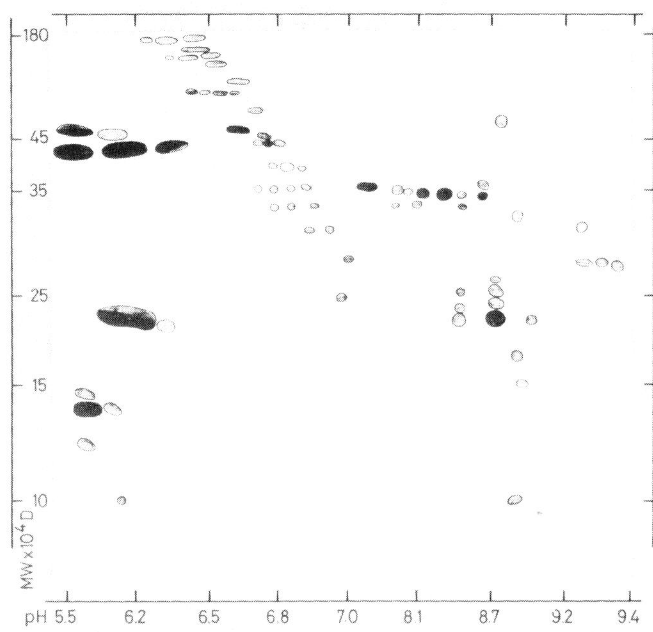

Fig. 3.22. Two-dimensional electropherogram of non-histone proteins of muscle cell nuclei. First direction: IEF: 120 × 2.5 mm polyacrylamide gel rod ($T=5\%$, $C=5\%$), with a urea concentration of 9.2 *M*, an Ampholin content of 2%, and an SDS content of 1%. Applied: 200 μg. Cathode electrolyte: 0.2 *M* NaOH. Anode electrolyte: 0.2 *M* H_3PO_4. Electrofocusing: 400 V, 20–0 mA (for 6 rods), for 12 h. Second direction: in a 260 × 150 × 3 mm polyacrylamide gel slab ($T=5\%$, $C=5\%$). Electrode buffer: 0.1 *M* TRIS–acetate, pH = 8.1. Electrophoresis: 125 V, 10 mA/cm^2, for 4 h. Fixing: with a solution of methanol–water–acetic acid (9 : 9 : 1). Staining: 0.25% Coomassie BB R-250. Destaining: with an aqueous solution of 7.5% acetic acid + 5% methanol. (By Ö. Takács, Department of Biochemistry, University Medical School, Szeged, Hungary.)

3.11 NON-ISOELECTRIC FOCUSING IN BUFFERS

Three extremely important features of IEF, viz. the instability in time of the pH gradient induced by the carrier ampholyte, the existence of a small current in the 'steady state', and the IEF of nucleic acids, led Nguyen *et al.* [238] to examine whether the multiphase buffer systems used in PAGE possess the ability to create a stable pH gradient without carrier ampholytes. Their experiments involved the use of amphoteric and non-amphoteric buffer mixtures, and by means of theoretical calculations recipes for very precisely designed buffer mixtures were compiled, by means of which a stable pH gradient, sharp buffer and protein zones, and a constant zone width could be reproduced in the cases examined.

3.12 ISOTACHOPHORESIS

In isotachophoretic separation, the sample ions to be separated (electrically charged and remaining in the ionic state throughout) exhibit different relative mobilities in an electric field. The ions of the same charge that can be separated by isotachophoresis have common counter-ions (e.g. TRIS), the mobilities of which in a constant electric field are smaller than those of the counter-ions of the previously leading electrolyte (e.g. the acetate ion), and larger than those of the ions of identical sign in the terminal electrolyte (e.g. the glycinate ion).

At the beginning of the separation, the sample ions are situated among both the leading and the terminal ions. In the separation, they become arranged in accordance with their relative mobilities (approximately in the sequence of their pI values) along the total length of the separation space (chamber, gel layer, column). After equilibrium has been established, the sample ions begin to migrate at the same rate towards the electrodes corresponding to their charges. In the course of this migration, at constant current, the electric field varies in a stepwise manner between the separated zones (which remain at a constant distance from one another); in this way, the potential is constant in each zone from the front to the start, but increases by jumps from one zone to another.

The separated substances, closely following one another at the same rate in narrow bands, can be kept at greater distances from one another by the addition to the sample of ampholytes with pH intervals corresponding to the pI values of the sample ions; hence, on completion of the separation, the ions of the sample, situated in a mobile pH-gradient, are differentiated from one another in sharp zones. The sample may be concentrated by a factor of as much as 1000–10 000.

The special fields of application of isotachophoresis are the separation of various peptides, metabolites, amino-acids and nucleotides.

Theoretical questions of isotachophoresis, its instrumentation and the scope of its application are discussed in detail by Everaerts, Beckers and Verheggen [239].

3.12.1 Analytical isotachophoresis

Any apparatus that may be employed for gel electrophoresis may also be used for analytical isotachophoresis. The proteins are first separated in an appropriate buffer without ampholyte, and then with the application of an ampholyte of the selected pH interval and suitable concentration.

Of the numerous gel compositions that may be employed for isotachophoresis, a gel with a 7.5% acrylamide content and the following composition has very wide application:

acrylamide solution ($T=30\%$, $C=3.33\%$)	1 volume
gel buffer, pH 8.2	1 volume
sucrose solution (25%)	1 volume
riboflavin solution, 0.004%	1 volume

If it is desired to work with a 5% or a 3.75% gel, 2 or 4 volumes, respectively, of distilled water are added to the composition above.

Compositions of the buffers:
Gel buffer, pH 6.1:

TRIS	5.2 g
1*M* phosphoric acid	39.0 ml
N, *N*, *N*′, *N*′-tetramethylethylenediamine (TEMED)	0.5 ml
H_2O	to 200.0 ml

Stock solution of leading buffer:

TRIS	5.2 g
1*M* phosphoric acid	39.0 ml
H_2O	to 200.0 ml

Before use it is necessary to dilute the stock solution of the leading buffer, to an extent depending on the composition of the gel:

for 7.5% gel: 1 vol. stock solution + 3 vols. water
for 5.0% gel: 1 vol. stock solution + 5 vols. water
for 3.75% gel: 1 vol. stock solution + 7 vols. water.

Terminal buffer, pH 8.2:

(*i*) β-alanine	35.8 g
H_2O	to 2000.0 ml
(*ii*) glycine	30.0 g
H_2O	to 2000.0 ml

The pH of the solutions must be adjusted to 8.2 with TRIS.

The gel solution is poured into 15 × 0.5 cm tubes to a height of 13 cm, gel buffer is layered onto its surface, and photo-polymerization is carried out for 1 h. After the polymerization, the gel tubes are placed in an apparatus that can be used for rod electrophoresis, and the protein sample, dissolved in terminal buffer, mixed with saccharose solution and ampholyte, is layered onto the surface of the gel. The upper bath is filled with terminal buffer, and the lower one (including the anode) with gel buffer. Isotachophoresis is carried out for 2–3 h at constant ionic strength and at 2 mA per gel. After completion, the gels are taken out, fixed and stained by the methods described in connection with IEF.

In isotachophoresis, the length of the protein zone is governed by the concentration of the protein, but is also influenced by the concentration of the ampholyte. A faster or slower separation can be attained, depending on the pH of the leading electrolyte. The concentration of the leading and terminal ions, and the distance between them, likewise influence the separation: a low concentration of the leading ions shortens the separation and the protein zones are longer. If the parameters are suitable, lengthening the separation does not impair its quality, since the possibility of diffusion is excluded.

3.12.2 Preparative isotachophoresis

Svendsen and Rose [240] used a preparative column for isotachophoresis. The current is conducted by the isotachophoretically migrating ions and the common counter-ions. Hence, the ions of the protein sample proceed one after another at the same rate. At a given rate the electric field is inversely proportional to the mobility, and thus, as in the analytical method, identical rates of the zones result in a stepwise increase of the field strength. The ions of the sample therefore cannot pass over into the neighbouring zone because of the lower or higher field.

Various forms of preparative isotachophoresis may be selected, and the loading may be about 1000 times that possible in preparative PAGE. In spite of this advantage, isotachophoresis has so far been used only to solve research problems, and has not yet gained ground among the routine methods.

PRACTICAL SUPPLEMENT

Two-dimensional electrophoresis according to O'Farrell [241]

Molarity or conc.	*Chemicals*	*Quantity*
	Soln. A. Lysis buffer	
9.5 *M*	Urea (ultra pure)	5.706 g
2% (w/v)	NP-40	200 mg
1.6% (w/v)	Ampholine LKB pH 5–7	0.400 ml
0.4% (w/v)	Ampholine LKB pH 3.5–10	0.100 ml
5%	2-Mercaptoethanol	0.500 ml

dilute to 10 ml with water (glass dist.)

	Soln. D. Acrylamide stock soln. for IEF	
28.38%	Acrylamide	28.38 g
1.62%	Bisacrylamide	1.62 g

water to 100 ml

Store at 4 °C!

	Soln. E. Nonidet P-40 stock soln.	
10% (w/v)	NP-40 (d = 1.06)	1.0 g

dilute to 10 ml with water

	Soln. G. Ammonium persulphate stock soln.	
10% (w/v)	Ammonium persulphate	1.0 g

dilute to 10 ml with water
Store at 4 °C, renew each week!

	Soln. H. Overlaying soln.	
8 *M*	Urea (ultra pure)	4.805 g

dilute to 10 ml with water

	Sol. I. Anode electrolyte	
10 m*M*	H_3PO_4 85%	0.675 ml

dilute to 1000 ml with water

	Soln. R. Running buffer with high conc. of SDS (2%)	
25 m*M*	TRIS	6.05 g
0.192 *mM*	Glycine	28.3 g
2.0%	SDS	4.00 g

dilute to 2000 ml with water and adjust pH as in the case of SOL.Q.

Soln. S. Bromophenol Blue

0.1% (w/v)	Bromophenol blue	25.0 mg

dilute to 25.0 ml with water

Soln. T. Staining solution

50% (w/v)	TCA	500 g
0.1% (w/v)	Coomassie Blue R	1.0 g

dilute to 1000 ml with water

Soln. U. Destaining solution

7% (w/v)	Acetic acid	

Protocols of two-dimensional electrophoresis

Tubes for IEF can be cut from any precision pipettes, or glass tubes with 2.3–2.5 mm ID. The appropriate length could be 12–20 cm. Gel tubes are washed in chromic acid, or special detergent solution overnight, rinsed with running tap water then with a number of changes of distilled water.

Before casting, the gel tubes are covered at one end with several layers of Parafilm; they are marked at 10–15 cm from the bottom, according to the length of gel column desired.

The IEF gel mixture is prepared and the tubes are filled from a very thin and long Pasteur pipette or a syringe with a long needle. The Parafilm is then removed and the water overlay is shaken off. The lower part of the apparatus is filled with the acid buffer and the upper part together with the tubes is placed in the apparatus. The top of the focusing gel is overlaid with 20 μl of soln. A from a Hamilton syringe, then filled with a cathode electrolyte.

Soln. L. Lower gel buffer pH 8.8

1.5 *M*	TRIS	18.15 g
0.4% (w/v)	SDS	400 mg

adjust pH to 6.8 with HCl and dilute to 100 ml with water
Store at 4 °C

Soln. M. Upper gel buffer pH 6.8

0.5 *M*	TRIS	6.05 g
0.4 (w/v)	SDS	400 mg

adjust pH to 6.8 with HCl and dilute to 100 ml with water

Soln. N. 30% Acrylamide stock soln. for SDS

29.2% (w/v)	Acrylamide	29.2 g
0.8% (w/v)	Bisacrylamide	0.8 g

dilute to 100 ml with water

Soln. O. SDS equilibration buffer

10% (w/v)	Glycerol	25.0 g
5% (w/v)	2-Mercaptoethanol	12.5 ml
2.3% (w/v)	SDS	5.75 g
62.5 m*M*	TRIS	1.892 g

adjust pH to 6.8 with HCl and dilute to 250 ml with water

Soln. P. Agarose gel in soln. O.

1% (w/v)	Agarose	1.0 g

dissolve in 100 ml of soln. O. and store at −20 °C in 5 ml aliquots

Soln. Q. Running buffer pH 8.3

25 m*M*	TRIS	6.05 g
0.192 *M*	Glycine	28.3 g
0.1%	SDS	2.0 g

dilute to 2000 ml with water, check pH, if necessary adjust with 1 *M* TRIS or 1 *M* glycine solution

Soln. J. Cathode electrolyte

20 m*M*	NaOH	0.800 g

dilute to 1000 ml with water

Soln. K. Overlay solution

9 *M*	Urea	2.702 g
0.8%	Ampholine pH 5–7	0.100 ml
0.2%	Ampholine pH 3.5–10	0.025 ml

dilute to 5 ml with water

Gel composition for isoelectric focusing

Urea (ultra pure)	5.50 g
Soln. D.	1.33 ml
Soln. E.	2.00 ml
Ampholine pH 5–7	0.40 ml
Ampholine pH 3.5–10	0.10 ml
Water	1.97 ml

Notes: The urea is weighed into a 125 ml Erlenmeyer flask with side-arm then the other components are added and the flask is gently swirled in a warm water bath to dissolve the urea. Once the urea is dissolved 10–15 μl of Soln. G are added. The solution is degassed by water pump vacuum and 10 μl of TEMED are added. The gel tubes have to be filled with the above solution. Gels are left to polymerize at room temperature for 2 h. Once the gel solution has been loaded, approximately 0.2 ml of dist. water can be overlaid to give a smooth surface during polymerization. The remainder of the electrolyte is added, filling the upper reservoir to about 2 cm above the gel tubes. The cover is fixed and the apparatus is connected to an appropriate power supply.

The pre-run is carried out at a constant voltage of 200 V for 15 min, then 300 V for 30 min and finally 400 V for 30 min. The power supply is turned off and the upper buffer is removed. The liquid in the gel tube is aspirated off carefully without sucking up the upper part of the gel. The samples can be loaded from a Gilson Pipetman or Hamilton syringe and then overlaid with 10–15 ml of Soln. K. The tubes are filled again with cathode buffer and the upper chamber is refilled. The focusing is performed with 400 V constant voltage overnight (16 h) then 800 V for 1 h.

The gels have to be removed carefully from the tubes as described [13]; care must be taken when gently pushing these into separate test tubes which are filled with Soln. O. Equilibration of gels for 30–45 min. is performed by setting the Parafilm-covered tubes horizontally in a shaker, agitated at 40–45 strokes/min. After this the second dimension may be started, or the gels frozen on dry ice and stored in an $-80\ ^\circ C$ freezer for several days.

Second dimension slab electrophoresis

Some equipment (e.g. Bio-Rad, Hoefer) is suitable for two dimensional electrophoresis. In general a pair of rectangular plates, the internal faces of both of them being ground obliquely to hold the focused gel. This is important because the slab gels are usually 0.8–1.2 mm thick, and the gel rods have 2.0–2.5 mm diameter. The glass plates and spacers are assembled and a separating gel (about 10 cm high) and a stacking gel (3.5–4% and 2.5 cm high) are cast. For the acrylamide/BIS ratios the following values are recommended.

Acrylamide	20%	17.5%	15%	12.5%
BIS	0.067%	0.07%	0.087%	0.10%

A typical gel protocol (12.5%) is:

1. Soln. L	7.5 ml
2. Soln. N	12.5 ml
3. Dist. H_2O	10.0 ml
4. Soln. G	50 μl
5. TEMED	50 μl
	30.0 ml

The gel volume can be varied according to thickness of the gel and the total number of the slabs. All the components are mixed except for the TEMED and degassed. The TEMED is then added and the solution loaded into the gel assembly by pipetting. The level to which the solution is added (10 cm) should be marked on the glass plates. Water-saturated isobutanol is then overlaid to produce a flat surface. Gels are left for a minimum of 1 h to polymerize. If the gels are to be run the same day, the overlay is aspirated and the gel surface is washed with several changes of dist. water, finally with 1:4 diluted Soln. S.

Composition of the stacking gel:

1. Soln. M.	2.5 ml
2. Soln. N.	1.5 ml
3. Dist. H_2O	5.5 ml
4. Soln. G.	0.5 ml
5. TEMED	10 μl
	10.0 ml

The components are mixed, degassed as described before, and the solution is pipetted over the separating gel, filling up to and slightly above the level of the notch (about 3 mm). Water-saturated butanol is overlaid again and the gels are left to polymerize at room temperature for 60 min. The stacking gel surface is rinsed with dist. water and excess liquid is shaken off. Before loading the focusing gels the Soln. P (agarose) should be warmed in a water bath to 100 °C. If the gels were thawed they have to be warmed up to room temperature and rinsed with dist. water. This could be performed with a U-shaped chemical spatula in which the gel is held by the fingers and the water added carefully from a squeeze bottle. The excess water is allowed to drain from the gel. The gel apparatus is inclined forward to create a V between the notched and rectangular plates. The hot agarose solution is taken up in a Pasteur pipette and transferred to the top of the stacking gel surface. While the hot solution runs into the V-shaped corner the focusing gel can be pushed down from the spatula and sealed to the surface of the stacking gel. Some agarose solution could be dropped over the focusing gel to firm it securely. Once the agarose sets the gels are ready to be run. Soln. Q (electrode buffer) is then poured in the lower and upper chambers. About 1 ml of Soln. S is added to the top buffer and mixed gently and the current is turned on. Gels can be run overnight at about 10 mA (60 V) or during the day (4–4.5 h) at 20 mA/slab (1.5 mm thick). The separation is finished when the blue front-line reaches the bottom of the slab. The gels can now be processed as desired for staining and/or autoradiography.

Preparation of exponent slab gels

Solutions

Soln. N/a

29.2% (w/v)	Acrylamide	29.2 g
0.8 (w/v)	Bisacrylamide	0.8 g

dissolve in final vol. of 100 ml 75% (w/v) glycerol

Soln. N/b

75% (w/v)	Glycerol	75.0 g

dilute to 100 ml with water

Composition of gradient gel (22.5%—5%)

A) Soln. N/a	30 ml
Soln. L	10 ml
Soln. G	40 μl

B) Soln. N	6.66 ml
Soln. L	10.0 ml
Water	13.0 ml
Soln. G	60 μl

Degas both solutions under vacuum, check the gradient-former system. Add 10μl of TEMED to Soln. A and 30 μl to Soln. B. Introduce the solutions into the gradient former, eliminate air bubbles, then start the pump and magnetic stirrer (in Soln. A).

Leave the gels to polymerize at room temperature for 2 h. Overlay with water-saturated butanol. After polymerization prepare the stacking gel as described previously.

Silver staining of proteins in polyacrylamide gels. From the several methods the modified procedures of Morrissey [242], Wray *et al.* [243], Poehling and Neuhoff [244] are recommended. See references for details.

REFERENCES

[1] Tiselius, A.: *Trans. Faraday Soc.*, **33,** 524 (1937).

[2] Hannig, K.: *Z. Anal. Chem.*, **181,** 244 (1961).

[3] Hannig, K.: *J. Chromatog.*, **159,** 183 (1978).

[4] Hannig, K.: *Jahrbuch der Max-Planck-Gesellschaft*, 1968, p. 117.

[5] Hannig, K.: *Separation of Cells and Particles by Continuous Free-Flow Electrophoresis*, in *Techn. Biochem. Biophys. Morph.*, Glick, D., Rosenbaum, R. M. (eds), Vol. 1., Wiley-Interscience, New York, 1972, p. 191.

[6] Hannig, K., Klofat, W., Endres, H.: *Z. Naturforsch.*, **19,** 1072 (1964).

[7] Grassmann, W., Hannig, K.: *Naturwiss.*, **37,** 496 (1950).

[8] Michaelis, L.: *Biochem. Z.*, **234,** 139 (1931).

[9] Grassmann, W., Hannig, K.: *Naturwiss.*, **37,** 397 (1950).

[10] Smithies, O.: *Biochem. J.*, **61,** 629 (1955).

[11] Smithies, O.: *Adv. Prot. Chem.*, **14,** 65 (1959).

[12] Scopes, R. K.: *Biochem. J.*, **107,** 139 (1968).

[13] Gordon, A. H.: *Electrophoresis of Proteins in Polyacrylamide and Starch Gels., Laboratory Techniques in Biochemistry and Molecular Biology*, Work, T. S., Work, E. (eds), North-Holland, Amsterdam, 1969.

[14] Becker, P. E. (ed.): *Humangenetik*, Band I/3, *Protein und Enzymvarianten*, Thieme, Stuttgart, 1975.

[15] Wieme, R. J.: *Agar Gel Electrophoresis*, Elsevier, Amsterdam, 1965.

[16] Grabar, P., Williams, C. A.: *Biochim. Biophys. Acta*, **20,** 193 (1953).

[17] Grabar, P., Williams, C. A.: *Immunoelectrophoretic Analysis*, Elsevier, Amsterdam, 1964.

[18] Scheidegger, J. J.: *Intern. Arch. Allergy*, **7,** 103 (1955).

[19] Backhausz, R.: *Immunodiffusion und Immunoelectrophorese*, Akadémiai Kiadó, Budapest, 1967.

[20] Allen, P. C., Hill, E. A., Stokes, A. M.: *Plasmaproteins, Analytical and Preparative Techniques*, Blackwell, Oxford, 1978.

[21] Wright, G. L., Mallmann, W. L.: *Proc. Soc. Exp. Biol. Med.*, **123,** 22 (1966).
[22] Ornstein, L.: *Ann. N. Y. Acad. Sci.*, **121,** 321 (1964); Davis, B. J.: *Ann. N. Y. Acad. Sci.*, **121,** 404 (1964).
[23] Shepherd, G. R., Gurley, L. R.: *Anal. Biochem.*, **14,** 356 (1966).
[24] Mandel, J. D.: *Dental Res.*, **45,** 634 (1976).
[25] Låås, T., Olsson, I. (Pharmacia Fine Chemicals), personal communication.
[26] King, E. E.: *J. Chromatog.*, **53,** 559 (1970).
[27] Jordan, E., Raymond, S.: *Anal. Biochem.*, **27,** 205 (1969).
[28] Richards, E. G., Coll, A., Gratzer, W. B.: *Anal. Biochem.*, **12,** 452 (1965).
[29] Aronson, T., Grönwall, A.: *Scand. J. Clin. Lab. Invest.*, **9,** 338 (1957).
[30] Blattler, D. T., Garner, F., Van Slyke, K., Bradley, A.: *J. Chromatog.*, **64,** 147 (1972).
[31] Hjertén, S.: *Arch. Biochem. Biophys.*, Suppl. **1,** 147 (1962).
[32] Hjertén, S.: *Anal. Biochem.*, **11,** 211 (1965).
[33] Michl, H., Pastuszyn, A.: *Mikrochim. Acta*, 880 (1963).
[34] Narayan, K. A., Vogel, M., Lawrence, J. M.: *Anal. Biochem.*, **12,** 526 (1965).
[35] Margolis, J., Kenrick, K. G.: *Anal. Biochem.*, **25,** 347 (1968).
[36] Fischbein, W. N.: *Anal. Biochem.*, **46,** 388 (1972).
[37] Ritchie, R. F.: *Maine Med. Assoc.*, **58,** 15 (1966).
[38] Homberg, H., Huttunen, E.: *Anal. Biochem.*, **55,** 620 (1973).
[39] Meyer, T. S., Lamberts, B. L.: *Biochim. Biophys. Acta*, **107,** 144 (1965).
[40] Chrambach, A.: *Anal. Biochem.*, **15,** 544 (1966).
[41] Weber, K., Osborn, M.: *J. Biol. Chem.*, **244,** 4406 (1969).
[42] Diezel, W., Kopperschläger, G., Hofmann, R.: *Anal. Biochem.*, **48,** 617 (1972).
[43] Griffith, I. B.: *Anal. Biochem.*, **46,** 402 (1972).
[44] Datyner, A., Finnimore, E.: *Anal. Biochem.*, **55,** 479 (1973).
[45] Gorovsky, M. A., Carlson, K., Rosenbaum, I. L.: *Anal. Biochem.*, **35,** 359 (1970).
[46] Fazekas, S., Webster, R. G., Datyner, A.: *Biochim. Biophys. Acta*, **71,** 377 (1963).
[47] Reisner, A. H., Nemes, O., Buchholtz, C.: *Anal. Biochem.*, **64,** 509 (1975).
[48] Kahn. R., Rubin, W.: *Anal. Biochem.*, **67,** 347 (1975).
[49] Traub, P., Boeckman, G.: *Z. Physiol. Chem.*, **359,** 571 (1978).
[50] Katovich-Hurley, C.: *Anal. Biochem.*, **80,** 624 (1977).
[51] Schwabe, C.: *Anal. Biochem.*, **27,** 201 (1966).
[52] Maurer, H. R.: *Z. Klin. Chem.*, **4,** 85 (1966).

[53] Clarke, J. T.: *Ann. N. Y. Acad. Sci.*, **121,** 428 (1964).
[54] Owen, J. A., Smith, H.: *Clin. Chim. Acta*, **6,** 441 (1961).
[55] Ribeiro, L. P., McDonald, H. K.: *J. Chromatog.*, **10,** 443 (1963).
[56] Ressler, N., Springgate, R., Kaufman, J.: *J. Chromatog.*, **6,** 409 (1961).
[57] Narayan, K. A., Narayan, S., Kummerow, F. A.: *Clin. Chim. Acta*, **14,** 227 (1966).
[58] Keyser, J. W.: *Anal. Biochem.*, **12,** 395 (1965).
[59] Kapitány, R. A., Zebrowski, E. J.: *Anal. Biochem.*, **56,** 361 (1973).
[60] Wardi, A. H., Michos, G. A.: *Anal. Biochem.*, **49,** 607 (1972).
[61] Segrest, J. P., Jackson, R. L.: *Molecular Weight Determination of Glycoproteins by Polyacrylamide Gel Electrophoresis*, in *Methods in Enzymology*, Vol. XXVIII B, Ginsburg V. (ed.), Academic Press, New York, 1972.
[62] McDevitt, C. A., Muir, H.: *Anal. Biochem.*, **44,** 612 (1971).
[63] Cutting, J. A., Roth, T. F.: *Anal. Biochem.*, **54,** 386 (1973).
[64] Grossbach, U., Weinstein, I. B.: *Anal. Biochem.*, **22,** 311 (1968).
[65] Peacock, A. C., Dingman, C. W.: *Biochemistry*, **6,** 1818 (1967).
[66] Kurnick, N. B.: *Exp. Cell. Res.*, **1,** 151 (1950).
[67] Radola, B. J., Delincée, H.: *J. Chromatog.*, **61,** 365 (1971).
[68] Borris, D. P., Aronson, J. M.: *Anal. Biochem.*, **32,** 273 (1969).
[69] Hartman, B. K., Udenfried, B.: *Anal. Biochem.*, **30,** 391 (1969).
[70] Nerenberg, S. T., Ganger, C., De Marco, L.: *Anal. Biochem.*, **43,** 56 (1971).
[71] Talbot, D. N., Yphantis, D. A.: *Anal. Biochem.*, **44,** 246 (1971).
[72] Fenner, C., Traut, R. T., Mason, D. T., Wikman-Coffelt, I.: *Anal. Biochem.*, **63,** 595 (1975).
[73] Takács, Ö., Pelle, T., Sohár, I., Guba, F.: unpublished data.
[74] Rodbard, D., Chrambach, A.: *Quantitative Polyacrylamide Gel Electrophoresis: Mathematical and Statistical Analysis of Data*, in *Electrophoresis and Isoelectric Focusing in Polyacrylamide Gels*, Allen, R. C., Maurer, H. R., (eds.), de Gruyter, New York, 1974.
[75] Hsieh, W.-C., Anderson, R. E.: *Anal. Biochem.*, **69,** 331 (1975).
[76] Aronson, I. N., Borris, D. F.: *Anal. Biochem.*, **14,** 27 (1967).
[77] Lewis, U. J., Clark, M. O.: *Anal. Biochem.*, **6,** 303 (1963).
[78] Maizel, I. V.: *Science*, **151,** 988 (1966).
[79] Maurer, H. R., Chalkley, G. R.: *Nachweis getrennter Substanzen durch Färbung*, in *Disk-Elektrophorese*, Maurer, H. R., (ed.), de Gruyter, New York, 1968.
[80] Sheu, C. W., Ries, J. J.: *Anal. Biochem.*, **49,** 23 (1972).
[81] Gilson, W., Gilson, R., Rueckert, R. R.: *Anal. Biochem.*, **47,** 321 (1972).

[82] Brade, W., Dietz, H.: *Anal. Biochem.*, **51,** 541 (1973).
[83] Hyden, H., Bjurstam, K., McEwen, B.: *Anal. Biochem.*, **17,** 1 (1966).
[84] Sai-Hung, J., Spring, T. G.: *Anal. Biochem.*, **85,** 287 (1978).
[85] Jovin, T., Chrambach, A., Naughton, M. A.: *Anal. Biochem.*, **9,** 351 (1964).
[86] Gordon, A. H., Lou[illegible] L. N.: *Anal. Biochem.*, **21,** 190 (1967).
[87] Jacobson, A., Lodish, H. F.: *Anal. Biochem.*, **54,** 513 (1973).
[88] Hjertén, S., Jerstedt, S., Tiselius, A.: *Anal. Biochem.*, **27,** 108 (1969).
[89] Van Jaarsweld, P. P., van der Walt, B. J., Le Roux, C. H.: *Anal. Biochem.*, **75,** 363 (1976).
[90] Boyde, T. R. C., Remtulla, N. A.: *Anal. Biochem.*, **55,** 492 (1973).
[91] Leboy, P. S., Cox, E. C., Flaks, J. G.: *Proc. Natl. Acad. Sci. U.S.*, **52,** 1367 (1964).
[92] Slater, G. G.: *Fed. Proc.*, **24,** 225 (1965).
[93] Margolis, J., Kenrick, K. G.: *Nature*, **214,** 1334 (1967).
[94] Allen, R. C.: *Polyacrylamide Gel Electrophoresis with Discontinuous Buffers at a Constant pH*, in *Electrophoresis and Isoelectric Focusing in Polyacrylamide Gel.* Allen, R. C., Maurer, H. R. (eds), de Gruyter, New York, 1974.
[95] Kopperschläger, G., Diezel, W., Biermayer, B., Hofman, E.: *FEBS Lett.*, **5,** 221 (1969).
[96] Kopperschläger, G., Stroch, H., Biermayer, G.: *Z. Med. Labor.-Diagn.*, **16,** 300 (1977).
[97] Rüchel, R., Jr.: *A Method for Continuous Gradient Gel Electrophoresis in Capillary Tubes*, in *Electrophoresis and Isoelectric Focusing in Polyacrylamide Gel*, Allen, R. C., Maurer, H. R. (eds.), de Gruyter, New York, 1974.
[98] Dames, W., Maurer, H. H.: *Simultaneous Preparation for Electrophoresis of a Large Number of Micro Polyacrylamide Gels with Continuous Concentration Gradients*, in *Electrophoresis and Isoelectric Focusing in Polyacrylamide Gel*, Allen, R. C., Maurer, H. R. (eds.), de Gruyter, New York, 1974.
[99] Foissy, H.: *J. Chromatog.*, **106,** 51 (1975).
[100] Wright, G. L. Jr., Farrell, K. B., Roberts, D. B.: *Clin. Chim. Acta*, **32,** 285 (1971).
[101] Wright, G. L. Jr., Schellhammer, P. F., Raulconer, R. L.: *Cancer Res.*, **37,** 4228 (1977).
[102] Lorentz, K.: *Anal. Biochem.*, **76,** 214 (1976).
[103] Tung, J. S., Knight, C. A.: *Anal. Biochem.*, **48,** 153 (1972).

[104] McDonagh, J., Messel, A., McDonagh, R. P., Murano, G., Blomback, B.: *Biochim. Biophys. Acta*, **257,** 135 (1972).
[105] Shapiro, A. L., Vinulea, E., Maizel, J. V.: *Biochem. Biophys. Res. Commun.*, **28,** 815 (1967).
[106] Patterson, B., Strohman, R.: *Biochemistry*, **9,** 4094 (1971).
[107] Talbot, D. N., Yphantis, D. A.: *Anal. Biochem.*, **44,** 246 (1971).
[108] Laemmli, U. K.: *Nature*, **227,** 680 (1970).
[109] Wyckoff, M., Rodbard, D., Chrambach, A.: *Anal. Biochem.*, **78,** 459 (1977).
[110] Porzio, M. A., Pearson, A. M.: *Biochim. Biophys. Acta*, **490,** 27 (1977).
[111] Bryan, J. K.: *Anal. Biochem.*, **78,** 513 (1977).
[112] Ferguson, K. A.: *Metabolism*, **13,** 985 (1964).
[113] Inouye, M.: *J. Biol. Chem.*, **246,** 4834 (1971).
[114] Catsimpoolas, N., Kenney, J.: *J. Chromatog.*, **64,** 190 (1972).
[115] Chiga, M., Brower, G. J., Lopez-Corella, E., Noelken, M. E.: *Clin. Chim. Acta*, **36,** 574 (1972).
[116] Williams, J. G., Gratzer, W. B.: *J. Chromatog.*, **57,** 121 (1971).
[117] Fessenden-Raden, J. M.: *Biochem. Biophys. Res. Commun.*, **46,** 1347 (1972).
[118] Lambin, P., Rochu, D., Fine, J. M.: *Anal. Biochem.*, **74,** 567 (1976).
[119] Tuszynski, G. P., Warren, L.: *Anal. Biochem.*, **67,** 55 (1975).
[120] Hearing, V. J. Klingler, W. G., Ekel, T. M., Montague, P. M.: *Anal. Biochem.*, **72,** 113 (1976).
[121] Rodbard, D., Chrambach, A.: *Anal. Biochem.*, **40,** 95 (1971).
[122] Takács, Ö.: (unpublished data).
[123] Alpers, D. H., Glickman, R.: *Anal. Biochem.*, **35,** 314 (1970).
[124] Leon, S. A., Bohrer, A. T.: *Anal. Biochem.*, **42,** 54 (1971).
[125] Paus, P. N.: *Anal. Biochem.*, **42,** 372 (1971).
[126] Tykva, R., Votruba, I.: *Anal. Biochem.*, **50,** 18 (1972).
[127] Fairbanks, G., Levinthal, G., Reeder, R. H.: *Biochem. Biophys. Res. Commun.*, **20,** 393 (1965).
[128] Uriel, J., Lavaille, C.: *Anal. Biochem.*, **42,** 509 (1971).
[129] Leinen, J. G., Witliff, J. L.: *Anal. Biochem.*, **86,** 279 (1978).
[130] Buisson, M., Reboud, A.-M., Reboud, I. P.: *Anal. Biochem.*, **75,** 656 (1976).
[131] Christoff, N., Anderson, P. I., Slotwiner, P., Song, S. K.: *Ann. N. Y. Acad. Sci.*, **135,** 150 (1966).
[132] Dietz, A. A., Lubrano, T.: *Anal. Biochem.*, **20,** 246 (1967).
[133] Hirschfeld, A.: *Acta Path. Microbiol.*, **47,** 169 (1959).

[134] Duspiva, F.: *Mikroskopisch-histochemische Enzym-Nachweise*, in *Methoden der Enzymatischen Analyse*, H.-U. Bergmeyer (ed.), Verlag Chemie, Weinheim, 1962, p. 920.
[135] Susor, W. A., Rutter, W. I.: *Anal. Biochem.*, **43,** 147 (1971).
[136] Taswell, H. F., Jeffer, D. M.: *Am. J. Clin. Pathol.*, **40,** 349 (1963).
[137] Anderson, P. J., Song, . K., Christoff, M.: *Proc. 4th Intern. Congr. Neuropath.*, Vol. I, p. 75.
[138] Pajdak, E., Pajdak, W.: *Anal. Biochem.*, **50,** 317 (1972).
[139] Scandalios, J. G.: *Biochem. Genet.*, **1,** 1 (1967).
[140] Ojala, K., Harmionen, A.: *Scand. J. Clin. Lab. Invest.*, **35,** 163 (1975).
[141] Harrison, R. A. P.: *Anal. Biochem.*, **61,** 500 (1974).
[142] Ward, C. W.: *Anal. Biochem.*, **74,** 242 (1976).
[143] März, L., Barna, J., Ebermann, R.: *J. Chromatog.*, **123,** 495 (1976).
[144] Hobson, G. E.: *Anal. Biochem.*, **75,** 637 (1976).
[145] Fernandez-Souza, J. M., Michelson, A. M.: *Biochem. Biophys. Res. Commun.*, **73,** 217 (1976).
[146] Rosenthal, A. L., Lacks, S. A.: *Anal. Biochem.*, **80,** 76 (1977).
[147] Louis-Ferdinand, R., Blatt, W. F.: *Clin. Chim. Acta*, **16,** 259 (1967).
[148] Makonkawkeyoon, S., Haque, R.-U.: *Anal. Biochem.*, **36,** 422 (1970).
[149] Olden, K., Yamada, K. M.: *Anal. Biochem.*, **78,** 483 (1977).
[150] Wright, G. L., Farrell, K. B., Roberts, D. B.: *Biochim. Biophys. Acta*, **295,** 396 (1973).
[151] Pun, J. Y., Lombrozo, L.: *Anal. Biochem.*, **9,** 9 (1964).
[152] Hyden, H., Bjurstam, K., McEwen, B.: *Anal. Biochem.*, **17,** 1 (1966).
[153] Burr, I. M., Balant, L., Stauffacher, W., Reynold, A. E.: *Anal. Biochem.*, **35,** 82 (1970).
[154] Neuhoff, V.: *Arzneumittelforsch.*, **18,** 35 (1967).
[155] Dames, W., Maurer, H. R., Neuhoff, W.: *Z. Physiol. Chem.*, **353,** 554 (1972).
[156] Been, A. C., Rasch, E. M.: *J. Histochem. Cytochem.*, **20,** 368 (1972).
[157] Reid, M. S., Bielski, R. L.: *Anal. Biochem.*, **22,** 374 (1968).
[158] Maurer, H. R., Dati, F. A.: *Anal. Biochem.*, **46,** 19 (1972).
[159] Goldberg, E.: *Ann. N. Y. Acad. Sci.*, **121,** 560 (1964).
[160] Goldberg, E.: *Science*, **139,** 602 (1963).
[161] Goldberg, E.: *Science*, **148,** 391 (1965).
[162] Amelunxen, R. E.: *Biochim. Biophys. Acta*, **122,** 175 (1966).
[163] Huang, R. C., Bonner, J.: *Proc. Natl. Acad. Sci. U.S.*, **54,** 960 (1966).
[164] Falaschi, A., Kornberg, A.: *J. Biol. Chem.*, **241,** 1478 (1966).
[165] Plummer, T. H., Hirsch, C. H. W.: *J. Biol. Chem.*, **238,** 1396 (1963).
[166] Herrick, H. E., Lawrence, J. M.: *Anal. Biochem.*, **12,** 400 (1956).

[167] Aronson, N. N., Davidson, E. A.: *J. Biol. Chem.*, **242,** 437 (1967).
[168] Reid, B. R., Cole, R. D.: *Proc. Natl. Acad. Sci. U.S.*, **51,** 1044 (1964).
[169] Reisfeld, R. A., Lewis, U. J., Williams, D. E.: *Nature*, **195,** 281 (1962).
[170] Neidle, A., Waelsch, H.: *Science*, **145,** 1059 (1964).
[171] Fambrough, D. M., Bonner, J.: *Biochemistry*, **5,** 2563 (1966).
[172] Gurley, L. R., Shepherd, G. R.: *Anal. Biochem.*, **14,** 364 (1966).
[173] McAllister, H. C., Wan, Y. C. Jr., Irvin, R. J.: *Anal. Biochem.*, **5,** 321 (1963).
[174] Marsh, W. H., Ord, M. G., Stocken, L. A.: *Biochem. J.*, **93,** 539 (1964).
[175] MacGillivray, A. J., Rickwood, D.: *Eur. J. Biochem.*, **41,** 181 (1974).
[176] Wada, H., Shell, E. E.: *Anal. Biochem.*, **46,** 548 (1972).
[177] Bray, D., Brownlee, S. M.: *Biochem. J.*, **55,** 213 (1973).
[178] Houston, L. L.: *Anal. Biochem.*, **44,** 81 (1971).
[179] Weiner, A. M., Platt, T., Weber, K.: *J. Biol. Chem.*, **247,** 3242 (1972).
[180] Takács, Ö., Sohár, I., Szilágyi, T., Guba, F.: *Acta Biol. Sci. Hung.*, **28,** 221 (1977).
[181] Takács, Ö., Scisłowski, P. W. D., Zydowo, M., Guba, F.: *Proc. 7th Eur. Conf. Motil., Warsaw*, 1978.
[182] Maurer, H. R., Allen, R. C.: *Clin. Chim. Acta*, **40,** 359 (1972).
[183] Hoffmeister, H.: *Potential Applications of Polyacrylamide Gel Electrophoresis for Clinical Diagnosis*, in *Electrophoresis and Isoelectric Focusing in Polyacrylamide Gel*, Allen, R. C., Maurer, H. R. (eds), de Gruyter, New York, 1974.
[184] Felgenhauer, K.: *Comparative Disc Electrophoresis of Serum and Cerebrospinal Fluid*, in *Electrophoresis and Isolectric Focusing in Polyacrylamide Gel*, Allen, R. C., Maurer, H. R. (eds.), de Gruyter, New York, 1974.
[185] Pastewka, J. V., Ness, A. T., Peacock, A. C.: *Clin. Chim. Acta*, **14,** 219 (1966).
[186] Narayan, K. A., Kummarow, F. A.: *Clin. Chim. Acta*, **13,** 532 (1966).
[187] Muniz, N.: *Clin. Chem.* **23,** 1826 (1977).
[188] Clements, R. L.: *Phytochemistry*, **5,** 243 (1966).
[189] Loeschcke, V., Stegemann, H.: *Phytochemistry*, **5,** 985 (1966); *Z. Naturforsch.*, **21,** 879 (1966).
[190] Barber, J. T., Wood, H. L., Steward, F. C.: *Canad. J. Botan.*, **45,** 5 (1967).
[191] Pauler, B.: *Z. Landwirtsch. Forsch.*, **25**/II. Sonderheft, 87 (1970).
[192] Wang, L. C.: *Cereal Chem.*, **48,** 229 (1971).
[193] Hall, T. C., McLeester, R. C., Bliss, F. A.: *Phytochemistry*, **11,** 647 (1972).
[194] Morita, Y., Horikoshi, M.: *Agr. Biol. Chem.*, **36,** 651 (1972).

[195] Nakadai, T., Nasuno, S., Iguchi, N.: *Agr. Biol. Chem.*, **37,** 757, 767, 775 (1973).
[196] Wright, D. J., Boulter, D.: *Phytochemistry*, **12,** 79 (1973).
[197] McLeester, R. C., Hall, T. C., Sun, S. M., Bliss, F. A.: *Phytochemistry*, **12,** 85 (1973).
[198] Kapoor, A. C., Desborough, S. L., Li, P. H.: *Potato Res.*, **18,** 469 (1975).
[199] Thanh, V. H., Okubo, K., Shibasaki, K.: *Tohoku J. Agr. Res.*, **25,** 41 (1975).
[200] Börner, T., Jahn, G., Hagemann, R.: *Biochem. Physiol. Pflanzen*, **169,** 179 (1976).
[201] Thoren, E.: *Var Föda*, **30,** 69 (1978).
[202] Smith, M. B., Back, J. F.: *Austral. J. Biol. Sci.*, **21,** 549 (1968); **23,** 1221 (1970).
[203] Coduri, R. J., Rand, A. G.: *J. Assoc. Offic. Anal. Chem.*, **55,** 461 (1972).
[204] Coduri, R. J., Rand, A. G.: *J. Assoc. Offic. Anal. Chem.*, **55,** 464 (1972).
[205] Parkinson, T. L.: *J. Sci. Food Agr.*, **23,** 659 (1972).
[206] Burley, R. W.: *Biochemistry*, **12,** 1464 (1973).
[207] Lopiekes, D. V., Dastoli, F. R., Price, S.: *J. Chromatog.*, **23,** 182 (1966).
[208] Ritchard, W. J.: *J. Chromatog.*, **16,** 327 (1964).
[209] Winston, S., Friedman, H., Schwartz, E. E.: *Anal. Biochem.*, **6,** 404 (1963).
[210] Bocci, V.: *J. Chromatog.*, **8,** 218 (1962).
[211] Tometsko, A. M., Delihas, N.: *Anal. Biochem.*, **18,** 72 (1967).
[212] Müller-Eberhard, H. J.: *Scand. J. Clin. Lab. Invest.*, **12,** 33 (1960).
[213] Kolin, A.: *Science*, **117,** 134 (1953).
[214] Vesterberg, O., Svensson, H.: *Acta Chem. Scand.*, **20,** 820 (1966).
[215] Svensson, H.: *Acta Chem. Scand.*, **15,** 325 (1961).
[216] Haglund, H.: *Sci. Tools (LKB)*, **14,** No. 2 (1967).
[217] Haglund, H.: *Isoelectric Focusing in pH Gradients*, in *Methods of Biochem. Anal.*, Vol. 19, Glick, D. (ed.), Wiley, London, 1971.
[218] Secchi, C.: *Anal. Biochem.*, **51,** 448 (1973).
[219] Osterman, L.: *Sci. Tools (LKB)*, **17,** 31 (1970).
[220] Godson, C. N.: *Anal. Biochem.*, **35,** 66 (1970).
[221] Finlayson, G. R., Chrambach, A.: *Anal. Biochem.*, **40,** 292 (1971).
[222] O'Brien, T. J. O., Liebke, H. H., Cheung, H. S., Johnson, L. K.: *Anal. Biochem.*, **72,** 38 (1976).
[223] Wrigley, C. W.: *Sci. Tools (LKB)*, **15,** 17 (1968).
[224] Fawcett, J. A.: *FEBS Lett.*, **1,** 81 (1968).
[225] Righetti, P., Drysdale, I. W.: *Biochim. Biophys. Acta*, **236,** 17 (1971).
[226] Wellner, D., Hayes, M. B.: *Ann. N. Y. Acad. Sci.*, **209,** 34 (1973).

[227] Gainer, H.: *Anal. Biochem.*, **51,** 646 (1973).
[228] Awdeh, Z. L., Williamson, A. R., Askonas, B. A.: *Nature*, **219,** 66 (1968).
[229] Leaback, D. H., Rutter, A. C.: *Biochem. Biophys. Res. Commun.*, **32,** 447 (1968).
[230] Bispink, G., Neuhoff, V.: *Z. Physiol. Chem.*, **357,** 991 (1976).
[231] Uriel, J., Berges, J.: in *Electrophoresis and Isoelectric Focusing in Polyacrylamide Gel*, Allen, R. C., Maurer, H. R. (eds.), de Gruyter, Berlin, 1974, p. 235.
[232] Graesslin, D., Weise, H. C., Rick, M.: *Anal. Biochem.*, **71,** 492 (1976).
[233] Dale, G., Latner, A. L.: *Clin. Chim. Acta*, **24,** 61 (1969).
[234] Scheele, G. A.: *J. Biol. Chem.*, **250,** 5375 (1971).
[235] Ferro-Luzzi Ames, G., Nikaido, K.: *Biochemistry*, **15,** 616 (1976).
[236] Strand, M., August, I. T.: *Cell*, **13,** 399 (1978).
[237] Anderson, L., Anderson, N. G.: *Proc. Natl. Acad. Sci. U. S.*, **74,** 5421 (1978).
[238] Nguyen, N. A., Rodbard, D., Svendsen, P. J., Chrambach, A.: *Anal. Biochem.*, **77,** 39 (1977).
[239] Everaerts, F. M., Beckers, J. L., Verheggen, P. E. M.: *Isotachophoresis: Theory, Instrumentation and Application*, Elsevier, Amsterdam, 1976.
[240] Svendsen, F. J., Rose, C.: *Sci. Tools (LKB)*, **17,** 13 (1970).
[241] O'Farrell, P. H.: *J. Biol. Chem.* **250,** 4007 (1975).
[242] Morrissey, J. H.: *Anal. Biochem.*, **117,** 107 (1981).
[243] Wray, W., Boulikas, T., Wray, W. P., Hancock, R.: *Anal. Biochem.*, **118,** 197 (1981).
[244] Poehling, H. M., Neuhoff, V.: *Electrophoresis*, **2,** 141 (1981).

CHAPTER 4

Column-chromatographic methods based on physical, chemical and biochemical binding

I. Kerese

4.1 GENERAL ASPECTS OF CHROMATOGRAPHY

Chromatography is a separation technique based on sorption and desorption. Chromatographic separation results when the substances bound on the thin initial zone of a porous sorbent column are made to move with different velocities by a liquid or gas flowing through the column [1, 2].

In the classical liquid–solid chromatographic methods used for the separation of proteins, peptides and amino-acids (adsorption, partition and ion-exchange chromatography), it must be taken into account that although the same regularities hold as for other ionic water-soluble materials, these regularities may be influenced in special ways by the molecular weights, stereo-structures, multivalent character, charges and solubilities of the individual proteins [3, 4]. These characteristics strongly limit the nature of the sorbents that may be employed in protein analysis. Fundamental requirements as regards the sorbents are as follows.

(a) The stationary phase, the sorbent column, should be insoluble in the mobile phase, the eluent.

(b) The proteins, peptides and amino-acids to be separated on the column should be bound with different strengths to the sorbent.

(c) The proteins, peptides and amino-acids to be separated should not be adsorbed irreversibly by the sorbent.

(d) The sorbent should not cause irreversible decomposition of the proteins, peptides and amino-acids to be separated, and should not interact with the eluent.

In gel chromatography based on the dimensions of the molecule and the pore size of the gel, the fundamental separation affect is molecular sieving; this is not appreciably affected by any adsorption phenomena that may occur (mainly with small molecules).

During the past 10 years, new chromatographic methods have been developed for use in protein analysis. They are due to the utilization of various sorption effects, and of the improved performances of sorbents and chromatographic instruments [5–7].

The application of new chromatographic sorbents has led to the development of ampholyte-displacement, covalent, affinity and hydrophobic-interaction chromatography. The use of these has in certain cases given rise to unique results in protein analysis.

With regard to some of the physical conditions of chromatography (the shortest duration, lowest pressure and lowest temperature of elution), in the interest of the best possible separation it is frequently necessary to come to a compromise in the chromatographic separation of proteins and of peptides containing different numbers of members. Which of these mutually conflicting interests is to be favoured in liquid–solid column chromatography depends primarily on the physico-chemical properties and mechanical strengths of the sorbents employed. Spherical sorbents with smaller particle sizes and higher strengths have led to the development of high-pressure liquid chromatography (HPLC), which has resulted in an increase in the flow-rate of the eluent (from 0.001–0.01 to 1 cm/s), making the separation very much faster. The advantages of HPLC may be summarized as follows: a smaller sorbent column and a quantity of sample are required; the much shorter working time increases the number of samples that can be handled; the high precision of the apparatus and the coupled integrator and computer evaluation give low limits of error.

A very extensive literature deals with the theoretical basis and various practical questions of HPLC [8–18], and accordingly an account of this technique will not be given in the present book.

In protein analysis the advantages of HPLC to some extent limit the usefulness of thin-layer chromatography (TLC). Experience to date shows that TLC will remain an excellent method for the screening of substances of unknown composition and for selection of a sorbent, and it will continue to be a good method for relatively low-performance, qualitative separations. However, the speed and accuracy of HPLC, and the possibility of further examination of the separated material, make it beyond doubt one of the fundamental chromatographic methods of the future [19].

A generally applicable method cannot be proposed for the optimum chromatography of proteins and of peptides of different sizes, because of their different primary and secondary structures, for these features fundamentally determine their physico-chemical properties, and consequently the conditions for their chromatography. A number of excellent monographs, technical books and reviews [20–24] provide accounts of the principles, the performance and the details of the chromatographic techniques, and abundantly satisfy the requirements for information that arise in protein analysis.

4.2 APPLICATIONS OF ADSORPTION COLUMN CHROMATOGRAPHY IN PROTEIN ANALYSIS

In the adsorption chromatographic separation of proteins and peptides, the stationary phases are primarily adsorbents possessing *hydrophilic groups*, such as hydroxyapatite, aluminium hydroxide, kaolin, celite and cellulose. The proteins and peptides are bound by van der Waals forces to the hydrophilic sites of these low-capacity sorbents. Proteins and peptides that are labile owing to their stereo-structure and multivalent character may be denatured by this binding and may even lose their biological activity. In spite of all this, there are some useful applications of adsorption chromatography in protein analysis [25]. For example, Tiselius succeeded in separating complex fractions from bovine plasma on hydroxyapatite, whereas this was not possible either by ion-exchange chromatography on cellulose or by electrophoresis [26, 27].

The work on the chromatography of proteins on hydroxyapatite is an excellent illustration of how to use adsorption column chromatography to advantage [27, 28] (see Table 4.1). The chromatography of various proteins dissolved in phosphate buffer was begun with 0.01–0.025 *M* phosphate buffer, and the molarity of the eluent was subsequently increased stepwise. The hold-up volume of the column was measured, and about 1.5 hold-up volumes of each buffer were used for initial fractionation, each fraction then being rechromatographed with a stepwise buffer gradient. The maximum pressure needed was 9.3–18.7 kPa (70–140 mmHg).

The industrial manufacture of hydroxyapatite sorbents with improved mechanical and physico-chemical properties has led to more extensive use of adsorption chromatography on a wider range of materials, and has assisted in deeper study of the theory and regularities of adsorption phenomena [29], especially with regard to the mechanism and types of substance that can be adsorbed (Table 4.1). Bernardi *et al.* [33, 34], for example, have disproved the idea that only peptides and proteins with molecular weights larger then 10^4 adsorb well on hydroxyapatite.

Stepwise elution is generally used in adsorption chromatography, but Skelton [35] used concentration gradient elution on a hydroxyapatite column to isolate two proteolytic enzymes with similar physical characteristics from a papaya latex extract, the pH 7 phosphate buffer concentration being raised from the initial 0.05 *M* to 0.2 *M*.

Table 4.1
Proteins that may be chromatographed on hydroxyapatite

Protein	Molarities of eluent phosphate buffers (pH = 6.8)	Ref.
Albumin (bovine plasma)	0.02, 0.04, 0.07, 0.11, 0.40	[27]
γ-Globulin	0.02–0.13 (pH = 7); constant concentration with 0.2 *M* NaCl	[27]
Chromoproteins from lipoprotein, egg-yolk	0.01, 0.02, 0.05	[27,30,31]
Lysozyme	0.02, 0.08, 0.11	[27]
Ovalbumin	0.025, 0.05, 0.07, 0.10	[27]
Serum protein	0.02, 0.03, 0.04, 0.05, 0.06, 0.08, 0.10, 0.15, 0.20, 0.30, 0.65	[32]

These high-quality hydroxyapatites have also permitted the chromatography, even in preparative quantities, of the skeletal proteins, which are otherwise difficult to separate. Mosersky *et al.* [36] chromatographed 200 mg of myosin on a 350 × 25.4 mm Bio-Gel HT column with 0.16, 0.23 and 0.4 *M* phosphate buffer of pH 6.8, containing mercaptoethanol and potassium chloride. In research into the microheterogeneity of tropocollagen, Kawasaki and Bernardi [37] successfully chromatographed this material on a hydroxyapatite column.

The disadvantage of hydroxyapatite columns, the slowness of elution, can be decreased by precipitating the hydroxyapatite onto cellulose. With water, this preparation (Serva) swells to six times its dry weight, and consequently the resistance pressure of the column is strongly diminished.

The types of adsorbent available for adsorption chromatography expanded with the appearance of the controlled-pore glass (CPG) sorbents. Crone *et al.* [38] chromatographed acetylcholinesterase with 5 m*M* TRIS buffer of pH 8, containing sodium chloride, on a 230 × 21.2 mm CPG column, and in the course of this also examined the regularities of the adsorption of proteins on the glass. Bock *et al.* [39] adsorbed membrane proteins and various soluble proteins on a CPG column with a surface area of

62–75 m^2/g and a pore size of 313–368 Å. The desired isolation was achieved by combining the eluent buffer with chaotropic salts* (e.g. LiBr), which allowed the proteins to be concentrated to a very great extent in one step. Bresler *et al.* [40] investigated the adsorption isotherms of influenza virus on a 200 × 10 mm column of CPG with a pore size of 850 Å and a particle diameter of 0.06–0.1 mm. This work also provided important theoretical data on the successful application of CPG to adsorption chromatography, e.g. the correlation between the pH of the eluent TRIS buffer and the p*K* of the SiOH group, which influences the sharpness of separation.

Another group of substances used as the solid phase in adsorption chromatography comprises the *apolar adsorbents*, such as active carbon, silanized silica gel, and a large number of industrial apolar preparations. The mobile phase in this reversed-phase chromatography is a polar solution. The reversed-phase technique is not suitable for the chromatography of native proteins and peptides with a polar character, but it is employed occasionally for the separation of peptides with low molecular weights. With a high-pressure technique involving an acetonitrile–water mixture on Phenyl-Corasil, Poragel PN and PS columns, Hansen *et al.* [41] achieved good separation up to decapeptides. Mönch and Dehnen [42] chromatographed a mixture of 10 dipeptides, 2 tetrapeptides, 1 pentapeptide, 1 heptapeptide and 1 octapeptide of various polarities, by means of a high-pressure technique with linear gradient elution on octadecylsilane. using methanol and a 0.05 *M* potassium dihydrogen phosphate buffer of pH 2.

It is to be expected that in the future the high-pressure technique will also have a favourable effect on the utilization of adsorption chromatography for protein analysis.

4.3 APPLICATIONS OF PARTITION COLUMN CHROMATOGRAPHY IN PROTEIN ANALYSIS

The basis of partition column chromatographic separation is the repeated counter-current partition of the material to be separated, between a stationary and a mobile liquid phase. The granular support carrying the stationary liquid phase takes over the role of the units of the counter-current extraction apparatus (Craig apparatus), and the tube-shaking is replaced by the diffusion between the stationary liquid phase and the mobile liquid phase flowing through it in the course of chromatography. The equilibrium

* A chaotropic salt increases the solubility of a hydrophobic species in water by causing extensive breaking of the water structure by hydration of its ions.

concentration partition coefficient between the phases may be modified by variation of the composition of the eluent. The characteristics of eluents have been studied in detail by Albertsson [43, 44] and Carpenter [45]. Normally an aqueous stationary phase is used, with an organic mobile phase, but reversed-phase methods are also used.

The classical support for partition column chromatography of proteins and peptides is Hyflo-Super-Cell. Some of the characteristic applications of this in protein analysis are listed in Table 4.2.

Table 4.2
Proteins separable by partition chromatography

Protein	Sorbent	Eluent solution	Ref.
Catalase	Hyflo-Supercel	ethanol–aqueous ammonium sulphate	[46]
Ribonuclease	Hyflo-Supercel	cellosolve–aqueous ammonium sulphate	[47]
Insulin	Hyflo-Supercel	ethyl butyl cellosolve–sodium phosphate solution, pH 3–7.6	[48]
Insulin	Hyflo-Supercel	butan-2-ol–0.5% dichloroacetic acid–0.1 *M* hydrochloric acid	[49]
γ-Globulin	Celite 545	ethyl butyl cellosolve–sodium phosphate solution, pH 3–7.6	[48]
Chymotrypsin, chymotrypsinogen			[50], [51]

More recently, gel filters have been employed as supports. Yamashiro [52] used Sephadex-25 in the separation of oxytocin, and eluted with n-butanol–benzene–pyridine–0.1% acetic acid (6 : 2 : 1 : 9). For the separation of a guinea-pig red blood cell hydrolysate, Anker [53] used 2% agarose gel, onto which Dextran T–70, Carbowax PEG–6000 and Blue Dextran had been adsorbed by shaking for 24 h. The mobile phase consisted of a solution of 10% Carbowax PEG–6000, 0.7% Dextran T–70, 0.01 *M* phosphate buffer (pH 7.4) and 0.15 *M* potassium chloride. The chromatography resulted in the same separation as counter-current extraction.

During the past 30 years, partition column chromatography has not attained such importance in protein research as in the study of other substances. The complications which arise in column chromatographic elution as a result of the interactions of the properties of the proteins and peptides, the stationary phase and the eluent, are sufficient to explain why first paper chromatography, and later thin-layer chromatography have been the preferred forms of partition chromatography for use in protein analysis.

4.4 APPLICATIONS OF ION-EXCHANGE COLUMN CHROMATOGRAPHY IN PROTEIN ANALYSIS

4.4.1 General aspects

The stationary phase in ion-exchange column chromatography is a substance with a polymeric matrix on which acidic or basic active groups capable of electrolytic dissociation are covalently bound. The polymeric matrix ensures that the exchanger is not water-soluble. The active (or functional) acidic groups fixed on a cation-exchange matrix dissociate to give protons, leaving on the matrix the anionic groups, which can then bind cations. The basic functional groups fixed on an anion-exchange matrix dissociate to give anions, leaving quaternary ammonium ions on the matrix, which are able to bind other anions. Thus the counter-ions neutralizing the fixed ions may be exchanged with other ions with the same charge-sign; the nature of the counter-ions is characteristic not of the ion-exchanger, but of its state. Accordingly, a cation-exchanger may be in the hydrogen form or the sodium form, for instance, depending on whether the counter-ion is H^+ or Na^+. In the same way, an anion-exchanger may be in the chloride or the acetate form, for example, depending on whether the counter-ion is Cl^- or $CH_3CO_2^-$.

The affinity of the ion-exchanger for ions increases with the effective charge on the ion, and with decreasing radius of the hydrated ion. The ion-exchange equilibrium also depends on the relative concentrations of the species involved.

In the ion-exchange chromatography of proteins and peptides existing as zwitterions, the physico-chemical processes of ion-exchange are much more complicated than in the case of simple ions. The proteins and peptides are dipolar macromolecules: polyelectrolytes with differentiated structures. Their sorption or desorption is based on electrostatic interactions with the coulombic forces on the active centres of the ion-exchange sorbent. In addition, however, other phenomena are manifested, such as apolar adsorption, binding by hydrogen-bonding, and partition as a consequence of solubility.

In ion-exchange chromatography, whether a cation- or an anion-exchange sorbent is selected is determined by the net charge of the molecules of the material to be separated (Fig. 4.1). The charges of proteins and peptides in solution are determined not only by the ratios of the acidic, neutral and basic amino-acids of which they are comprised, but also by the pH of the solution. In a solution with a pH lower than its isoelectric point, the dissolved protein and peptide molecules are positively charged, while in a solution of higher pH than the isoelectric point they are negatively charged.

Accordingly, a cation-exchanger must be used for the electrostatic sorption of positively-charged substances, and an anion-exchanger for that of negatively-charged ones. At a pH between that of its isoelectric point and 1 unit above, a protein can be chromatographed on either a cation- or an anion-exchanger. However, neither type of exchanger binds the zwitterion at its isoelectric point. The bound substances may be eluted separately by variation

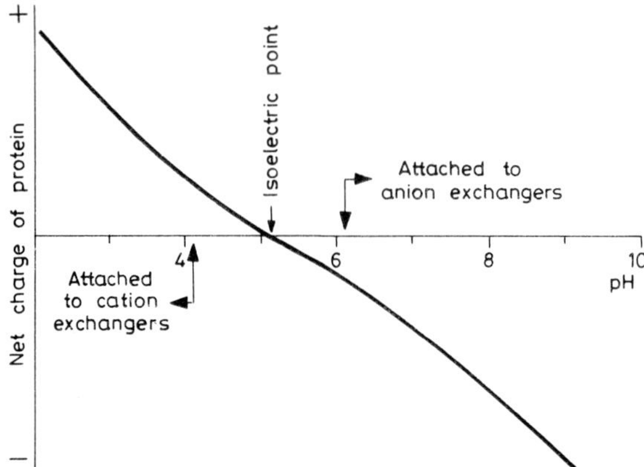

Fig. 4.1. The net charge on a model protein, as a function of pH (by kind permission of Pharmacia Fine Chemicals AB, Sweden)

of the pH and the ionic strength of the eluent. The pH of the sample solvent and of the eluent buffer must be chosen so that the proteins remain stable and do not undergo decomposition.

Selection of an appropriate ion-exchanger is facilitated if we know the isoelectric point of the substance to be separated, and also the pH range of stability of active materials. In research work, these data may often be unknown to the analyst. The correct selection of the sorbent is then much easier if the isoelectric point of the unknown substance can be established. If the substance is soluble in water or a neutral salt solution, the isoelectric point may be determined by titration [54], pH measurement, or isoelectric focusing. An empirical choice can also be made by a procedure based on the principle of the method used by Hirs *et al.* [55] to establish the capacity of an ion-exchanger, with the difference that the same quantities of cation- or anion-exchange sorbent, the same volumes of buffers with pH appropriate for the character of the sorbents, and the same amounts of examination material must be used throughout. After the completion of binding by the sorbents, accelerated by shaking, protein determination on the supernatant solutions shows the extent of sorption: the solution with the lowest concentration corresponds to the greatest degree of sorption. As a check on irreversible binding in the case of enzymes, the amount of enzyme that can be desorbed from the ion-exchanger must also be determined.

The capacity of the selected ion-exchanger for the material under examination may be determined by the procedure of Hirs *et al.* [55]: identical volumes of solutions with known, different concentrations of the substance to be separated are added to constant amounts of the chosen ion-exchange sorbent. Binding by the sorbent can be accelerated by shaking. The results of protein determinations on clear supernatant samples may then be used to calculate how much protein was bound by 1 g of ion-exchanger from the solutions of different concentrations, and a saturation curve may be constructed, which gives the capacity of the ion-exchanger with respect to the substance to be chromatographed.

Factors decisively influencing the ion-exchange separation are temperature, and the ionic strength and pH of the eluent solution. The components of the sample, bound with different affinities, can be fractionally eluted off the ion-exchanger by varying the pH of the eluent towards the isoelectric points of the bound components, or by increasing the initially low ionic strength of the eluent. The pH and ionic strength may be varied continuously or stepwise.

Cellulose-, agarose- and dextran-based ion-exchange gels are mainly used for the column chromatographic separation of proteins and peptides. Synthetic ion-exchange resins are used to a smaller extent.

4.4.2 Column chromatography of proteins and peptides on synthetic ion-exchange resins

The newer ion-exchange sorbents give better chances for separation of proteins than synthetic ion-exchange resins do. Depending on the properties of the ion-exchange resins used, certain proteins with high molecular weights and strongly differentiated stereo-structures may readily be denatured as a consequence of the sorption, and other proteins may be bound irreversibly on the functional groups of the ion-exchange resin. Various synthetic resins are utilized for ion-exchange column chromatography in protein analysis: a cation-exchanger with a matrix of polymethacrylic acid (Amberlite IRC 50) or of sulphonated polystyrene cross-linked to various extents with divinylbenzene (Dowex 50, Amberlite IR 120); anion-exchangers with a matrix of polystyrene containing the quaternary ammonium ion (Dowex 1, Dowex 2). Table 4.3 lists some characteristic examples of proteins and peptides successfully chromatographed on the various ion-exchange resins.

Table 4.3

Proteins and peptides that may be chromatographed on an ion-exchange resin column

Proteins and peptides separated on synthetic cation-exchange resins

Proteins separated on resins with a polymethacrylate matrix	
ribonuclease	[55]
cytochrome C	[56]
desoxyribonuclease	[57]
lysozyme	[58]
insulin	[59]
prolactin	[60]
growth hormone	[61]
histones	[62]
CO-haemoglobin	[56]
chymotrypsin, chymotrypsinogen	[63]
trypsinogen	[64]
papain	[65]
cellulase	[66]
pectinase	[67]

Peptides separated on resins with a sulphonated polystyrene matrix	
acidic peptides of lysozyme	[68]
tryptic peptides of human haemoglobin	[69]
tryptic peptides of oxidized ribonuclease	[70]
peptides of thermally-denatured collagen	[71]
tryptic peptides of starch phosphoglycerate kinase	[72]
peptides of carboxymethylated protein-avidin	[73]
peptides of soya hydrolysate	[74]
tryptic peptides of extracellular ribonuclease	[75]
lysozyme	[76]
apomyoglobin	[76]
glycosyl-Asn derivatives	[77]
tryptic peptides of haemoglobin, lysozyme, ribonuclease	[78]
Peptides separated on resins with a polymethacrylate matrix	
tryptic peptides of ribonuclease	[79]
peptic peptides of ribonuclease	[80]
decomposition products of ribonuclease after *IEF*	[81]

Peptides chromatographed on anion-exchange resins

Peptides separated on a resin with a polystyrene matrix containing the quaternary ammonium group	
tryptic peptides of tobacco mosaic virus	[82]
tryptic peptides of liver catalase	[83]
aminoethylated γ-chain tryptic peptides of foetal haemoglobin	[83]
tryptic peptides of human globin	[84]
tryptic peptides of collagen	[85]
Witte peptone	[86]

The separations given in this Table include some interesting examples of exploitation of composition of the eluent [69, 71, 84], automation [72, 76, 78], and the combination of cation- and anion-exchange columns [74].

In spite of these examples, it must not be forgotten that, because of their strong binding and non-specific adsorption, proteins and peptides cannot be quantitatively desorbed from ion-exchange resins. Hence the ion-exchange resins are more limited in scope than ion-exchangers with a matrix of cellulose, agarose or dextran.

4.4.3 Column chromatography of amino-acids on synthetic ion-exchange resins

By ion-exchange column chromatography, the amino-acid composition of a protein or the free amino-acids in a physiological solution may be determined in an extremely short time and with high accuracy. This possibility arose from the research of the group of Moore and Stein. In 1948 the qualitative reaction of ninhydrin with amino-acids, which had been known for 30 years, was developed into a quantitative determination [87], and then in 1951 a method was published for the ion-exchange column chromatography of the amino-acids. These two methods are the basis for automated amino-acid analysis.

Before chromatographic determination of the amino-acids, the samples must be appropriately pretreated. The amino-acids must be liberated from their peptide binding by hydrolysis, and a physiological sample must be suitably deproteinated. The correct performance of these operations is an indispensable precondition for accurate results to be obtained. These operations will therefore now be surveyed.

Hydrolysis of the sample

Before the hydrolysis, it is advisable to determine the nitrogen content and the dry-matter content of the substance being examined, to establish the weight of sample to take. If the sample contains water in excess of the air-dried moisture content, this water will dilute the acid or base used for the hydrolysis, and such samples must be dried before hydrolysis.

Most amino-acids undergo various degrees of hydrolysis, depending on the other substances present and on the manner of hydrolysis. For total amino-acid determination, therefore, it is practical to subject the samples to both acidic and alkaline hydrolysis for various lengths of time.

The different parameters of acidic hydrolysis have been investigated by many research workers; accordingly, data are available on electrostatic [89] and steric effects in the splitting of various types of peptide bonds, on the reasons for the special stability of certain peptide bonds [90], and on the influence of the nature and concentration of the acid, the temperature, and the duration of the hydrolysis.

Various *acids for hydrolysis* are recommended in the literature. Until fairly recently, hydrochloric acid was most generally used. Linderstrøm–Lang [91] employed the azeotropic acid (21.88%, w/v; ~6 *M*). The degree of decomposition of the amino-acids during hydrolysis with hydrochloric acid depends on the duration of the hydrolysis, but other important factors are the presence of oxygen (i.e. the degree of evacuation), the purity of the hydrochloric acid, and the temperature of the hydrolysis. A number of authors stress the importance of using hydrochloric acid redistilled from glass apparatus, because metal ions may catalyse the destruction of several amino-acids in the course of hydrolysis [92]. Linderstrøm–Lang [91] draws attention in particular to the danger of contamination of the hydrochloric acid with iron. In non-isolated protein preparations, and especially in agricultural products, the quantities of microelements exceed the metal ion impurities in the hydrochloric acid of analytical purity used for their hydrolysis; hence, analytical-purity hydrochloric acid of reliable manufacture (with Fe < 0.00005%) may be employed even without distillation [93, 94].

Table 4.4
Amino-acids decomposing during hydrolysis with 6 *M* hydrochloric acid

Amino-acids decomposed	Substances promoting decomposition	Refs.
Trp	O_2, carbohydrates, edestin, Ser	[102–104]
Asn, Gln	Cys, pyruvic acid, Zn	[105]
Thr, Ser		[105–111]
Cys, Cys-Cys	O_2, carbohydrates	[93, 111]
Cys	carbohydrates	[112]
Met	O_2	[93, 113]
Tyr, Glu, Pro, Asp		[111, 114]

An exact description of the correct performance of the hydrolysis is given by Moore and Stein [93]. The most important points are as follows: the hydrolysis should be done in glass (e.g. Pyrex, Corning Cat. No. 9860) not containing substances that can be leached by hydrochloric acid; before the hydrolysis apparatus is sealed off, the air-space above the hydrochloric acid frozen out in it must be evacuated to 7.9 Pa (0.06 mmHg); on every sample, one hydrolysis for 20 h and one for 70 h must be performed, and from the results of these, by extrapolation to zero time, those amino-acids that may be determined by hydrolysis with hydrochloric acid are obtained.

A number of authors [95–101] agree that, of the other substances that may be present, the carbohydrates are of special importance in affecting the course of the hydrolysis. For this reason, the method of hydrolysis in the case of carbohydrate-containing agricultural products differs from that for

Table 4.5
Recommended amounts of test substance and 6 *M* hydrochloric acid for hydrolysis

Test substance	Amounts of substance and 6 *M* HCl	Hydrolysis temperature °C	Hydrolysis time, h	Refs.
ribonuclease	5 mg + 1 ml of 6 *M* HCl	110	22 and 70	[116]
various carbohydrate-containing substances	1 : 500 – 1 : 5000*			[117]
foodstuffs, fodders	100 mg + 100 ml of 6 *M* HCl	137	20	[118, 119]

* "... amounts of 6 *M* HCl equal to 500 to 5000 times the weight of the protein " [117].

isolated proteins. Table 4.4 details the destructive amino-acid reactions which can occur during hydrolysis with 6 *M* hydrochloric acid.

The results for Val and Ile will be low if the hydrolysis time is too short, because certain valyl and isoleucyl bonds are very stable, and hydrolysis for 96 h is necessary to liberate these amino-acids from certain peptide bindings [115].

Numerous modifications are given in the literature, to decrease humin formation and the accompanying amino-acid destruction, during hydrolysis with hydrochloric acid; some of these involve the ratio of substance to be hydrolysed to the amount of hydrochloric acid (Table 4.5).

Other authors add reducing agents to the 6 *M* hydrochloric acid (Table 4.6).

The more recent efforts to diminish the destruction of the amino-acids are based on use of acids other than hydrochloric. Liu and Chang [126] hydrolysed 2–3 mg of protein for 22, 48 and 72 h at 110 °C with 1 ml of 3 *M* *p*-toluenesulphonic acid containing 0.2% tryptamine, the ampoule being evacuated to 2.7–4.0 Pa 0.02–0.03 mmHg before sealing. Liu [127] subsequently performed the hydrolysis with tryptamine in 4 *M* methanesulphonic acid in place of the *p*-toluenesulphonic acid. Penke *et al.* [128] report a procedure for hydrolysis with 3 *M* mercaptosulphonic acid.

Table 4.6
Hydrolyses with hydrochloric acid containing additives

Test substance	Composition of acid used for hydrolysis	Refs.
plant samples	6 *M* HCl + phenol	[120–122]
	6 *M* HCl + 0.5% 1,4-butanediol	[123]
	6 *M* HCl + 2–4% thioglycollic acid	[124, 125]
lysozyme, fetuin	6 *M* HCl + mercaptopropionic acid	[125]
	6 *M* HCl + mercaptosuccinic acid	[125]
	6 *M* HCl + oxalic acid	[125, 126]

The performic acid oxidation procedure developed by Hirs [129] can be usefully employed for the reliable determination of Cys and Met; the oxidized products, Cys(O_3H) and Met(O), are subjected to chromatography.

Alkaline hydrolysis must be used for the determination of Trp. As for acidic hydrolysis, a number of variants and modifications are to be found in the literature (see Table 4.7). The study by Noltmann *et al.* [130] deserves special attention; their basic findings may be regarded as guidelines for

Table 4.7
Media for alkaline hydrolysis of test substances

Hydrolysis medium	Refs.
4 *M* $Ba(OH)_2$	[130–132]
4 *M* NaOH	[133, 134]
5 *M* NaOH	[135]
6 *M* NaOH	[136]
4.2 *M* NaOH + partly hydrolysed starch	[137]

correct alkaline hydrolysis: silicon-glass subject to minimal leaching is used (Vycor, Corning Cat. No. 7900, with 96% Si content), and before sealing-off, the hydrolysis tube is evacuated to 13.3 Pa (0.1 mmHg). Both acidic and neutral amino-acids can be separated by chromatography of the alkaline hydrolysate. If the Leu results obtained are compared with the Leu value found by acid hydrolysis, conclusions may be drawn as to the Trp losses originating from the operations of the alkaline determination.

The methodological problems involved in Trp determination are very well reflected by many publications, and these have been surveyed (96 references) by Friedman and Finley [138].

Individual hydrolysis methods can be recommended only on the basis of some definite aspect of the analysis (e.g. the absence or presence of carbohydrate) and none seems universally applicable.

Deproteination

For the chromatography of the free amino-acids in physiological solutions, tissue fluids and organ extracts, the sample solution must be freed not only from the inorganic species, but also from the proteins and peptides in the solution: these would otherwise clog the column as contaminants on the resin, and also the ninhydrin-positive peptides eluted from the column after different times would give peaks of various sizes on the chromatogram, thereby hampering its evaluation. Three methods are generally employed for this process (deproteination) of freeing the sample solution from proteins and high molecular-weight peptides: precipitation with chemicals; purification by ion-exchange chromatography; ultrafiltration.

Compounds frequently utilized for deproteination by chemical precipitation include picric acid, sulphosalicylic acid, trichloroacetic acid and tungstic acid.

In deproteination with 1% picric acid [139], the Trp is decomposed and the entire procedure is a little complicated.

There is a deterioration in the chromatographic separation of Asp, Met (O_2), Thr and Ser from serum deproteinated with 3% sulphosalicylic acid; the use of a larger amount of sulphosalicylic acid leads to a poorer resolution throughout the whole of the chromatographic process [140]. Mondino *et al.* [141] eliminated this problem by eluting with a lithium citrate buffer. When deproteination with 3% sulphosalicylic acid was used for 7 parallel samples of a given plasma, the coefficients of variation for the amino-acids determined varied between 1.5% (Arg) and 22.2% (Asp). In the comparative investigations by Block *et al.* [142], 5–48% more free amino-acid was obtained if the plasma was deproteinated with sulphosalicylic acid than if picric acid was used.

According to Bhatty [143], deproteination with trichloroacetic acid may not be complete and different materials require different concentrations of the reagent.

In a study of deproteination with tungstic acid, Bitó and Dawson [144] found that the loss of free amino-acid depends on the relative concentration of the protein and on the precipitant. The losses of the individual amino-acids varied between 50% (Arg) and 1.5% (neutral amino-acids).

Deproteination on a cation-exchange resin (Chromobead Type A, H^+ form) was examined by Reid [145]. The coefficients of variation in 7 parallel analyses on a given serum lay between 2.7% (Lys) and 13.6% (Asp).

The best procedure for deproteination before the chromatography of free amino-acids is ultrafiltration. This has the advantages over the above-mentioned procedures that there is no amino-acid loss as a consequence of contamination, and that it requires only a short time, so decomposition is avoided.

All the modifications of ultrafiltration are simple, and a wide choice of suitable filtering devices is available commercially, together with filter membranes with well-defined pore sizes, of excellent quality (see Chapter 2). Of the various types of ultrafiltration apparatus available, the low-volume ones are of special importance. Gressner [146] gives the design of a small ultrafiltration apparatus that may be home-made from plastic, and is suitable for the deproteination of 20–100 μl of physiological solution. In this apparatus, 20 μl of plasma are forced through the Amicon UM-2 membrane by nitrogen at about 2.5 atm in 10 min; this method is therefore of particular significance in the analysis of free amino-acids in the serum of infants.

Chromatography of amino-acids

The amino-acids are bound by various forces (coulombic and van der Waals) to a sulphonated polystyrene resin in the sodium (or lithium) form, 8% cross-linked with divinylbenzene. In the course of the sorption and elution of the amino-acid molecules bound on the ion-exchanger, the characteristic charge, pK, molecular weight and side-chain composition (polar or apolar side-chains) of the individual amino-acids exert their effects with different intensities; the sequence of desorption is the overall resultant of all these features.

The commercial resins available in sodium form for amino-acid determination are nowadays all of the spherical type; for 'single-column' chromatography with a duration of 4 h, the particle size is 17 μm, while for high-pressure chromatography with a duration of 45–70 min it is 8 μm. In the event of the fairly strong contamination of a resin, or the necessity to change from the sodium to the lithium form, a good method is washing with 6 M nitric acid (the resin being removed from the chromatography column for this purpose), followed by reconversion into the sodium form or conversion into the lithium form by the procedure of Moore and Stein. The eluent buffers are citrate solutions. Of the numerous variants, the sodium-salt buffers used most often for the 'single-column' procedure are as follows:

(*i*) pH 2.2 sodium citrate buffer, 0.2 *N* in sodium, for dissolution of the hydrolysate;

(*ii*) pH 3.25 sodium citrate buffer, 0.2 *N* in sodium, for elution of acidic amino-acids;

(*iii*) pH 4.25 sodium citrate buffer, 0.2 *N* in sodium, for elution of neutral amino-acids;

(*iv*) pH 6.45 sodium citrate buffer, 1.2 *N* in sodium, for single-column elution of basic amino-acids.

With buffers (*ii*)–(*iv*), elution is done at 52 °C.

Many publications have appeared in connection with the lithium elution of the amino-acids [147–154]. The buffers used in the method of Kedenburg [150] are as follows:

(*v*) pH 2.2 lithium citrate buffer, 0.3 *N* in lithium, for dissolution of free amino-acids;

(*vi*) pH 2.8 lithium citrate buffer, 0.3 *N* in lithium, for elution at 39 °C up to Cys;

(*vii*) pH 4.1 lithium citrate buffer, 1.2 *N* in lithium, for elution at 60 °C of the other amino-acids.

After completion of the chromatography, the contaminating substances still on the column are dissolved off with 0.4 *M* lithium hydroxide, and the column must then be equilibrated with the starting buffers.

The preparation of the ninhydrin solution for determination of the eluted amino-acids is described in detail by Spackman *et al.* [155]; the $E_{1\,\mathrm{cm}}^{1\%}$ values for the product should be 0.280 at 430 nm; 0.040 at 460 nm; 0.006 at 570 nm. Numerous modified preparations are to be found in the literature, with the aim of avoiding the aging of the ninhydrin solution and the accompanying decreased colour formation [156–160]. To avoid the drawbacks of methyl-cellosolve, in 1968 Moore [161] used dimethylsulphoxide for dissolution of the ninhydrin.

In part the difficulties of preparing the ninhydrin reagent, and the changes occurring in it on storage, and in part the necessity for a more sensitive reagent for the detection of amino-acids in quantities less than the nmole level, which is sufficient for high-pressure amino-acid analysers, led to the introduction of fluorescence reagents that interact with primary amino groups. Lustenberger *et al.* [162, 163] prepared the pyridoxyl derivatives of the amino-acids in the eluate, and then measured these fluorometrically. Alternatively, they converted the pyridoxil amino-acids into Schiff's bases with $NaBH_4$, and determined the radioactivity of the eluate; this latter step improved the sensitivity of the amino-acid determination to 10^{-13} mole, with a limit of error of $\pm 5\%$. Udenfriend *et al.* [164, 165] investigated in great detail the reactions of amino-acids (and proteins) with fluorescamine, and also published a design for the construction of an amino-acid analyser based on these reactions [166]. This method has the advantage that the reaction with the eluate takes place within 20 s at room temperature; its drawbacks are that Pro does not react directly with fluorescamine (for this the addition of 0.1 *M* *N*-chlorosuccinimide is required), and that the solution of fluorescamine in acetone strongly corrodes the packings of the pumps. The quantity of hydrolysate that can be analysed falls within the limits 1–100 μg. Benson and Hare [167] use fluorescamine in the presence of 2-mercaptoethanol to transform primary amines into fluorescent substances. Roth and Hampaï [168] make the amino-acids fluorescent by reacting them with *o*-phthalaldehyde, which is readily soluble in water. This reaction, which yields a bluish fluorescence, is 50 times more sensitive than that with ninhydrin. The lower limit of detection is 0.5 nmole of amino-acid. *o*-phthalaldehyde, which reacts only with α-amino groups, forms a fluorescent compound with Pro after oxidative decarboxylation of the latter with chloramine-T, *N*-bromosuccinimide or hypochlorite.

The method of detecting the eluted amino-acids is of great importance in designing the amino-acid analyser: the fluorescence method requires the substantially more complex and expensive fluorometer in place of the simple photometer used with the ninhydrin reaction.

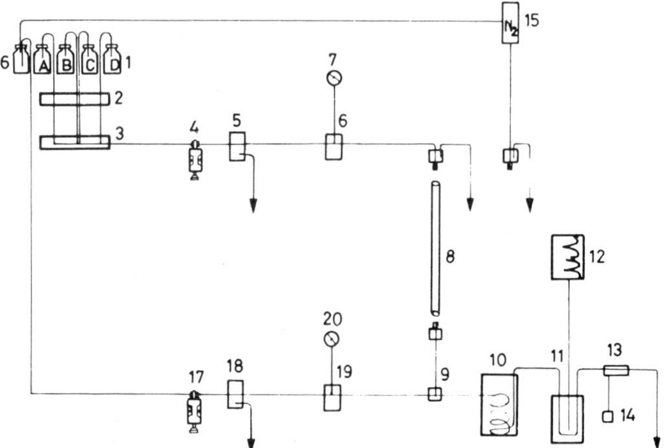

Fig. 4.2. Schematic diagram of automatic amino-acid analysers: 1, buffer reservoirs; 2, buffer valves; 3, collecting block; 4, buffer pump; 5, deaerator; 6, filter; 7, manometer; 8, chromatographic column; 9, mixing block; 10, reactor; 11, photometer: 12, chart-recorder; 13, rotameter; 14, injector; 15, nitrogen reservoir; 16, ninhydrin reservoir; 17, ninhydrin pump; 18, deaerator; 19, filter; 20, manometer

The automatic amino-acid analysers developed by Spackman *et al.* [155] involved the use of two columns: on one column, 150 cm in length, only the acidic and neutral amino-acids were chromatographed, while on the other, 15 cm in length, the basic amino-acids were eluted. Since this publication in 1958, amino-acid analysers have undergone very considerable technical development, but even the most modern HPLC amino-acid analyser is still based on the same principle. Figure 4.2 illustrates schematically the construction of automatic amino-acid analysers.

The main developments over the past 20 years are as follows.

(*i*) The manufacture of smaller, spherical resin particles, which permit chromatography with a higher pressure (15–20 atm) the elution thereby being shortened to 4 h.

(*ii*) Development of the 'single-column procedure', i.e. continuous determination of the basic amino-acids on the same column after elution of the acidic and neutral amino-acids.

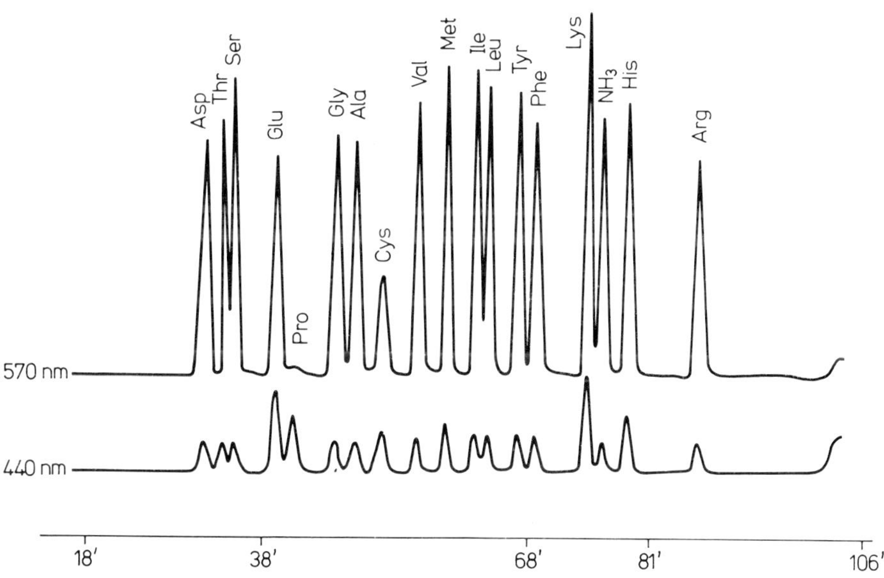

Fig. 4.3. Chromatogram of standard amino-acids, by single-column HPLC. Sample: each amino-acid 1.5×10^{-10} *M*. Amino-acid analyser: Liquimat I. Resin column: 230 × 4 mm Durrum DC-4A. Flow-rate of eluent 20 ml/h; addition of ninhydrin solution 10 ml/h. Back-pressure of the column 45–55 atm. Addition of eluent buffers: pH 3.25 sodium citrate buffer, 0.2 *N* in sodium, up to 18th min; pH 3.50 sodium citrate buffer, 0.7 *N*, up to 38th min; pH 3.65 sodium citrate buffer, 1.6 *N* in sodium, up to 68th min. Regeneration: 0.2 *M* NaOH, up to 81st min; equilibration with 0.2 *N* Na citrate, pH 3.25, up to 106th min. Temperature: 35 °C up to 11th min, 65 °C up to 38th min, 75 °C up to 68th min; cooling to 35 °C up to 85th min

(*iii*) The appearance in 1972 of the high-pressure amino-acid analysers, the very sensitive detectors of which allow the determination of amino-acids at even the nmole level.

The degree of automation and the programming possibilities may differ from analyser to analyser, but this does not essentially change the accuracy of the determination. Hence, having a good buffer- and ninhydrin-pump, an ion-exchange resin of uniform particle size and good quality, and a precisely-working recorder, a specialist experienced in the field of amino-acid chromatography can 'home-build' not only an analyser with a chromatography time of 4 h, but even a high-pressure one, as demonstrated by Scott *et al.* [169].

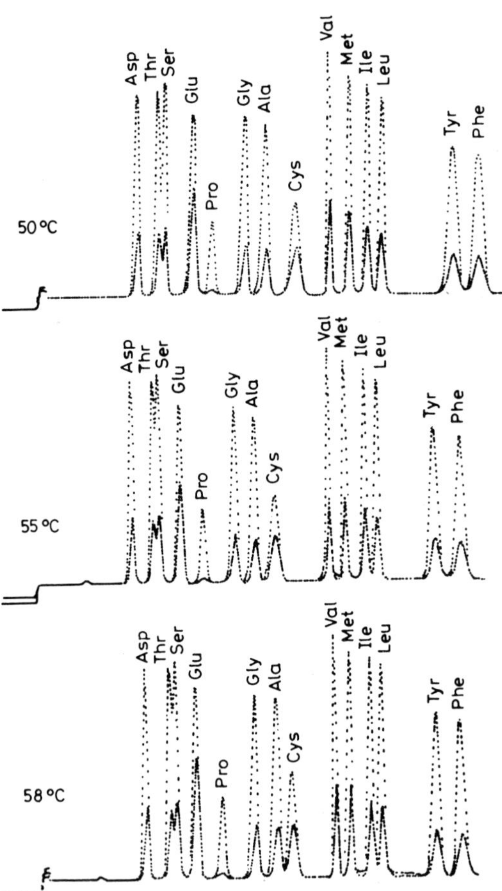

Fig. 4.4. Effect of temperature on the elution of amino-acids (BC-200 manual; by kind permission of LKB Instruments GmbH)

Figure 4.3 presents a chromatogram obtained with a 75-min, single-column method, the elution data also being reported.

The ion-exchange chromatographic separation of the amino-acids, and the sequence of elution, can be influenced by variation of the temperature, and the pH and cation concentration of the eluent citrate buffer. With elevation of

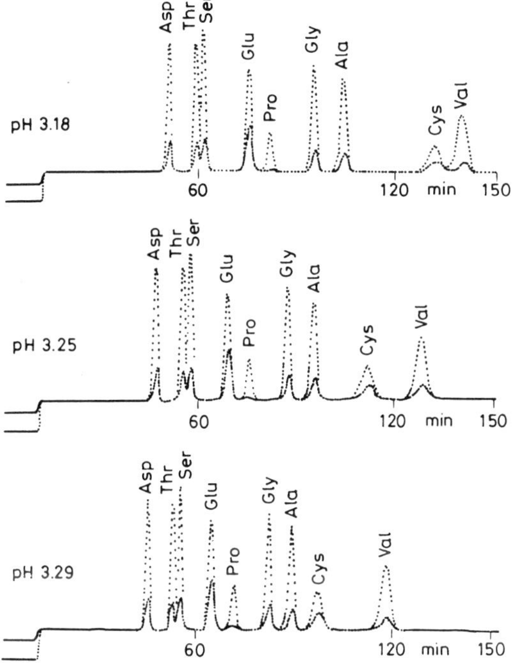

Fig. 4.5. Effect of pH of eluent buffer on the elution of amino-acids (BC-200 manual; by kind permission of LKB Instruments GmbH)

the temperature of elution, the chromatographic separation is accelerated, but above a certain temperature the selectivity deteriorates (Fig. 4.4).

The sharpness of the separation depends to a large extent on the pH of the first eluent buffer. With sodium citrate buffer of pH 3.25, Cys is situated midway between Ala and Val on the chromatogram. With buffer of higher pH, the Cys is eluted earlier, closer to the Ala peak (Fig. 4.5). For hydrolysates that also contain Hyp (e.g. from collagen), it is advisable to use a buffer with pH lower than 3.25, so that a separate peak is obtained for Hyp before the Asp; however, in this case the Cys lies nearer to the Val peak, as may be seen in the chromatogram obtained with a buffer of pH 3.18.

With modern analysers, the chromatograms are evaluated by means of a coupled integrator and computer [170–175]. Chromatograms recorded with a chart recorder in the earlier analysers without integrators were evaluated by means of the product of the peak-height and the width at the peak half-height. Such an evaluation necessitates the periodic chromatography of standard solutions of known concentration, the absolute quantities of the separated amino-acids being obtained by reference to this calibration. Calculation of the absolute amount of protein is facilitated by the addition of internal standards to the sample before hydrolysis; suitable standards are norleucine (eluted after Leu) [176], ε-aminocaproic acid [177], *S*-β-(4-pyridylethyl)-L-cysteine [178], and *S*-β-(4-pyridylethyl)-DL-penicillinamine [179].

Special chromatographic methods

The development of protein analysis is reflected by the fact, among others, that besides the chromatography of protein hydrolysates becoming a routine procedure, other requirements have arisen, such as the selective chromatography of the total free amino-acids of physiological solutions, the separation and determination of non-protein amino-acids, and the application of shortened chromatographic procedures for definite tasks.

One difficulty in the chromatography of the free amino-acids in physiological solutions is caused, for instance, by the overlap of the Ser, Asn, Gln and Glu peaks. Although computerized peak-integration can evaluate even overlapping peaks, the chromatographic separation of these amino-acids may be improved by starting the elution at a lower temperature, lowering the pH of the eluent buffer for the acidic amino-acids, decreasing the flow-rate of the buffer, and employing lithium citrate buffers in place of sodium citrate [147–154].

In the determination of the non-protein amino-acids, the analyst is faced with two problems: knowledge of the retention times of the amino-acids in question, i.e. the positions of their peaks in the chromatogram, and knowledge of the specific absorptivities of the reaction products of the amino-acids with ninhydrin. Both can readily be solved empirically, even in the absence of literature data, if individual amino-acids of chromatographic purity are available.

In the past few years, numerous publications have appeared, describing the chromatography of individual members of the very significant group of non-protein amino-acids, or the group of methylated amino-acids. The more important of these are listed in Table 4.8.

The requirement for shortened procedures arises when only one amino-acid in the sample needs to be determined. Chromatography with a shortened programme can be achieved by appropriate variation of the ion strengths and pH values of the buffers, and sometimes the temperature.

For the diagnosis of phenylketonuria, Phe may be chromatographed within 29 min on a resin column 22 cm high, with 0.35 *M* sodium acetate

Table 4.8
Methylated, non-protein amino-acids separated by chromatography

Amino-acids	Ref.
N^{ε}-methylated lysines	[180]
N^{ε}-methyl-lysine	[181]
N^{G},N^{G}-dimethylarginine, N^{G},N'^{G}-dimethylarginine, N^{ε}-mono-, di- and trimethyl-lysine	[182]
Methylated basic amino-acids	[183]
Guanidino-methylated arginines, *N*-methylated lysines	[184]
N^{ε}-monomethyl-lysine	[185]
N^{G}-monomethylarginine, N^{G},N^{G}-dimethylarginine, N^{G},N'^{G}-dimethylarginine, N^{ε}-monomethyl-lysine, N^{ε}-dimethyl-lysine, N^{ε}-trimethyl-lysine	[186]
all known methylated basic amino-acids	[187]
methylated amino-acids from meat	[188]
methylated basic amino-acids	[189]
radioactive N^{ε}-methylated lysines	[190]

buffer of pH 5.28 or 6.5, at 57 °C. For the diagnosis of hypervalinaemia and leucinosis, Val and Leu can be chromatographed within 90 min, on a similar 22-cm resin column with 0.35 *M* sodium citrate buffer of pH 3.7, at 57 °C.

In the procedure of Dévényi [191], the two main amino-acids in protein-feed research and plant-breeding, Met and Lys, can be determined in 62 min on a 14-cm resin column, with pH 3.28 sodium citrate buffer, 0.2 *N* in sodium and pH 4.25 sodium citrate buffer, 0.8 *N* in sodium, at 50 °C. Villegas *et al.* [192] developed an extremely original, shortened chromatographic procedure for determination of the Lys contents of a large number of samples; this permits 36 Lys determinations daily on a single amino-acid analyser.

The theoretical basis and methodology of automatic ion-exchange chromatography of the amino-acids are discussed in some excellent reviews and monographs [93, 104, 117, 193, 194]; in these the protein analyst can find answers to the questions arising in the course of his practical work, and can also discover data for the solution of various problems.

4.4.4 Chromatography of proteins and peptides on ion-exchange gels

Ion-exchangers with a polysaccharide matrix have proved essentially more useful sorbents than the ion-exchange resins: non-specific adsorption does not occur, and elution of the material is nearly quantitative. The availability of ion-exchangers with different polysaccharide matrices makes it possible to choose the appropriate sorbent in accordance with the properties of the protein to be separated. Fundamental factors in the choice of the ion-exchangers are the size and pH-stability of the protein molecule.

It is practical to select the ion-exchanger on the basis of molecular size as follows:

(*i*) for substances with molecular weights lower than 10^4 a dextran matrix;

(*ii*) for substances with molecular weights in the range 10^4–5×10^5 a cellulose and a moderate pore-sized dextran matrix;

(*iii*) for proteins with molecular weights in excess of 5×10^5 a cellulose or an agarose matrix.

The nature of the sample determines whether the separation should be performed on a cation- or an anion-exchange sorbent (see Fig. 4.1), while the pH of the initial elution buffer is selected on the basis of the pH-stability of the sample. It is practical to elute the bound substances by stepwise or continuous increase of the ionic strength, and in addition by continuous variation of the pH of the eluent buffer towards the isoelectric point of the sample. Continuous-gradient elution provides the possibility of rechromatography of multicomponent fractions originating from a preliminary separation (e.g. a gel filtration), for these may be applied together to the ion-exchange sorbent, without concentration. In such cases, however, the starting buffer must be selected so that the component to be isolated is not eluted by it. If the component to be separated is eluted from the applied sample by the starting buffer, then the quantity of sample to be added to the column must be dissolved in a small volume.

It is advisable to use a quantity of sample binding only about 5–10% of the total capacity of the sorbent, and the column should already be in equilibrium with the starting buffer; thus, in choosing the parameters of the chromatography it is necessary to start either from the volume and capacity of the column, or from the amount of material to be separated.

The simplest means of checking the equilibration of ion-exchange gel columns, and of monitoring the ionic strength of the eluate, is by conductivity measurement.

The first step in the regeneration of a gel ion-exchanger after chromatography is to use a salt solution of high strength ($I=2$) to remove the substances remaining sorbed after the elution. If the sorbed substances are lipids, or other contaminants that are not dissolved off by a salt solution with such a high concentration, then 0.1 *M* sodium hydroxide or non-ionic detergents should be used. Following this dissolution step, the ion-exchange gel is washed with a 0.5–1 *M* salt solution containing the counter-ion of the starting buffer, and is next equilibrated with the starting buffer. Cellulose- and dextran-based ion-exchangers are best removed from the column for regeneration, as they undergo volume changes depending on the concentration of the salt solution used.

If an ion-exchange gel is to be out of use for a prolonged period, its further utility can be ensured by storage in suitable media. After use, the gel column is well washed, and then a 0.003% toluene solution or a 0.05% trichlorobutanol ("Chloreton") solution is passed through it. For cation-exchangers it is possible to use 0.02% sodium azide solutions, or 0.005% Merthiolate solution, slightly acidified, and for anion-exchangers 0.001–0.01% phenylmercury acetate, borate or nitrate solutions, or 0.002% chlorohexidine ('Hibitane') solution.

Chromatography with cellulose-based ion-exchangers

The active groups of ion-exchanger cellulose derivatives are situated on the surfaces of the cellulose fibres. However, the hydroxyl groups of the cellulose molecule also sorb the differentiated-structure polyelectrolyte proteins and high molecular-weight peptides, without these being exposed to the danger of denaturing. Their utilization became widespread following the work of Peterson and Sober [195], who used ion-exchange cellulose preparations involving fibre modifications. The ion-exchange cellulose gels employed nowadays are bead-shaped or microgranular, with greater strength, a higher capacity, and a better resolving ability.

More than 90% of the chromatography done on the commercially-available cation- and anion-exchange cellulose derivatives involves the use of CM and DEAE cellulose. Cation-exchange phosphorylated and anion-exchange AE and ECTEOLA cellulose preparations can be employed within wider pH limits, the technique being similar to that already described, but with these extremely strong ion-exchangers it is always necessary to consider whether the proteins and peptides to be chromatographed at extremely low or high pH will be denatured.

Table 4.9
Capacity data on ion-exchange cellulose gels

Anion-exchange cellulose gels	
DEAE Sephacel	150 mg of serum albumin per ml of gel (0.04 *M* TRIS-phosphate, pH 8.4)
DEAE microgranular cellulose, DE-32 Whatman	660 mg of serum albumin per g of DE-32 (0.01 *M* Na phosphate, pH 8.5)
Cation-exchange cellulose gel	
CM microgranular cellulose, CM-32 Whatman	400 mg of 7S γ-globulin per g of CM-32 (0.08 *M* Na phosphate, pH 3.5)

The inorganic ion capacities of ion-exchange celluloses are low, and their protein-binding capacities vary from preparation to preparation, depending on the nature of the ion exchanger and on the buffer used (Table 4.9).

Because of the hydroxyl groups of the cellulose molecule and the fibrous structure of the cellulose, ion-exchange cellulose derivatives are able to take up extremely large amounts of water, partly as water of hydration stabilized by hydrogen-bonds, and partly as water of imbibition, filling the spaces between the fibres; these together can come to 7–10 times the weight of the cellulose derivative. In dry cellulose preparations some of the hydrogen-bonds formed between adjacent hydroxyl groups are particularly stable; in water or in buffer solution these do not break, and this results in a decreased water-uptake and swelling. After their hydration, swelling and sedimentation, the first step in the production of dry cellulose preparations is the removal of the residual suspended particles. However, the maximum swelling (ensuring optimum chromatographic separation) of the ion-exchange cellulose preparations available commercially in the dry state can be attained only by a procedure consisting of several steps.

In the following, the measured weight of sorbent will be taken as unit volume. DEAE cellulose is suspended in 15 volumes of 0.5 *M* hydrochloric acid (CM cellulose in 15 volumes of 0.5 *M* sodium hydroxide) with careful stirring. Then, after sedimentation for 30 min, the liquid is decanted. The sedimented swollen sorbent is next suspended in distilled water, transferred to a Buchner funnel, and washed with distilled water to remove the acid (alkali in the case of CM) until the effluent attains a pH of 4 (8 for CM). The washed sorbent is suspended for 30 min in 15 volumes of 0.5 *M* sodium hydroxide (15 volumes of 0.5 *M* hydrochloric acid for CM) and the suspension is transferred to a Buchner funnel and washed with distilled water until the effluent is neutral. The DEAE thus prepared in OH^- form (H^+ form for CM) must be equilibrated with the first buffer to be used in the chromatography.

The DEAE-Sephacel preparation has the advantage that it need not be submitted to this time-consuming swelling, sedimentation and activation procedure, for the commercial product is available in a gel consistency and in an activated state, and only the equilibration with the first buffer must be performed. A typical separation on Sephacel is shown in Fig. 4.6.

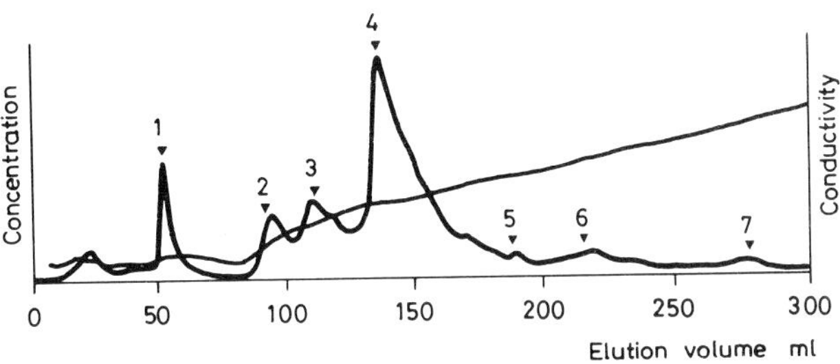

Fig. 4.6. Fractionation of serum proteins on DEAE Sephacel. Sample, human serum; column diameter, 1.6 cm; bed height, 10 cm; eluent, TRIS-phosphate (0.04 *M*, pH 8.4 with concave gradient to 0.5 *M*, pH 3.4); flow rate, 0.25 ml/min. (By kind permission of Pharmacia Fine Chemicals AB, Sweden.)

Before the DEAE cellulose preparations suspended in the first buffer are packed into the column, it is advisable to free them from air; the CO_2 dissolved in the buffer solution from the air would otherwise bind as carbonate on the anion-exchanger, which would decrease the subsequent ion-exchange during chromatography.

Ion-exchange cellulose preparations have been employed to solve virtually all of the technical problems of ion-exchange separation encountered in the isolation of proteins. For this reason, we shall not attempt to compile a list (which inevitably would be incomplete) and will say only that assistance towards the solution of a particular protein-separation problem may be obtained from the isolation data to be found in the relevant literature [195].

Chromatography on agarose-based ion-exchangers

The DEAE and CM agarose derivatives are excellently suitable for the ion-exchange chromatography of proteins, membrane components and other substances of high molecular weight. For chromatography, the DEAE and CM Sepharose CL–6B gels cross-linked with covalent bonds are verv advantageous. They may be used in the pH range 3–10, they display high resolution and high liquid-permeability, and the volume of the gel column decreases by only about 2% between pH 10 and pH 4, which permits regeneration of the gel bed in the chromatograph tube itself, without emptying. Some data characterizing the capacities of Sepharose ion-exchangers follow:

DEAE Sepharose CL–6B: 10 g of haemoglobin/100 ml of gel, using 0.1 *M* TRIS–HCl buffer of pH 8.

CM Sepharose CL–6B: 10 g of haemoglobin/100 ml of gel, using 0.1*M* sodium acetate–acetic acid buffer of pH 5.

The possibility of high resolution ensures a much faster elution than with the other gel ion-exchangers, as is demonstrated by the chromatograms of human serum, obtained with DEAE Sepharose CL–6B, shown in Fig. 4.7.

The utility of Sepharose-based ion-exchange gels for the separation of membrane proteins is well exemplified by the work of Janson [196], who used chromatography on DEAE Sepharose CL–6B to separate the soluble membrane components of *Cytophaga johnsonii*, and as a result succeeded in isolating dextranase. Shannon *et al.* [197] isolated isoenzymes of horse-radish root peroxidase on a CM Sepharose CL–6B column, with a linear-gradient sodium acetate–acetic acid buffer of pH 3.8, at a flow-rate of 60 ml/h.

Sepharose-based gels also allow unusual elutions, such as the chromatography of various insulin derivatives (arginyl-insulin, arginyl-insulin ethyl and methyl esters, desaminated insulin) on a DEAE Sepharose CL–6B column, with a pH 8.3 solution of 0.1 *M* ammonium chloride containing 60% ethanol.

In chromatography with agarose-based ion-exchangers cross-linked with covalent bonds, the technical pretreatment (packing and equilibration of the column, preparation and application of the sample) and the elution are identical with those employed for other ion-exchange gels.

Chromatography with dextran-based ion-exchangers

Cross-linked dextran-based ion-exchangers have the interesting property that, although they retain their gel-filter characteristics, the chromatographic separation is essentially based on ion-exchange. Thus, there are two technical conditions in the application of dextran-based ion-exchangers: as ion-exchange sorbents, they are similar to the other ion-exchange gels, while with regard to their matrix they are similar to dextran gel filters. As is the case for non-ion-exchange gel filters, in water or in aqueous solutions with low ionic strength they swell very considerably, the degree of swelling depending on the pH and ionic strength of the solution, and the nature of the counter-ion.

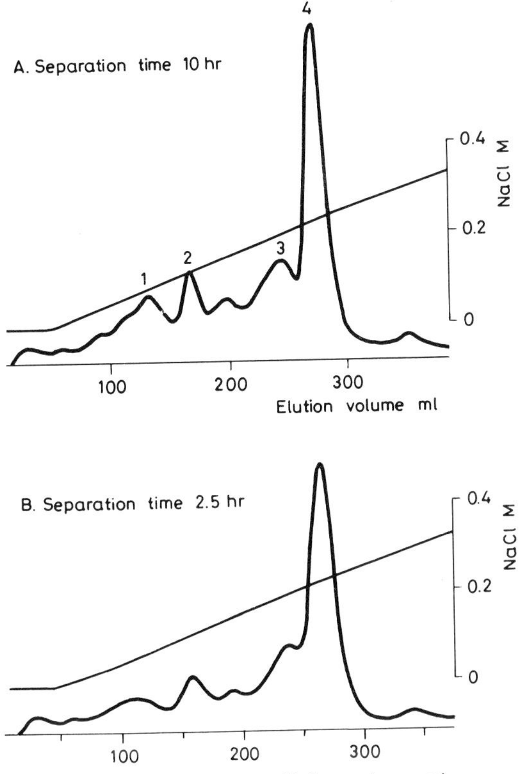

Fig. 4.7. Fractionation of whole serum on DEAE Sepharose CL-6B. Column diameter, 1.6 cm; bed volume, 30 ml; eluent, 0.05 *M* TRIS–HCl buffer at pH 8.6 with a linear gradient of 0–0.5 *M* NaCl (total volume of gradient: 500 ml). *A*, Flow rate: A, 37.5 ml/h; B, 145 ml/h. (Reproduced by kind permission of Pharmacia Fine Chemicals AB, Sweden.)

Increase of the ionic strength of the eluent during the chromatography leads to dehydration of the gel, thereby diminishing the swelling, which may cause shortening of the sorbent column. In their application, therefore, the principles and procedures relating to the use of both the other ion-exchange gels and Sephadex should be taken into consideration.

Characteristic data on some of the most generally-used dextran-based ion-exchange preparations are listed in Table 4.10.

Although substances with molecular weights higher than 2×10^5 cannot penetrate into the A-50 and C-50 pores, it is nevertheless more advantageous to use the A-25 and C-25 exchangers for these. Since these are less hydrated, the density of the ion-exchange groups on the surface of the sorbent is higher;

Table 4.10
Characteristic data on ion-exchange Sephadexes

Sephadex type	Haemoglobin capacity, g/g	Recommended pH	Molecular weights of proteins that can be chromatographed
DEAE Sephadex A-25	0.6	2–9	$<3\times10^4$, $>2\times10^5$
A-50	5.0	2–9	$3\times10^4-2\times10^5$
QAE Sephadex A-25	0.4	2–10	$<3\times10^4$, $>2\times10^5$
A-50	6.0	2–10	$3\times10^4-2\times10^5$
CM Sephadex C-25	0.4	6–10	$<3\times10^4$, $>2\times10^5$
C-50	9.0	6–10	$3\times10^4-2\times10^5$
SP Sephadex C-25	0.2	2–10	$<3\times10^4$, $>2\times10^5$
C-50	7.0	2–10	$3\times10^4-2\times10^5$

The haemoglobin capacities of the DEAE and QAE Sephadexes were obtained with the use of a TRIS-HCl buffer, pH 8, and $I=0.01$, and those of the CM and SP Sephadexes at pH 5 and $I=0.01$

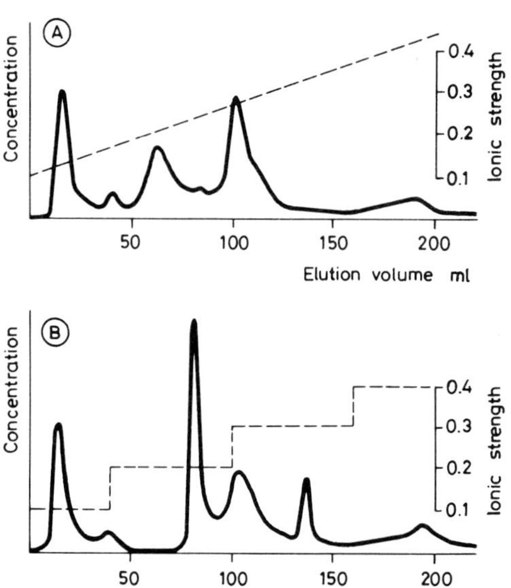

Fig. 4.8. Elution characteristics of bovine serum in continuous and stepwise gradient elution on QAE Sephadex A-50. Gel column 1.5 × 26 cm. Sample 4 ml of 3% w/v lyophilized bovine serum. Starting buffer 0.1 *M* TRIS–HCl, pH 6.5 (0.5 *M* NaCl was used for the gradient), flow rate 12 ml/h; A, continuous gradient; B, stepwise NaCl gradient. In both chromatograms the first peak is given by IgG. Albumin gives peak 4 with the continuous gradient, and peaks 4 and 5 with the stepwise gradient. (Reproduced by kind permission of Pharmacia Fine Chemicals AB, Sweden.)

in addition, as a consequence of the lower swelling accompanying the lower hydration, the mechanical strength of the sorbent particles is also higher than that of the A-50 or C-50 preparations.

The chromatograms of bovine serum chromatographed by continuous and stepwise gradient elution on the strongly basic anion-exchanger QAE-Sephadex A-50 (Fig. 4.8) demonstrate the possibilities of application of dextran-based ion-exchange chromatography.

The literature data indicate that dextran-based ion-exchangers have been successfully employed for the purification of virtually all proteins in the past 12 years.

4.5 APPLICATIONS OF AMPHOLYTE DISPLACEMENT CHROMATOGRAPHY IN PROTEIN ANALYSIS

Ampholyte displacement chromatography can be regarded as a modification of ion-exchange chromatography, in which the fractionating elution of the substances bound on the ion-exchanger is not achieved by variation of the pH of the eluent buffer solution, or of its ionic strength, but occurs by the action of a carrier dissolved in the eluent. The method was first described by Leaback and Robinson [198], who isolated isoenzymes of beta-*N*-acetyl-D-hexosamidase on a CM cellulose column with a 40-g/l solution of Ampholine of pH 8–10. By means of displacement chromatography, Young and Webb [199] separated α-foetoprotein and albumin, as well as other serum proteins, on a 10 × 0.9 cm DEAE cellulose column; the non-adsorbed proteins were first eluted with a 10 m *M* sodium phosphate buffer of pH 7.8, at a flow-rate of 13.5 ml/h, then albumin and foetoprotein with a 40-g/1. Ampholine phosphate buffer of pH 4 6, and subsequently the proteins still bound to the ampholyte, with 0.2–0.3 *M* sodium chloride. Young and Webb state that the elution is a consequence of the interaction between the protein and the carrier ampholyte.

4.6 APPLICATIONS OF COVALENT CHROMATOGRAPHY IN PROTEIN ANALYSIS

Whereas the basis of the chromatographic effect in ion-exchange chromatography is the ionic bonding occurring between certain ions of the stationary and mobile phases, the basis of covalent chromatography is that thiol-containing proteins in the mobile phase are covalently bound on the 2-mercaptopyridyl groups of the stationary phase, from which they can be eluted with specific reducing agents.

The method was developed by Brocklehurst *et al.* [200]. The stationary phase consisted of a glutathione spacer bound on an agarose matrix activated with bromine cyanide, and into which a 2-pyridyl group is incorporated. By means of the free thiol group on the surface, peptides and proteins are reversibly bound on the stationary phase in such a way that a mixed disulphide bridge is formed with its —S–S— bond, a 2-thiopyridone molecule being split off the stationary phase. The immobilized proteins may be dissolved off with buffer containing a reducing agent.

For the covalent chromatography of thiol-containing proteins and enzymes, it is convenient to use Activated Thiol-Sepharose 4B. The spacer of this is comprised of one molecule of 2,2′-dipyridyl disulphide and one molecule of glutathione, coupled to CNBr-activated Sepharose 4B. The hydroxyphilic glutathione spacer group decreases the steric limiting error of the disulphide reaction. The Activated Thiol-Sepharose 4B has a covalent binding capacity of 1.25–2.5 mg of protein/ml of gel. At the start of the chromatography, the non-bound protein is washed out with TRIS, phosphate or acetate buffer prepared in 0.1–0.5 *M* sodium chloride. Immobilized molecules are released and eluted with a buffer of pH 8 containing a low concentration of reducing agent (5–20 m*M* L-cystine).

Covalent chromatography of low molecular-weight peptides containing the thiol group can be carried out very conveniently with Thiopropyl-Sepharose 6B; in the procedure of Axén *et al.* [201], a 2-hydroxypropyl

Table 4.11
Proteins separated by covalent chromatography

	Refs.
Isolation of active papain from dry papaya latex	[200]
Isolation of bovine mercaptalbumin	[202]
Separation of thiol-containing peptide from high molecular-weight proteins	[203]
Partial fractionation of whole casein	[204]
Purification of ficin (from *Ficus glabrata*) and characterization of its active centres	[205]
Isolation of high-activity urease from *Canavalia ensiformis*	[206]
Separation of two soft-tissue collagens	[207]
Purification of prealbumin	[208]
Separation of plasma proteins	[209]
Synthesis of metallothionein	[210]
Study of primary structure of allergen M (from cod)	[211]
Study of ferroxidase	[212, 213]
Jack-bean urease preparation	[214]
Specific isolation of cysteine peptides	[215]

radical is bound as spacer to the Sepharose 6B in a stable ether bond. A 2-mercaptopyridyl group is linked by a disulphide bond to the terminal end of the spacer. The covalently-bound substance to be isolated is recovered from the mixed disulphide bridge with a buffer containing dithiothreitol or 2-mercaptoethanol.

Covalent chromatography is particularly convenient for the solutions of problems arising in the course of protein-structure research. The use of covalent chromatography to isolate thiol-containing fragments of biologically-active molecules, and to purify the products of peptide syntheses, has the advantage, among others, that the biologically-active peptides isolated in this way do not lose their activities.

Table 4.11 presents some of the applications of covalent chromatography.

4.7 APPLICATIONS OF AFFINITY CHROMATOGRAPHY IN PROTEIN ANALYSIS

In contrast with the traditional chromatographic methods, which separate materials on the basis of their physico-chemical properties, affinity chromatography (first described in 1968 [216]) utilizes their biochemical properties for the separation of the component proteins of the sample. Affinity chromatography can be regarded in effect as a biospecific adsorption chromatography, whereby a biologically-active substance may be isolated directly, in a single step, on the basis of its biological affinity.

The stationary phase in affinity chromatography is a sorbent, to the insoluble matrix of which a spacer is linked by a covalent bond; to the free end of this a biologically active molecule may be attached in a ligand bond. This molecule bound to the spacer is a specific biosorbent, which can interact with an appropriate biologically-active compound (an enzyme with its inhibitor, an antibody with the antigen).

If a specifically biologically-active molecule from the mobile phase is bound on the sorbent, the binding may be reversible. Such a bond may be reversible. Such a bond may be broken by increasing the ionic strength of the eluent and sometimes by variation of the pH. With reversible binding, if the stationary phase is an enzyme bound on the matrix, the enzyme inhibitor can be obtained, e.g. the trypsin inhibitor of physiological organs on immobilized trypsin; similarly, with an antigen bound to the spacer a high-titre immunoglobulin may be obtained.

The purpose of biosorption in a given case, however, may also be the catalysis of a biochemical process, in the course of which a biochemical molecule in the mobile phase undergoes chemical transformation. In this case the biologically-active substance bound to the spacer, e.g. trypsin, hydrolyses the peptide bonds of the proteins and peptides dissolved in the mobile phase. Trypsin bound in this way is not released from its binding during the reaction. This phenomenon has led to the development of a preparative-scale methodology for the immobilized enzymes [217].

Sorbents with very different properties have proved suitable as the base-materials of biosorbents used for affinity chromatography, e.g. Sephadex, CM cellulose, cross-linked Sepharose, polyacrylamide, maleic anhydride copolymer, and porous glass preparations (with active ester, thiol and carboxyl bonds). The simplest matrix for affinity chromatography is CNBr-activated Sepharose 4B, which is suitable directly for the covalent binding of proteins, macromolecules containing amino groups, and spacers of various natures and lengths [218].

$$\begin{matrix}\text{—O}\\ \text{—O}\end{matrix}\!\!>\!C{=}NH + H_2NR \rightarrow \begin{matrix} & NH \\ & \| \\ \text{—O—} & C\text{—NHR} \\ \text{—OH} & \end{matrix}$$

The possibility of binding to CNBr-activated Sepharose 4B exists in the pH range 6–10, with the optimum at pH 8–9. The binding capacity differs for different proteins and at different pH values; typically 1 ml of gel will bind 8 mg of chymotrypsinogen at pH 8.

An overall picture of the chemical process of Sepharose activation with CNBr is given by Cuatrecasas [219].

The macrobead form of Sepharose activated with CNBr, CNBr-activated Sepharose 6 MB (particle diameter 200–300 μm), may be used for the fractionation and purification of various cells. The gel is suitable for the rapid, easy and non-damaging immobilization of cells through the primary amine groups of their specific ligands.

As in chromatography with ion-exchange gels, here too only about 5% of the sorption capacity of the column may be utilized, and the material taken for chromatography must be bound on the top 1–2 cm of the short (15–30 cm) column. Because of this, the volume of the sample applied to the column does not play a primary role. For the first elution buffer it is practical to use a solution with a fairly high ionic strength, e.g. 0.5 *M* sodium cloride, which washes the non-specifically bound substances from the other active groups of the sorbent. The buffer solutions suitable for the elution of biospecifically-bound molecules differ considerably in both their pH and their ionic composition: e.g. a Gly/HCl buffer of pH 2.8 is used for an albumin–

immunoglobulin complex; IgE is eluted from an anti-IgE–Sepharose 4B column with sodium thiocyanate solution [220]; collagenase from a collagen–Sepharose column by the addition of 1 *M* sodium chloride to the starting buffer [221]; and lipoprotein lipase from the other proteins on a heparin–Sepharose column with an increasing ionic-strength gradient [222].

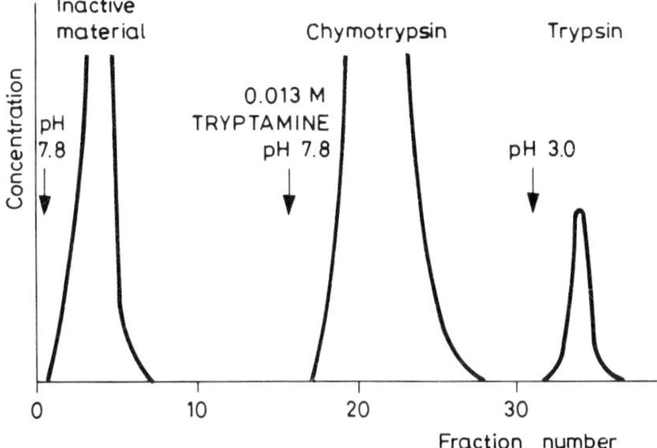

Fig. 4.9. Purification of trypsin and chymotrypsin on soya-bean trypsin inhibitor coupled to CNBr-activated Sepharose 4B. Crude pancreatic extract was applied at pH 7.8. Tryptamine (0.013 *M*) was added to the buffer to elute chymotrypsin, and the elution pH was changed to pH 3.0 to elute trypsin. (Reproduced by kind permission of Pharmacia Fine Chemicals AB, Sweden.)

A very illustrative example of specific elution is given by the purification of trypsin and chymotrypsin from a crude pancreatic extract, by means of binding to soya-bean trypsin inhibitor bound on CNBr-Sepharose 4B (Fig. 4.9).

For the dissociation of the chymotrypsin bound on the soya-bean trypsin inhibitor, a tryptamine solution of pH 7.8 is necessary; this is a specifically reversible inhibitor of chymotrypsin. Then, the binding of trypsin can be broken by decreasing the pH of the eluent solution to pH 3.

Various Sepharose-based matrixes are available to the research worker for the separation of different biologically-active products by affinity chromatography. A survey of the applications and some characteristic data relating to these is presented in Table 4.12.

A number of reviews [219, 223–225] and monographs [226, 227] provide detailed information regarding the selection and preparation of the matrix and biosorbent most appropriate for the purpose, and also the technical solutions of the chromatography. Here, therefore, it need only be stated that with this rapid and elegant procedure separations can be achieved that yield biologically intact preparations.

Table 4.12
Affinity chromatographic utilization of Sepharose-based matrixes

Preparation	Spacer	Optimum pH for binding	Separated materials
CH Sepharose 4B	6-aminohexanoic acid with terminal COOH group		primary amino groups of proteins, peptides, amino-acids, with carbodi-imide reaction
Activated CH Sepharose 4B	6-aminohexanoic acid with COOH group esterified with *N*-hydroxysuccinimide	8	primary amino groups of free COOH-containing proteins, carboxylic acids, amino-acids, mainly for small ligands
AH Sepharose 4B	1,6-diaminohexane with free amino group	4.5–6	free COOH-containing amino-acids, carboxylic acids, keto-acids, with carbodi-imide reaction
Blue Sepharose CL-6B	Cibacon Blue FsG-A bound to Sepharose CL-6B with tyrazine binding method	7	various proteins, albumin, blood-clotting factors, kinases, adenylate cyclase, DNA polymerase, nitrate reductase, succinyl CoA transferase; their specific elution with coenzymes
5′-AMP Sepharose 4B	N^6-(6-aminohexyl)-5′-AMP bound to CNBr-Sepharose 4B	7	NAD^+-dependent dehydrogenase, ATP-dependent kinases, various other enzymes
2′,5′-ADP Sepharose 4B	N^6-(6-aminohexyl)adenosine-2′,5′-diphosphate covalently bound to Sepharose 4B	7	$NADP^+$-dependent dehydrogenases, various enzymes showing affinity with $NADP^+$
Con A Sepharose	Concanavaline A covalently bound to CNBr-activated Sepharose 4B	6	glycoproteins, carbohydrate-containing biopolymers, membrane fragments
Protein A Sepharose CL-4B	Protein A (isolated cell wall of *Staphylococcus aureus*) covalently bound to CNBr-activated Sepharose 4B	7	IgG-type antibodies and their fragments which contain the F_c region from F_{ab} fragments
Wheat-germ Lectin Sepharose 6 MB	Wheat-germ lectin covalently bound to CNBr-activated Sepharose 6 MB		glycoproteins, lymphocytes, other cell types, cell surfaces

4.8 APPLICATIONS OF HYDROPHOBIC INTERACTION CHROMATOGRAPHY IN PROTEIN ANALYSIS

In the chromatographic procedures used in protein analysis, the sorption and desorption are based on the hydrophilic properties (and the utilization of these) of the active groups of the sorbent and the hydrophilic sites situated on the surface of the protein molecules. From knowledge of the colloid-physics laws relating to the surface phenomena involved in the content of different phases, it is obvious that the proteins may also bind to hydrophobic substances through the uncharged hydrophobic moieties on their surfaces.

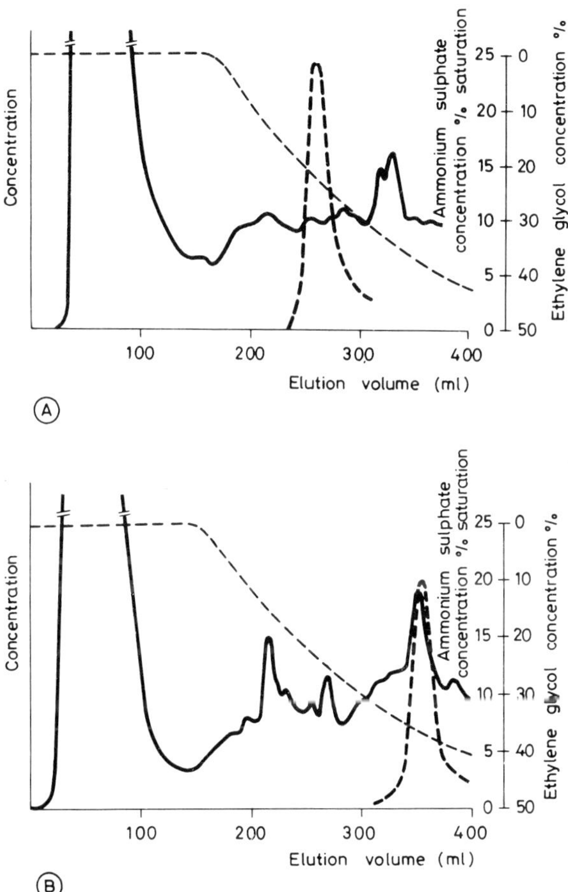

Fig. 4.10. Purification of β-amylase from barley meal by hydrophobic interaction chromatography on A, Octyl-Sepharose CL-4B and B, Phenyl-Sepharose CL-4B. Sample 40 ml of barley meal extract in 0.01 *M* sodium phosphate buffer, pH 6.8, 25% saturated with ammonium sulphate (sample buffer). Column diameter, 1.6 cm; bed volume, 30 ml; flow rate: 25 ml/h. After sample addition, elution was continued with 85 ml of sample buffer, followed by a gradient of decreasing ammonium sulphate concentration and increasing ethylene glycol concentration (final concentrations: 0% and 50%, respectively). (Reproduced by kind permission of Pharmacia Fine Chemicals AB, Sweden.)

The hydrophobic interaction binding of protein molecules may occur when, as a consequence of the high salt concentration of the solvent, the hydrate sheath linked to the dissolved protein molecule by van der Waals forces is bound with lower energy than that of the interaction between the hydrophobic sorbent and the hydrophobic surface of the solute. Hydrophobic interaction chromatography is therefore based on the interactions between the hydrophobic matrix, the hydrophobic surface parts of the solute to be separated, and the aqueous eluent solution. The interaction, and consequently the chromatographic effect, is strongly influenced by temperature, pH, ionic strength, and the chaotropic (see footnote, p. 163) properties of the dissolved ions [228]. The chaotropic properties of the ions of salts affecting the hydrophobic interaction increase roughly in accordance with the Hofmeister lyotropic series.

An advantageous feature of hydrophobic interaction chromatography is that homogenized organ extracts and extract solutions prepurified by ammonium sulphate precipitation, but containing a high concentration of

Table 4.13
Proteins separated by hydrophobic interaction chromatography

Proteins	Ref.
aspartate transcarbamylase	[229]
glycogen phosphorylase	[230]
alkaline phosphatase	[231]
potato phosphorylase	[232]
muscle enzymes	[233]
α-isopropyl malate isomerase	[234]
protein mixtures	[235]
human IgA	[236]
wheat-germ aspartate transcarbamylase	[237]
chloramphenicol acetyltransferases	[238]
lipoamide dehydrogenase	[239]
human serum albumin	[240]
ovalbumin	[240]
N-acetyl-*β*-D-hexosaminidase	[241]

ammonium sulphate, can be added directly to sorbents such as Octyl-Sepharose CL-4B. The elution begins with a salt solution of high concentration that dissolves the sample; this washes out the non-bound proteins from the column. The adsorbed proteins are dissolved off in the decreasing sequence of their hydrophobic properties, by continuous lowering of the ammonium sulphate concentration of the eluent and by the addition of ethylene glycol in continuously increasing concentration.

The most frequently employed solid phase for the hydrophobic interaction chromatography of proteins and peptides is a cross-linked agarose gel, to the monosaccharide units of which octyl or phenyl groups are linked in stable ether bonds (e.g. Octyl-Sepharose CL-4B and Phenyl-Sepharose CL-4B). Besides their other favourable properties (a high degree of chemical stability; absence of charged groups; heat-resistance), they are characterized by high capacities: under optimum conditions, 1 ml of gel is capable of retaining 15–20 mg of human serum albumin in hydrophobic binding.

For the hydrophobic interaction chromatography of proteins that are bound by particularly strong hydrophobic forces to Octyl-Sepharose CL-4B (e.g. haemoglobin, β-lactoglobulin), and are therefore difficult to elute without being denatured, it is better to use Phenyl-Sepharose CL-4B. The differences between the hydrophobic chromatographic effects of the two sorbents are illustrated by the two chromatograms in Fig. 4.10.

Table 4.13 lists possibilities of application of hydrophobic interaction chromatography.

The theoretical basis of hydrophobic interaction chromatography is reviewed in the publications by Hjertén *et al.* [240, 242, 243].

REFERENCES

[1] Mikeš, O. (ed.): *Laboratory Handbook of Chromatographic and Allied Methods*, Horwood, Chichester, 1979.

[2] Heftman, E.: *Chromatography*, 3rd Ed., Reinhold, New York, 1975.

[3] Alexander, P., Block, R. J.: *Analytical Methods of Protein Chemistry*, Pergamon Press, London, 1960.

[4] Morris, C. J. O. R., Morris, P.: *Separation Methods in Biochemistry*, Pitman, London, 1963.

[5] Haschemeyer, R. H. H., Haschemeyer, A. E. V.: *Proteins: A Guide to Study by Physical and Chemical Methods*, Wiley-Interscience, New York, 1973.

[6] Dévényi, T., Gergely, J.: *Amino Acids, Peptides and Proteins*, Elsevier, Amsterdam, 1974.

[7] Kirkland, J. J.: *Modern Practice of Liquid Chromatography*, Wiley-Interscience, New York, 1971.

[8] Scott, R. P. W.: *Contemporary Liquid Chromatography*, Wiley, London, 1976.

[9] Engelhardt, H.: *Hochdruck-Flüssigkeits-Chromatographie*, Springer, Berlin, 1975.

[10] Halász, I.: *Liquid Chromatography: An Overview*, in Kirkland, J. J.: *Modern Practice of Liquid Chromatography*, Wiley-Interscience, London, 1971, p. 325.

[11] Horváth Cs. (ed.): *High-Performance Liquid Chromatography—Advances and Perspectives*, Vol. 1, Academic Press, New York, 1980.

[12] Kaiser, R. E., Oelrich, E.: *Optimization in HPLC*, Hüthig, Heidelberg, 1981.

[13] Davankov, V. A., Bochkov, A. S., Belov, Yu. P.: *J. Chromatog.*, **218,** 547 (1981).

[14] Kurganov, A. A., Davankov, V. A.: *J. Chromatog.*, **218,** 559 (1981).

[15] Snyder, L. R., Kirkland, J. J.: *Introduction to Modern Liquid Chromatography*, Wiley, New York, 2nd Ed., 1979.

[16] Dixon, P. F., Gray, C. H., Linn, C. K., Stoll, M. S. (eds): *High-Pressure Liquid Chromatography in Clinical Chemistry*, Academic Press, London, 1976.

[17] Hearn, M. T. W., Grego, B.: *J. Chromatog.*, **218,** 497 (1981).

[18] Hawk, G. L. (ed.): *Biological/Biochemical Applications of Liquid Chromatography* III. Dekker, Basel, 1982.

[19] Done, J. N., Kennedy, G. J., Knox, J. H.: *Nature*, **237,** 77 (1973).

[20] Bailey, J. L.: *Techniques in Protein Chemistry*, 2nd Ed., Elsevier, Amsterdam, 1967.

[21] Eastoe, J. E., Courts, A.: *Practical Analytical Methods for Connective Tissue Proteins*, Spon, London, 1963.

[22] Calmon, C., Kressman, T. R. E.: *Ion-Exchangers in Organic Chemistry and Biochemistry*, Wiley-Interscience, London, 1957.

[23] Neilands, J. B., Stumpf, P. K.: *Outlines of Enzyme Chemistry*, 2nd Ed., Wiley, London, 1958.

[24] Parris, N. A.: *Instrumental Liquid Chromatography*, Elsevier, Amsterdam, 1976.

[25] Snyder, L. R.: *Principles of Adsorption Chromatography*, Dekker, New York, 1968.

[26] Tiselius, A.: *Arkiv Kemi*, **7,** 443 (1954).

[27] Tiselius, A., Hjertén, S., Levin, Ö.: *Arch. Biochem. Biophys.*, **65,** 132 (1956).

[28] Johansson, B.: *Nature*, **181,** 996 (1958).

[29] Kawasaki, T.: *J. Chromatog.*, **82,** 167, 191, 219, 237, 241 (1973).

[30] Hoxo, F., O'Eocha, C., Norris, P.: *Arch. Biochem. Biophys.*, **54,** 162 (1955).

[31] Bernardi, G., Cook, W. H.: *Biochim. Biophys. Acta*, **44,** 96 (1960).
[32] Hjertén, S.: *Biochim. Biophys. Acta*, **31,** 216 (1959).
[33] Bernardi, G., Kawasaki, T.: *Biochim. Biophys. Acta*, **160,** 301 (1968).
[34] Bernardi, G., Giro, M. G., Gaillard, C.: *Biochim. Biophys. Acta*, **278,** 409 (1972).
[35] Skelton, S.: *J. Chromatog.*, **35,** 283 (1968).
[36] Mosersky, S. M., Gugger, R. E., Pettani, J. D., Kolman, S. D.: *J. Chromatog.*, **100,** 192 (1974).
[37] Kawasaki, T., Bernardi, G.: *Biopolymers*, **9,** 269 (1970).
[38] Crone, H. C., Dawson, R. M., Smith, E. S.: *J. Chromatog.*, **103,** 71 (1975).
[39] Bock, H. G., Skene, P., Fleischer, S., Cassidy, P., Harshman, S.: *Science*, **191,** 380 (1976).
[40] Bresler, S. E., Katushkina, N. V., Kolikov, V. M., Potokin, J. L., Vinogradskaya, G. N.: *J. Chromatog.*, **130,** 275 (1977).
[41] Hansen, J. J., Greibrokk, T., Currie, B. L., Johansson, K. N.-G., Folkers, K.: *J. Chromatog.*, **135,** 155 (1977).
[42] Mönch, W., Dehnen, W.: *J. Chromatog.*, **140,** 260 (1977).
[43] Albertsson, P. A.: *Partition of Cell Particles and Macromolecules*, Chapter 5, *Countercurrent distribution*, Wiley, Almqvist, Wickell, New York and Stockholm, 1960.
[44] Albertsson, P. A.: *Partition of Cell Particles and Macromolecules in* Polymer Two-Phase Systems, *Adv. Protein Chem.*, **24,** 309 (1970).
[45] Carpenter, F. H.: *Arch. Biochem. Biophys.*, **78,** 539 (1958).
[46] Herbert, D., Pinsent, A. J.: *Biochem. J.*, **43,** 193 (1948).
[47] Martin, A. J. P., Porter, R. R.: *Biochem. J.*, **49,** 215 (1951).
[48] Porter, R. R.: *Biochem. J.*, **53,** 320 (1953).
[49] Carpenter, F. H., Hess, G. P.: *J. Am. Chem. Soc.*, **78,** 3351 (1956).
[50] Porter, R. R.: *The Chemical Structure of Proteins*, p. 31, Wostenholme, G. E. W., Cameron, M. P. (eds), Churchill, London, 1953.
[51] Porter, R. R.: *Brit. Med. Bull.*, **10,** 237 (1954).
[51] Yamashiro, D.: *Nature*, **201,** 76 (1964).
[53] Anker, H. S.: *Biochim. Biophys. Acta*, **229,** 290 (1971).
[54] Greenberg, D. M.: *Amino Acids and Proteins*, Thomas, Springfield, 1951, p. 449.
[55] Hirs, C. H. W., Moore, S., Stein, W. H.: *J. Biol. Chem.*, **200,** 493 (1953).
[56] Boordman, N. K., Partridge, S. M.: *Biochem. J.*, **59,** 543 (1955).
[57] Koszalka, T. R., Falkenstein, R., Altmann, K. I.: *Biochim. Biophys. Acta*, **23,** 648 (1957).
[58] Tallan, H. H., Stein, W. H.: *J. Biol. Chem.*, **200,** 507 (1953).
[59] Cole, R. D.: *J. Biol. Chem.*, **235,** 2294 (1960).

[60] Cole, R. D.: *J. Biol. Chem.*, **236,** 1369 (1961).
[61] Papkoff, H., Li, C. H.: *Biochim. Biophys. Acta*, **29,** 145 (1958).
[62] Crampton, C. F., Stein, W. H., Moore, S.: *J. Biol. Chem.*, **225,** 363 (1957).
[63] Hirs, C. H. W.: *J. Am. Chem. Soc.*, **77,** 5743 (1955).
[64] Tallan, H. H.: *Biochim. Biophys. Acta*, **27,** 407 (1958).
[65] Finkle, B. J., Smith, E. L.: *J. Biol. Chem.*, **230,** 669 (1958).
[66] Myers, F. L., Northcote, O. H.: *Biochem. J.*, **71,** 749 (1959).
[67] Talboys, P. W.: *Nature*, **166,** 1077 (1950).
[68] Thompson, A. R.: *Biochem. J.*, **61,** 253 (1955).
[69] Schroeder, W. A., Jones, R. T., Cormick, J., McCalla, K.: *Anal. Chem.*, **34,** 1570 (1962).
[70] Catravas, G. N.: *Anal. Chem.*, **36,** 1146 (1964).
[71] Pikkarainen, J., Kulonen, E.: *Acta Chem. Scand.*, **22,** 1267 (1968).
[72] Machleidt, W., Kerner, W., Otto, J.: *Z. Anal. Chem.*, **252,** 151 (1970).
[73] Huang, T. S., De Lange, R. J.: *J. Biol. Chem.*, **246,** 686 (1971).
[74] Ozawa, Y., Suzuki, K., Tajima, O., Mogi, K.: *Agr. Biol. Chem.*, **36,** 1371 (1972).
[75] Hartley, R. W.: *Anal. Biochem.*, **46,** 676 (1972).
[76] Atassi, M. Z., Perlstein, M. T., Rosemblatt, M. C., Rocek, P.: *Anal. Biochem.*, **49,** 164 (1972).
[77] Kimmel, M. T., Plummer, T. H. Jr.: *Anal. Biochem.*, **49,** 267 (1972).
[78] Benson, J. V. Jr., Jones, R. T., Cormick, J., Patterson, J. A.: *Anal. Biochem.*, **16,** 91 (1966).
[79] Redfield, R. R., Anfinsen, C. B.: *J. Biol. Chem.*, **221,** 385 (1956).
[80] Anfinsen, C. B.: *J. Biol. Chem.*, **221,** 405 (1956).
[81] Jacobs, S.: *Analyst*, **98,** 25 (1973).
[82] Wittmann, H. G., Braunitzer, G.: *Virology*, **9,** 726 (1959).
[83] Schroeder, W. A., Robberson, B.: *Anal. Chem.*, **37,** 1583 (1965).
[84] Rudloff, V., Braunitzer, G.: *Z. Physiol. Chem.*, **323,** 129 (1961).
[85] Grassmann, W., Hannig, K., Schleyer, M.: *Z. Physiol. Chem.*, **322,** 71 (1960).
[86] Wieland, Th., Determann, H., Albrecht, E.: *Liebig's Ann.*, **633,** 185 (1960).
[87] Moore, S., Stein, W. H.: *J. Biol. Chem.*, **176,** 367 (1948).
[88] Moore, S., Stein, W. H.: *J. Biol. Chem.*, **192,** 663 (1951).
[89] Long, D. A., Truscott, T. G.: *Trans. Faraday Soc.*, **59,** 918 (1963).
[90] Synge, R. L. M.: *Biochem. J.*, **39,** 351 (1945).
[91] Linderstrøm-Lang, K.: (private communication) in Tristram, G. R., Smith, R. H.: *Adv. Protein Chemistry*, **18,** 331 (1963).
[92] Hamilton, P. B.: *Ann. N. Y. Acad. Sci.*, **102,** 55 (1962).

[93] Moore, S., Stein, W. H.: *Methods in Enzymology*, Vol. **VI,** 1963, 819.
[94] Kirsten, E., Kirsten, R.: *Biochem. Z.*, **339,** 287 (1964).
[95] Kofrányi, E.: *Z. Physiol. Chem.*, **287,** 170 (1951).
[96] Dustin, J. P., Czaikoska, C., Moore, S., Bigwood, E. J.: *Anal. Chim. Acta*, **9,** 256 (1953).
[97] Pirie, N. W.: *Proteins*, in *Moderne Methoden der Pflanzenanalyse*, Vol. 4, Paech, K., Tracey, M. V. (eds.), Springer-Verlag, Berlin, 1955, p. 23.
[98] Krampitz, G.: *Z. Tierphysiol. Tierernähr. Futtermittelkunde*, **15,** 76, 227 (1960).
[99] Wilson, R. F., Tilley, J. M. A.: *Sci. Food Agric.*, **16,** 174 (1965).
[100] Dornseifer, T. P., El'ode, K. E., Keith, S. E., Powers, J. J.: *J. Food Sci.*, **31,** 351 (1966).
[101] Rooney, L. W., Salem, A., Johnson, J. A.: *Cereal Chem.*, **44,** 539 (1967).
[102] Moore, S., Stein, W. H.: *J. Biol. Chem.*, **211,** 893 (1954).
[103] Tristram, G. R.: *Techniques in Amino Acid Analysis*, Technicon, Frankfurt 1966, p. 61.
[104] Hill, R. L.: *Adv. Protein Chem.*, **20,** 37 (1965).
[105] Rees, M. W.: *Biochem. J.*, **40,** 632 (1946).
[106] Smith, E. L., Stockell, A., Kimmel, J. R.: *J. Biol. Chem.*, **207,** 501, 551 (1954).
[107] Kassel, B., Laskowski, M.: *J. Biol. Chem.*, **236,** 1996 (1961).
[108] Stein, E. A., Junge, J. M., Fischer, E. H.: *J. Biol. Chem.*, **235,** 371 (1960).
[109] Wilcox, P. E., Cohen, E., Tan, W.: *J. Biol. Chem.*, **228,** 999 (1957).
[110] Wallenfels, K., Arens, A.: *Biochem. Z.*, **332,** 217, 247 (1960).
[111] Crampton, C. F., Moore, S., Stein, W. H.: *J. Biol. Chem.*, **215,** 787 (1955).
[112] Yoritaka, T., Ono, T.: *Nagasaki Igakkai Zasshi*, **29,** 400 (1954).
[113] Osono, K., Mukai, J., Tominaga, F.: *Nagasaki Igakkai Zasshi*, **30,** 156 (1955).
[114] Mahowald, T. A., Noltmann, E. A., Kuby, S. A.: *J. Biol. Chem.*, **237,** 1138 (1962).
[115] Harfenist, E.: *J. Am. Chem. Soc.*, **75,** 5528 (1953).
[116] Hirs, C. H. W., Moore, S., Stein, W. H.: *J. Biol. Chem.*, **211,** 941 (1954).
[117] Tristram, G. R., Smith, R. H.: *Adv. Protein Chem.*, **18,** 227 (1963).
[118] Davies, M. G.: *Shandon Instr. Appl.*, No. 26, 1968, p. 26.
[119] Mondino, A., Bongiovanni, G.: *J. Chromatog.*, **52,** 405 (1970).
[120] Drawert, F., Reuther, K. H.: *Angew. Chem.*, **75,** 169 (1963).
[121] Sanger, F., Thompson, E. O.: *Biochim. Biophys. Acta*, **71,** 468 (1963).
[122] Benisek, W. F., Raftery, M. A., Cole, R. D.: *Biochemistry*, **6,** 3780 (1967).
[123] Freedlender, E. F., Haber, E.: *Biochemistry*, **11,** 2362 (1972).

[124] Matsubara, H., Sasaki, M.: *Biochem. Biophys. Res. Commun.*, **35,** 175 (1969).

[125] James, L. B.: *J. Chromatog.*, **68,** 123 (1972).

[126] Liu, T. Y., Chang, Y. H.: *J. Biol. Chem.*, **246,** 2842 (1971).

[127] Liu, T. Y.: in Moore, S.: *The Precision and Sensitivity of Amino Acid Analyses, in Chemistry and Biology of Peptides*, Meienhofer, J. (ed.), Ann Arbor, 1972, p. 629.

[128] Penke, B., Ferenczi, R., Kovács, K.: *Anal. Biochem.*, **60,** 45 (1974).

[129] Hirs, C. H. W.: *J. Biol. Chem.*, **219,** 611 (1956).

[130] Noltmann, E. A., Mahowald, T. A., Kuby, S. A.: *J. Biol. Chem.*, **237,** 1146 (1962).

[131] Homer, A.: *J. Biol. Chem.*, **22,** 369 (1915).

[132] Dréze, A., Reith, W. S.: *Biochem. J.*, **62,** 3P (1956).

[133] Kuiken, K. A., Lyman, C. M., Hale, F.: *J. Biol. Chem.*, **171,** 551 (1947).

[134] Dévényi, T., Báti, J., Fábián, F.: *Acta Biochim. Biophys. Acad. Sci. Hung.*, **6,** 133 (1971).

[135] Jorpes, E.: *Biochem. J.*, **26,** 1488 (1932).

[136] Oelshlcgcl, F. J. Jr., Schroeder, J. R., Stahmann, M. A.: *Anal. Biochem.*, **34,** 331 (1970).

[137] Hugli, T. E., Moore, S.: *J. Biol. Chem.*, **247,** 2828 (1972).

[138] Friedman, M., Finley, J. W.: *Agric. Food Chem.*, **19,** 626 (1971).

[139] Stein, W. H., Moore, S.: *J. Biol. Chem.*, **211,** 915 (1954).

[140] London, R.: *Aminosäuren-Säulenchromatographie*, Technicon, Frankfurt, 1965, p. 105.

[141] Mondino, A., Bongiovanni, G., Fumero, S. I., Rossi, L.: *J. Chromatog.*, **74,** 255 (1972).

[142] Block, W. D., Markow, M. E., Steele, B. F.: *Proc. Soc. Exp. Biol. Med.*, **122,** 1089 (1966).

[143] Bhatty, R. S.: *Cereal Chem.*, **49,** 729 (1972).

[144] Bito, L. Z., Dawson, J.: *Anal. Biochem.*, **23,** 95 (1969).

[145] Reid, R. H. P.: *Techniques in Amino Acid Analysis*, Technicon, Frankfurt 1966, p. 43.

[146] Gressner, A. M.: *Anal. Biochem.*, **56,** 532 (1973).

[147] Benson, J. V. Jr., Gordon, M. J., Petterson, J. A.: *Anal. Biochem.*, **18,** 228 (1967).

[148] Peters, J. H., Berridge, B. J., Cummings, J. G., Liu, S. C.: *Anal. Biochem.*, **23,** 459 (1968).

[149] Vega, A., Nunn, P. B.: *Anal. Biochem.*, **32,** 446 (1969).

[150] Kedenburg, C. P.: *Anal. Biochem.*, **40,** 35 (1971).

[151] Lorenz, H.: *Phytochemistry*, **10,** 63 (1971).

[152] Houpert, Y., Tarallo, P., Siest, G.: *J. Chromatog.*, **115,** 33 (1975).
[153] Tutschek, R., Meier, K. D., Grönöng, F., Stubba, W.: *J. Chromatog.*, **139,** 211 (1977).
[154] Adriaens, P., Meesschaert, B., Wuyts, W., Vanderhaeghe, H., Eyssen, J.: *J. Chromatog.*, **140,** 103 (1977).
[155] Spackman, D. H., Stein, W. H., Moore, S.: *Anal. Chem.*, **30,** 1190 (1958).
[156] Rosen, H.: *Arch. Biochem. Biophys.*, **67,** 10 (1957).
[157] Rothwell, R.: *Technicon Research Bulletin,* No. 20. (1960).
[158] Yamamoto, M., Young, J. L.: *J. Chromatog.*, **57,** 152 (1971).
[159] Langer, H. J.: *Technicon Symposium* No. 976 (1971).
[160] James, L. B.: *J. Chromatog.*, **59,** 178 (1971).
[161] Moore, S.: *J. Biol. Chem.*, **243,** 6281 (1968).
[162] Lustenberger, N., Lange, H. W.: *Angew. Chem. Intern. Ed.*, **11,** 227 (1972).
[163] Lange, H. W., Lustenberger, N., Hempel, K.: *Z. Anal. Chem.*, **261,** 337 (1972).
[164] Samejima, K., Dairman, W., Udenfriend, S.: *Anal. Biochem.*, **42,** 222 (1971).
[165] Samejima, K., Dairman, W., Stone, J., Udenfriend, S.: *Anal. Biochem.*, **42,** 237 (1971).
[166] Stein, S., Böhlen, P., Stone, J., Dairman, W., Udenfriend, S.: *Arch. Biochem. Biophys.*, **155,** 202 (1973).
[167] Benson, J. R., Hare, P. E.: *Proc. Natl. Acad. Sci. U. S.*, **72,** 619 (1975).
[168] Roth, M., Hampaï, A.: *J. Chromatog.*, **83,** 353 (1973).
[169] Scott, C. D., Jolley, R. L., Pitt, W. W., Johnson, W. F.: *Am. J. Clin. Path.*, **53,** 701 (1970).
[170] Fishman, M. L., Landgraff, L. M., Burdick, D.: *J. Chromatog.*, **86,** 37 (1973).
[171] Lamaziere, J.: *J. Chromatog.*, **106,** 191 (1975).
[172] Scott, K., Cannell, G. R., Zerner, B.: *Anal. Biochem.*, **69,** 474 (1975).
[173] Owen, J. M., Dale, A. D.: *J. Chromatog.*, **107,** 207 (1975).
[174] Brown, J. H., Walker, S., Casto, L., Howell, R.: *J. Chromatog.*, **116,** 293 (1976).
[175] van der Walt, S. J.: *Anal. Biochem.*, **75,** 301 (1976).
[176] Walsh, K. A., Brown, J. R.: *Biochim. Biophys. Acta*, **58,** 596 (1962).
[177] Bates, L. S.: *Anal. Biochem.*, **41,** 158 (1971).
[178] Cavins, J. F., Friedman, M.: *Anal. Biochem.*, **35,** 489 (1970).
[179] Friedman, M., Noma, A. T., Masri, M. S.: *Anal. Biochem.*, **51,** 280 (1973).
[180] Hempel, K., Lange, H. W.: *Z. Physiol. Chem.*, **349,** 603 (1968).
[181] Seely, J. H., Edatell, R., Benoiton, N. L.: *J. Chromatog.*, **44,** 618 (1969).

[182] Kakimoto, Q., Akazawa, S.: *J. Biol. Chem.*, **245,** 5751 (1970).
[183] Deibler, G. E., Martenson, R. E.: *J. Biol. Chem.*, **248,** 2387 (1973).
[184] Tyihák, E., Ferenczi, S., Hazai, I., Zoltán, S.: *J. Chromatog.*, **102,** 257 (1974).
[185] Beckerton, A., Buttery, P. J.: *J. Chromatog.*, **104,** 170 (1970).
[186] Lange, H. W., Hempel, K.: *J. Chromatog.*, **107,** 389 (1975).
[187] Zakardas, C. G.: *Canad. J. Biochem.*, **53,** 96 (1975).
[188] Rangeley, W. R. D., Lawrie, R. A.: *J. Food Techn.*, **12,** 9 (1977).
[189] Hehn, R., Vančíková, O., Macek, K., Deyl, Z.: *J. Chromatog.*, **133,** 390 (1977).
[190] Paik, W. K., Dimaria, P., Pearson, E., Kim, S.: *Anal. Biochem.*, **90,** 262 (1978).
[191] Dévényi, T.: *Acta Biochim. Biophys. Acad. Sci. Hung.*, **6,** 129 (1971).
[192] Villegas, E., McDonald, C. E., Gilley, K. A.: *Cereal Chem.*, **45,** 432 (1968).
[193] Moore, S.: *The Precision and Sensitivity of Amino Acid Analysis*, in *Chemistry and Biology of Peptides*, Meienhofer, J. (ed.), Ann Arbor 1972.
[194] Blackburn, S.: *Amino Acid Determination*, Arnold, London, 1968.
[195] Peterson, E. A., Sober, H. A.: *Cellulosic Ion Exchangers*, Elsevier, Amsterdam, 1975.
[196] Janson, J.-C.: *J. Gen. Microbiol.*, **88,** 209 (1975).
[197] Shannon, L. M., Kay, E., Lew, J. Y.: *J. Biol. Chem.*, **241,** 2166 (1966).
[198] Leaback, D. H., Robinson, H. K.: *Biochem. Biophys. Res. Commun.* **67,** 248 (1975).
[199] Young, J. L., Webb, A. B.: *Anal. Biochem.*, **88,** 619 (1978).
[200] Brocklehurst, K., Carlsson, J., Kierstan, M. P. J., Crook, E. M.: *Biochem. J.*, **133,** 573 (1973).
[201] Axén, R., Drevin, H., Carlsson, J.: *Acta Chem. Scand.*, **B 29,** 471 (1975).
[202] Carlsson, J., Svenson, A.: *FEBS Lett.*, **42,** 183 (1974).
[203] Egorov, T. A., Svenson, A., Rydén, L.: *Proc. Natl. Acad. Sci. U.S.*, **72,** 3029 (1975).
[204] Nijhuis, H., Klostermeyer, H.: *Milchwissenschaft*, **30,** 530 (1975).
[205] Malthouse, J. P. G., Brocklehurst, K.: *Biochem. J.*, **159,** 221 (1976).
[206] Norris, R., Brocklehurst, K.: *Biochem. J.*, **159,** 245 (1976).
[207] Sykes, B. C.: *FEBS Lett.*, **61,** 180 (1976).
[208] Fex, G., Lindgren, R.: *Biochim. Biophys. Acta*, **493,** 410 (1977).
[209] Laurell, C.-B., Thulin, E., Bywater, R. P.: *Anal. Biochem.*, **81,** 336 (1977).
[210] Squibb, K. S., Cousins, R. J.: *Biochem. Biophys. Res. Commun.*, **75,** 806 (1977).
[211] Elsayed, S., Bennich, H.: *Scand. J. Immunol.*, **4,** 203 (1975).

[212] Rydén, L., Björk, I.: *Biochemistry*, **15,** 3411 (1976).
[213] Rydén, L., Eaker, D.: *FEBS Lett.*, **53,** 279 (1975).
[214] Carlsson, J., Olsson, I., Axén, R.: *Acta Chem. Scand.*, **B 30,** 180 (1976).
[215] Svenson, A., Carlsson, J., Eaker, D.: *FEBS Lett.*, **73,** 171 (1977).
[216] Cuatrecasas, P., Wilchek, M., Anfinsen, C. B.: *Proc. Natl. Acad. Sci. U.S.*, **61,** 636 (1968).
[217] Wingard, L. B. Jr., Katchalski-Katzir, E.: *Immobilized Enzyme Principles*, Academic Press, New York, 1976.
[218] Axén, R., Porath, J., Ernbach, S.: *Nature*, **214,** 1302 (1967).
[219] Cuatrecasas, P.: *J. Biol. Chem.*, **245,** 3059 (1970).
[220] Bennich, H., Johansson, S. G. O.: *Adv. Immunol.*, **13,** 1 (1971).
[221] Bauer, E. A., Jeffrey, J. J., Eisen, A. Z.: *Biochem. Biophys. Res. Commun.*, **44,** 813 (1971).
[222] Olivesrona, T., Egelrud, T.: *Biochem. Biophys. Res. Commun.*, **43,** 524 (1971).
[223] Cuatrecasas, P., Anfinsen, C. B.: *Ann. Rev. Biochem.*, **40,** 259 (1971).
[224] Cuatrecasas, P.: *Agric. Food Chem.*, **19,** 600 (1971).
[225] Jakoby, W. B., Wilchek, M.: *Affinity Techniques of Enzyme Purification*, in *Methods in Enzymology*, Vol. XXXIV, Academic Press, New York, 1974.
[226] Lowe, C. R., Dean, P. D. G.: *Affinity Chromatography*, Wiley, London, 1974.
[227] Turková, J.: *Affinity Chromatography*, Elsevier, Amsterdam, 1978.
[228] Hippel, P. H., Schleich, T.: in *Structure and Stability of Biological Macromolecules*, Timasheff, S. N., Fasman, G. D., (eds), Dekker, New York, 1969, p. 417.
[229] Yon, R. J.: *Biochem. J.*, **137,** 127 (1974).
[230] Er-el, Z., Shaltiek, S.: *FEBS Lett.*, **40,** 142 (1974).
[231] Doellgast, G. J., Fishman, W. H.: *Biochem. J.*, **141,** 103 (1974).
[232] Hokse, H.: *Carbohydr. Res.*, **37,** 390 (1974).
[233] Jennissen, H. P., Heilmeyer, L. M. G.: *Biochemistry*, **14,** 754 (1975).
[234] Bigelis, R., Umbarger, H. E.: *J. Biol. Chem.*, **250,** 4315 (1975).
[235] Hofstee, B. H. J.: *Prep. Biochem.*, **5,** 7 (1975).
[236] Doellgast, G. J., Plaut, A. G.: *Immunochemistry*, **13,** 135 (1975).
[237] Yon, R. J., Simmonds, R. J.: *Biochem. J.*, **151,** 281 (1975).
[238] Zaidenzaig, Y., Shaw, W. V.: *FEBS Lett.*, **62,** 266 (1975).
[239] Lowe, C. R.: *Biochem. Soc. Trans.*, **5,** 253 (1977).
[240] Pahlman, S., Rosengren, J., Hjertén, S.: *J. Chromatog.*, **131,** 99 (1977).
[241] Artyukov, A. A., Moloatsov, N. V.: *J. Chromatog.*, **130,** 451 (1977).
[242] Hjertén, S.: *J. Chromatog.*, **87,** 325 (1973).
[243] Hjertén, S., Rosengren, J., Pahlman, S.: *J. Chromatog.*, **101,** 281 (1974).

CHAPTER 5

Thin-layer chromatography of proteins, peptides and amino-acids

E. Tyihák

5.1 BASIC PRINCIPLES OF THIN-LAYER CHROMATOGRAPHY

Chromatographic methods can be divided into two main groups: *column systems* such as classical column chromatography, high-performance liquid chromatography, gel filtration and gas chromatography, which are suitable for quantitative work, and *planar systems* such as paper chromatography and thin-layer chromatography, the main advantage of which is the ease of identification of the separated substances.

During the past 15–20 years, thin-layer chromatography has found widespread use. A number of efforts led to the emergence of the variants of the method known today [1, 2]; its present basic form was developed by Kirchner *et al.* [3], while the rapid spread of the method began with the publication by Stahl *et al.* [4] which appeared in 1956. In that paper, the authors described a standardized method; that is, they succeeded in solving the problem of layering the sorbent (e.g. silica gel), previously used only in the form of a column, onto a glass plate. This became the basis of modern thin-layer chromatography, providing reproducible results.

There is a close connection in principle between column and thin-layer chromatography, the two systems of liquid chromatography. The resolving power of column chromatography is considerably greater than that of thin-layer chromatography, where the number of theoretical plates is limited. The maximum attainable theoretical plate number in classical thin-layer chromatography is around 2000; in contrast, in column chromatography, and particularly in modern HPLC, it is 10^5 [5]. In theory, therefore, column chromatography should be able to deal with all the separation problems soluble by liquid chromatography. In spite of this, there seem to be no limits to the spread of thin-layer chromatography, and the reasons for this arise from its advantages listed below.

(*a*) The chromatograms can be evaluated visually by use of selective or specific colour reactions so individual compounds may be detected specifically in the presence of many others.

(*b*) In column chromatography certain components (possibly the biologically most active) of the sample may be bound irreversibly on the column, and hence may not be identified. In contrast, in thin-layer chromatography separation a picture of the entire sample is obtained.

(*c*) In thin-layer chromatography, selection of the chromatographic conditions (sorbent, solvent, etc.) requires a comparatively short time (5–50 min).

(*d*) The latest patterns of separation chamber [6–8] permit comparison of a number of separation conditions on a single thin-layer plate.

(*e*) The method is cheap, simple and fast and can deal with even quite complex mixtures.

In classical variants of thin-layer chromatography (TLC), the sorbent (silica gel, aluminium oxide, polyamide powder, cellulose powder, ion-exchange resin, etc.), generally slurried in water or a polar organic solvent, is spread by hand or with an appropriate device, in a thin (0.25–1.0 mm) layer on a glass plate or a plastic sheet or foil. Most devices for making the thin-layers have a slurry-reservoir or trough, and a spreading-edge at the appropriate distance above the surface of the glass plate; by movement of either the spreader or the plate, the slurry poured onto the plate is smoothed into a uniform layer. These spread layers are used after drying or activation by heat-treatment at 50–150 °C. The prepared plates may be stored in a suitable holder or a desiccator.

There is an ever growing choice and use of precoated plates marketed by various firms; glass, plastic or aluminium foil is used as the support in these plates. Precoated plates have a number of advantages: (*i*) a highly uniform layer; (*ii*) simple documentation; (*iii*) the foil type may readily be cut into slices; (*iv*) simple storage; (*v*) good reproducibility.

A solution of the substance to be examined is added dropwise to a point or a band on the layer, from a suitable micropipette or capillary; a current of air is used to accelerate the drying of the spots or bands. The quantity of sample may vary between wide limits, but if too much is applied, there may be distortion of the separated zones. After addition of the sample, the thin-layer plate is placed in a development chamber fitted with a ground-class cover, the bottom of the chamber being filled with the development solvent mixture. After the solvent mixture has migrated an appropriate distance, the plate is removed and dried and the separated substances are detected by spraying with a suitable reagent, with or without heating. Some substances can be located without a reagent, on the basis of their colour, fluorescence or absorption of ultraviolet radiation.

In most forms of thin-layer chromatography, consideration must be given to the vapour in the space above the sorbent layer. Its composition depends on the nature of the development mixture and the chamber system. The components in the vapour space are adsorbed to different degrees on the dry layer, and therefore the vapour space is of particular importance as regards the effectiveness of separation [8].

When in contact with a dry support, a multicomponent solvent mixture undergoes demixing during flow. The theory of polyzonal thin-layer chromatography is based on the phenomenon of solvent demixing [9, 10]. The theory may be illustrated by considering duozonal thin-layer chromatography. The theory states that the more polar component (β-front) falls behind the less polar solvent (α-front), the rate of migration of the β-front being lower than that of the α-front by a constant retarding factor, k_β. The sandwich (S) [11] and the horizontally-arranged BN chamber [12] are the most convenient for zonal thin-layer chromatographic separations, since the solvent equilibration is relatively slight in these chambers. The use of unsaturated chambers, or the presaturation of the thin-layer with the vapours of one or more solvents, may give valuable results in some cases (vapour-programmed thin-layer chromatography) [13].

An ultramicro (UM) chamber may be used to clarify the role of the solvent vapour; the essence of this is that the sorbent layer is covered completely by a cover-plate, so that it is not in contact with the surface of the solvent, and thus the disturbances originating from capillary action are eliminated [14, 15].

In gradient thin-layer chromatography, better separation is attained by variation of the composition of the support or the pH of the development mixture [16].

One of the great advantages of TLC over the other chromatographic methods is the wide choice of detection methods. Differential thermal analysis has been used to identify dicarboxylic acids by means of their melting-point endotherms [17]. In functional thin-layer chromatography [18], the reaction used proceeds before the chromatography, on the plate or in a separate vessel. In reaction thin-layer chromatography, the layer participates in the reaction directly [19]. The various types of sensitizing reactions in the identification also belong to reaction thin-layer chromatography [20].

In TLC [21, 22], impregnation of the layers and the use of additives are becoming increasingly more common, because better resolution may be achieved.

Mini thin-layer chromatography considerable reduces the dimensions of the TLC apparatus [23], and gives cheap and rapid analysis.

Two-dimensional work is commonly used, thin-layer electrophoresis frequently being used in one of the dimensions (fingerprint technique). For this purpose various firms have designed appropriate instruments, characterized by the incorporation of electrophoretic cells with small vapour spaces. The electrode compartments and the sorbent layer may be connected with filter paper. A cooling surface and a buffer of low ionic strength are used in order to avoid heat effects.

A sharp division cannot be made between the classical and modern variants of TLC. The modern variants mainly use a higher flow-rate and give a better resolution.

One of the differences between conventional (classical) TLC and instrumentalized high-performance TLC (HPTLC) lies in the particle size of the adsorbent. Conventional TLC chromatograms are generally developed linearly, whereas three development modes, viz. linear, circular and anticircular, may be employed in HPTLC [24–26].

HPTLC offers distinct advantages when (a) a comparatively large number of samples have to be analysed; (b) when only some fractions of a complex sample need to be quantified; (c) when a short development distance is enough for analysis; (d) when the samples contain only some components; (e) when the absence of the mobile phase during detection reduces sensitivity problems. It is shown that the total analysis time per individual sample lies in the region of 1.5–2.0 minutes; thus the speed is comparable with that of HPLC [27].

In HPTLC, three steps of the procedure—sample application, development and quantitative assessment—are instrumentalized [26].

Reversed-phase HPTLC is used for the separation of many classes of both polar and non-polar organic compounds [26]. In some cases, the reversed-phase HPTLC owing to the application of polar solvents or their mixtures has a long separation time, so diffusion is important but mainly over longer development distances. This major problem in reversed-phase HPTLC has been adequately solved for at least one chemically bonded stationary-phase material, viz. by the addition of 3% NaCl to the mobile phase [28]. Quantitative evaluation of HPTLC chromatoplates is based on absorption or fluorescence measurements in the reflectance, transmission, or simultaneous reflectance/transmission mode [26].

Application of samples to the layer in a concentrating zone is a special, fairly recent development. Use is made of a double layer, prepared by spreading side by side a wider, active (e.g. silica gel) layer and a narrower, inactive (e.g. kieselguhr) layer [29]. The sample is applied dropwise to the inactive zone; the solvent mixture then causes the substances to accumulate in a sharp band, without separation, on the border with the active zone; further development then leads to better resolution.

The development of overpressured thin-layer chromatography (OPTLC) is an important stage in planar liquid chromatography because this technique combines the advantages of classical TLC, modern HPTLC using fine-particle sorbent layers and HPLC [30–34]. In this technique the separation proceeds in a pressurized ultramicro chamber (PUM chamber), where the sorbent layer is completely covered by a plate held in contact with it by external pressure, while the solvent mixture is forced through under pressure. It thus corresponds to HPLC on a column with a very thin but wide cross-section.

In the PUM chamber the external pressure on the flexible cover membrane must always be higher than the input pressure of the solvent. The input pressure of the solvent increases linearly with increasing solvent migration distance. An increase in the solvent flow velocity always results in higher input pressures, which must be taken into account in choosing an appropriate external pressure on the membrane.

The linear migration of the solvent front in a PUM chamber can be carried out by impregnating the sides of the sorbent layer and placing a narrow plastic sheet on the layer or making a narrow channel in the layer before the position of the solvent inlet.

OPTLC is especially attractive when a large number of samples is to be investigated because the time required for analysis is short compared to an HPLC separation. OPTLC is still more attractive when used in the two-dimensional mode.

The outstanding properties of chemically bonded phases on various chromatoplates can be used in OPTLC for the separation of different substance groups. The poorer wettability of reversed-phase plates does not appear a problem in OPTLC as the solvent is admitted into the PUM chamber by means of a pump system.

The quadratic law of TLC development is not valid in linear OPTLC using different plates and solvents. The migration of the solvent front in linear OPTLC is described by a simple equation [34]:

$$k \sim \frac{Z_f}{t}$$

where the velocity constant (k) is a function of the flow velocity of the mobile phase, the quality of the sorbent and the dimensions of the bed particles. This relationship shows that the linear OPTLC technique really approaches column chromatographic conditions.

5.2 THIN-LAYER CHROMATOGRAPHY OF PROTEINS

5.2.1 Separation of proteins on inorganic and organic layers

Bauer [35] attempted the separation of various protein preparations (protamines, histones, etc.) on silica gel and cellulose layers, using several aqueous development mixtures in succession at intervals of 15 min (e.g. 1 *M*; 0.25 *M*; 0.1 *M*; 0.01 *M* hydrochloric acid; 50% acetic acid). Others also tried to separate known enzymes (pepsin, trypsin, etc.) on a silica gel support, but their work is now mainly of historical interest [36].

Kibardin and Lasurkina [37] separated proteins fairly successfully in phosphate buffer (pH = 6.8) on a hydroxyapatite layer. With this method they

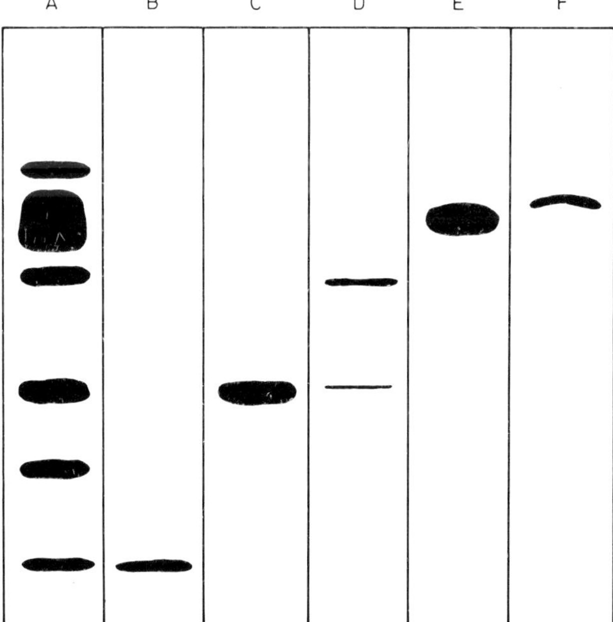

Fig. 5.1. Rechromatography on a hydroxyapatite thin layer of the water-soluble protein fraction obtained from egg-yolk [38]. *A:* total fraction; *B:* fraction WSF-1; *C:* fraction WSF-3; *D:* fraction WSF-4; *E–F:* amino-acids. (Reproduced from *Canad. J. Biochem.*, **49**, 44 (1971) by permission of the National Research Council of Canada.)

were later able to measure the quantitative changes occurring in various proteins on denaturation with various agents (e.g. cupric chloride). The deviation coefficient (q) was introduced to characterize the changes in the mobilities of the control and modified proteins (R_{fC} and R_{fM}).

Thin-layer chromatography can be used effectively for the study of lipoproteins, because of the favourable chromatographic characteristics of the non-protein parts. Gornall and Kuksis [38, 39] examined the lipoproteins of egg-yolk on a layer prepared from hydroxyapatite.

Figure 5.1 shows the results of chromatography (on a hydroxyapatite layer) of water-soluble fractions obtained from egg-yolk.

Blaton *et al.* [40] investigated the lipoproteins of hyperlipidaemic patients of types II, III and IV by chromatography on silica gel with petroleum ether–ether–acetic acid mixture (150:20:0.5) as developing agent; phosphoric acid–copper acetate reagent was employed to detect the lipoproteins, with heating at 190 °C for 30 min. Trautmann [41] also used silica gel in his study of the carrier proteins and special lipoproteins of ^{3}H-labelled compounds with juvenile hormone effects, obtained from the glands of *Tenebrio molitor* L. larvae, the developing mixtures being cyclohexane–ethanol (1:1), petroleum ether–ethyl acetate (9:1), etc.

5.2.2 Separation of proteins on a thin-layer gel

Thin-layer gel chromatography can be regarded as a limiting case of liquid–liquid (partition) chromatography, in which the same solvent is used for the stationary and mobile phases.

Dextran-based semisynthetic, and acrylamide-based synthetic gel-formers are used in general for preparation of the gel layer. Most methods reported in the literature utilize the finest particle-size (superfine) Sephadex for preparation of the gel layer. Table 5.1 details the water-uptake data for the Sephadex gels used to prepare the gel layer, and also the swelling times.

The gel layers are generally prepared on glass plates, with the usual spreaders. The optimum thickness of the gel layer is in the range 0.4–1.0 mm (Table 5.2). It is advisable to let the spread layers stand freely in air for 15–20 min, after which they may be used for thin-layer gel chromatography. The spread layers may be stored for fairly long periods in a wet chamber, but care must be taken to eliminate infection with micro-organisms. The spread gel layer may even be dried if it is not to be used immediately. The surface of the thin layer may undergo cracking, even with very mild drying, but G-25, G-50 and G-75 layers can easily be regenerated by spraying with buffer solution.

As a consequence of the nature of the thin-layer gel, in practice a wet layer is always used for the chromatography. Before application of the sample the

Table 5.1
Minimum swelling times and water-uptakes of Sephadex gels of superfine quality*

Sephadex type	Water-uptake g of H_2O/g of dry gel	Minimum swelling time, h	
		at room temp.	on a boiling water-bath
G-50	5.0	3	1
G-75	7.5	24	2
G-100	10.0	72	3
G-200	20.0	72	3

* From the Pharmacia Fine Chemicals AB prospectus "Thin-layer gel filtration with the Pharmacia TLG apparatus".

Table 5.2
Approximate quantities of gel suspension necessary for the spreading of gel thin-layers of various thicknesses*

Layer thickness, mm	Volume of gel suspension, ml	
	20 × 20 cm plate	20 × 40 cm plate
1.0	55	110
0.8	45	90
0.6	35	70
0.4	25	50

* From the Pharmacia Fine Chemicals AB prospectus "Thin-layer gel filtration with the Pharmacia TLG apparatus".

gel layer must be equilibrated for 10–15 h in the development chamber with the buffer it is planned to use.

The protein sample (1–5 μl) may be applied to the gel layer at a point or in a band in the customary manner, with a capillary or a micropipette, simultaneously with reference compounds (authentic proteins) of known molecular weight. Chromatography of coloured substances of high molecular weight (e.g. Blue Dextran 2000, cytochrome C, haemoglobin) along with the sample is particularly advantageous for following the migration of the buffer solution.

A development chamber suitable for separations on a thin-layer gel is shown in Fig. 5.2; this is a very simple apparatus, that can readily be home-made, possibly from plastic. The glass plate supporting the gel layer is inclined at an angle of 10–20 °C to the vessel containing the buffer solution, and the buffer is led onto the layer with a strip of filter paper (e.g. Whatman No. 3) as wide as the layer, and adhering to it over a length of 1–1.5 cm. If the filter paper strip at the lower end of the layer is only just moistened, this inhibits the departure of the solution excess from the bottom of the plate, but at the same

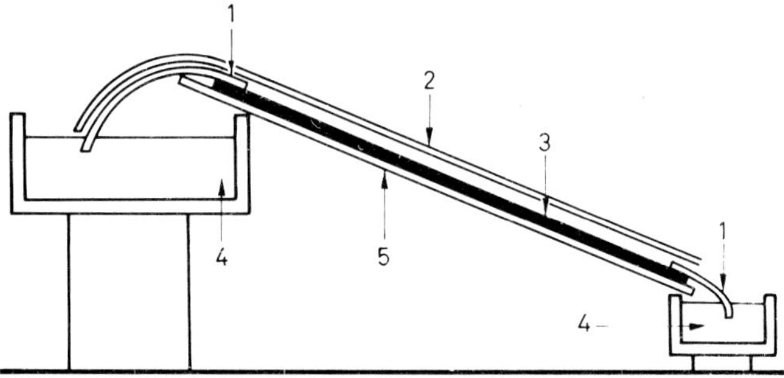

Fig. 5.2. Scheme of simple thin-layer gel chromatographic chamber [45]: *1*, filter paper; *2*, cover plate; *3*, protein; *4*, solution; *5*, glass plate. (Reproduced from *J. Chromatog.*, **16,** 167 (1964), with the permission of Elsevier Scientific Publishing Company.)

time it ensures a good connection with the gel layer. In general, the optimum rate of migration is 1–2 cm/h. Thus, in the case of a standard 20-cm layer plate, roughly one working day is needed for optimum development.

The buffer solutions used for thin-layer gel chromatography are practically the same as those used for gel chromatography, so the thin-layer system can well be used in the selection of gel chromatographic buffer systems.

After development, the gel layer plate is removed from the chamber and evaluated. For this purpose coloured and fluorescent substances are inspected directly on the wet plate, otherwise colour reactions or stains are applied.

As with sorbent TLC, the gel plates may be dried for staining. Determann [42] states that the gel layer may be dried effectively by placing a moist filter paper above the layer. It is also practical to place the plate on an electric hot-plate (at 50–70 °C), which results in drying under mild conditions. The dry gel layers are mechanically stable and can readily be handled.

The most effective method for detection of the separated proteins on the gel layer, however, is to transfer the proteins to filter paper (Whatman 3 MM or Schleicher and Schüll 2043/b). The gel layer is covered with dry filter paper (bubble formation is to be avoided), which is carefully smoothed down over it.

Next the paper is dried together with the layer at 120 °C, or is carefully peeled off the layer and dried at 110 °C. The proteins on the paper may be detected by the procedures used in paper chromatography.

A dried gel layer is pretreated for 10 min in the rinsing bath (750 ml of methanol, 200 ml of water and 50 ml of glacial acetic acid), then immersed for 1 h in the staining bath (a saturated solution of Amido Black 10B in a mixture of 750 ml of methanol, 200 ml of water and 50 ml of glacial acetic acid), and finally destained for 2 h.

A dried paper is immersed for 15 min in a staining bath (450 ml of methanol or ethanol, 450 ml of water, 100 ml of glacial acetic acid and 0.6 g of Amido Black 10B), and then destained in three destaining baths (1% acetic acid solution or tap-water), for 30 min in each. Alternatively the paper is stained for 5 min in a 1% solution of Bromophenol Blue, saturated with mercuric chloride, and then destained in five 0.5% acetic acid baths (30 min in each).

Thin-layer gel chromatography can be utilized to great advantage for the determination of the approximate molecular weights of peptides and

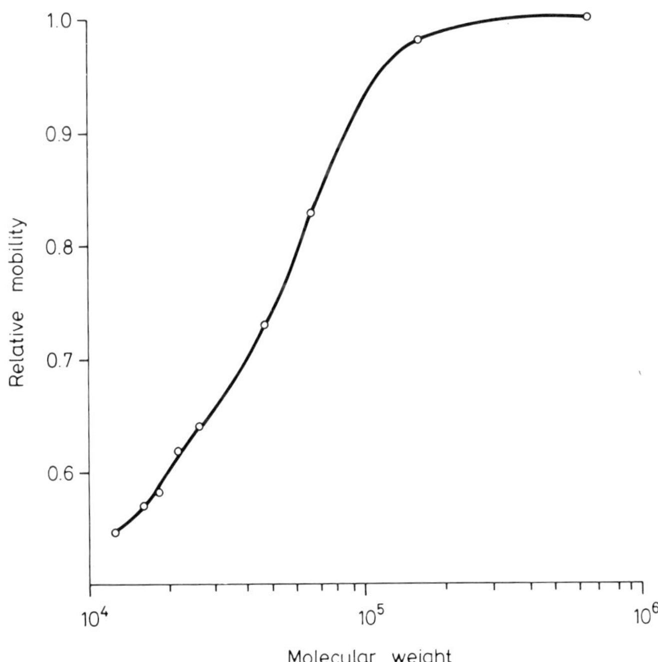

Fig. 5.3. Plot of correlation between molecular weights of proteins and their chromatographic mobilities [44]. (By kind permission of *Biochemical Journal.*)

proteins. Even the first publication on it [43] dealt with a simple molecular weight determination for proteins. Andrews [44] found a close correlation between the logarithm of the molecular weight of a protein and its chromatographic behaviour (the distance covered in constant time on a given layer) (Fig. 5.3). On this basis, the molecular weights of unknown proteins can be determined approximately, by thin-layer gel chromatography in the

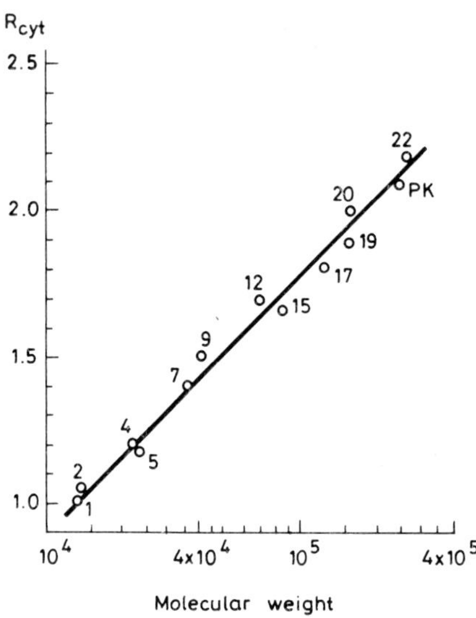

Fig. 5.4. Standard plot for determination of the molecular weights of proteins by thin-layer gel chromatography [46]. Sephadex G-200 (superfine) gel layer: *1*, cytochrome C; *2*, ribonuclease; *4*, soya-bean trypsin inhibitor; *5*, α-chymotrypsin; *7*, pepsin; *9*, peroxidase; *12*, serum albumin; *15*, creatine phosphate kinase; *17*, glyceraldehyde phosphate dehydrogenase; *19*, aldolase; *20*, alcohol dehydrogenase; *22*, catalase; *PK*, phosphocreatine kinase; *XOD*, xanthine oxidase. (Reproduced from *J. Chromatog.*, **25**, 303 (1966), with the permission of Elsevier Scientific Publishing Company.)

presence of proteins of known molecular weights. The behaviour of proteins on a thin-layer gel and in a gel column was examined by Morris [45], who found that there is a linear correlation between the *R* value measured on the thin-layer gel and the *K* value measured in the gel column. Determann and Michel [46] also reported a standard curve, obtained with the use of a large number of known proteins, for the determination of molecular weights (Fig. 5.4). For determination of the molecular weights of polypeptide chains, Heinz and Prosch [47] recommend thin-layer gel chromatography in 6 *M* guanidine hydrochloride. Table 5.3 gives some data obtained with this method for proteins of known molecular weights [47].

Table 5.3
Measurement of molecular weights of polypeptide chains by thin-layer gel chromatography, in 6 *M* guanidine hydrochloride [47]*

Protein	M.W.	No. of polypeptide chains	Polypeptide M.W.	Own measurement	
Lysozyme	14,500	1	14,500	X	17.500
				S.D.	500
				n	15
Lactic acid dehydrogenase	126,000	4	31,500	X	31,000
				S.D.	500
				n	15
Glyceraldehyde phosphate dehydrogenase (muscle)	140,000	4	53,000	X	35,000
				S.D.	1,200
				n	15
Alkaline phosphatase	41,000	1	41,000	X	41,500
				S.D.	800
				n	15
Phosphorylase b	188,000	2	94,000	X	98,000
				S.D.	500
				n	15
β-Galactosidase	135,000	1	135,000	X	132,000
				S.D.	400
				n	15

X = mean; n = number of runs; S.D. = standard deviation
* Reproduced by kind permission of *Anal. Biochem.*

In the systematic examinations by Radola [48], a linear correlation was observed between the logarithm of the radius of a protein molecule and its R_m value (the R_m value referred to myoglobin). The following equations were derived on the basis of the calibration curves prepared from the data measured on various Sephadex layers:

G-75: $\log r = 0.627 R_m - 0.376$
G-100: $\log r = 0.588 R_m - 0.329$
G-200: $\log r = 0.479 R_m - 0.202$

Radola [48] similarly found a quantitative correlation between the logarithm of the molecular weight of a globular protein and its R_m value:

$$\text{G-75: } \log M = 1.402 R_m + 2.776$$
$$\text{G-100: } \log M = 1.281 R_m + 2.896$$
$$\text{G-200: } \log M = 1.172 R_m + 3.015$$

Whereas Radola [48] calculated R_m values referred to myoglobin, Morris [45] worked with R_{Hb} values, referred to haemoglobin. However, he was able to establish a correlation between the R_{Hb} value and the logarithm of the molecular weight of the protein only in the case of Sephadex G-200:

$$\text{G-200: } \log M = 1.47 R_{Hb} + 3.0$$

Radola [48] also compared Sephadex G-200 and Bio-Gel P-300; the latter gives a better separation than the former does, for ferritin. Sample separation is exemplified in Fig. 5.5. Details of gel preparation: 8 g of Sephadex G-75, 6 g of Sephadex G-100, 4 g of Sephadex G-200, 5.4 g of Bio-Gel P-60 or 3.5 g of Bio-Gel P-300 swollen at 4 °C for 2–3 days with 100 ml of 0.02 *M* phosphate buffer (pH 7.2–7.4) that is 0.5 *M* in sodium chloride.

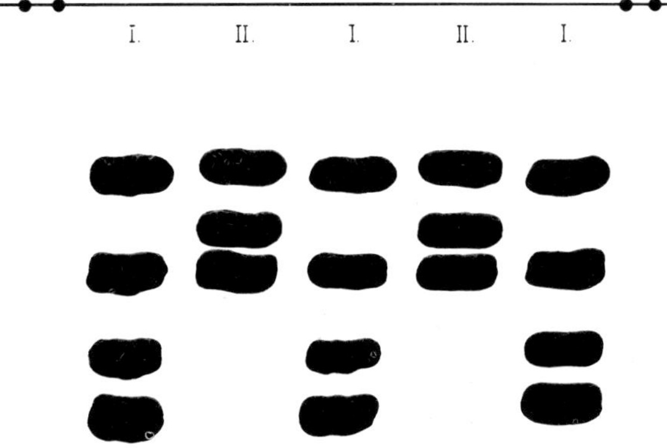

Fig. 5.5. Gel filtration on a Sephadex G-200 thin layer [48]. *I:* myoglobin, calf serum albumin, calf γ-globulin and ferritin; *II:* cytochrome C, ovalbumin and calf serum albumin. Quantity applied: 10 μl of a 1% solution onto a 20 × 20 cm layer. Separation time: 6.6 h. Staining: Amido Black 10B solution. (Reproduced from *J. Chromatog.*, **38**, 61 (1968), with the permission of Elsevier Scientific Publishing Company.)

Further significant applications of thin-layer gel chromatography are the characterization of body fluids, and the detection of pathologic proteins. The reliable study of such complex mixtures is possible only with a two-dimensional method: it is practical to use gel filtration in the first direction, and electrophoresis or immunodiffusion in the second. Hanson *et al.* [49, 50] were able to characterize the paraproteins of the serum and other biological fluids from healthy and ill persons by a combination of thin-layer gel chromatography and thin-layer gel electrophoresis on a Sephadex G-200 layer with a veronal buffer (pH 8.4). For this purpose an apparatus was constructed in which the layer plate could be situated at the necessary oblique angle during the thin-layer gel chromatography, and horizontally during electrophoresis. Some authors have attempted to use agar-agar in the form of a thin gel layer; this is very suitable for immunodiffusion after thin-layer gel chromatography [51, 52]. Agostini *et al.* [53] transferred the proteins from a gel layer to a cellulose acetate membrane strip, where the performance of immunodiffusion does not cause a problem. Peeters and Vantrappen [54] described a simple method using a Sephadex G-200 layer for a screening test on human serum for the detection of macroamylasaemia. The buffer used was saturated with starch, and iodine was employed for visualization.

Some of the most important results in the preparative gel chromatography of proteins are those of Maier [55], who added a Lissamine–Rhodamine B-200 mixture to the eluent buffer for the detection of proteins.

Table 5.4
Study of enzymes on a thin-layer gel

Enzyme	Origin	Layer	Detection	Ref.
Oestrogen sulphotransferase	calf tissue	Sephadex G-200	"paperprint"	[56]
Glyoxalate reductase	Pseudomonas	Sephadex G-200	"paperprint"	[57]
Kojic acid oxidase	plant tissue	Sephadex G-200	"paperprint"	[58]
Alkaline phosphatase	milk	Sephadex G-200	"paperprint"	[59]
Ornithine carbamoyl transferase	Staphylococcus	Sephadex G-75	"paperprint"	[60]
Acid phosphatase	equine muscle	Sephadex G-75	hydroxylamine-$FeCl_3$	[61]
Lactic acid dehydrogenase	human urine	Sephadex G-200	"paperprint"	[62]

Table 5.4 lists the most significant results of the thin-layer gel chromatographic study of enzymes. It is evident and noteworthy that the detection reactions for enzymes can be carried out most easily on filter (chromatography) paper, after transfer of the proteins from the gel layer.

Ramsey [63] separated esterases on a starch gel layer in 0.025 *M* borate buffer (pH 8.55) with a potential gradient of 12 V/cm. 1-Naphthyl butyrate was used as the substrate for the detection reaction, and bis-diazotized *o*-dianisidine as the coupling compound. Maruna [64] in-

Table 5.5
Chromatographic study of various peptides on a silica gel thin layer

Peptide type	Origin	Development mixture	Detection	Ref.
Angiotensin		n-butanol–acetic acid–water, 4:1:1 n-butanol–pyridine–water, 30:20:6	ninhydrin	[65]
Angiotensin antagonists		$CHCl_3$–acetone, 5:1 n-butanol–acetic acid–water, 4:1:1	ninhydrin	[66]
Cyclotripeptide	*Aspergillus ochraceus*	various (5)	Cl_2–tolidine $KMnO_4$	[67]
Cytochrome C peptide fragment syntheses		$CHCl_3$–methanol–acetone, 16:1:1 $CHCl_3$–methanol, 16:1	m.p. and elemental analysis	[68]
Bradykinin		propan-2-ol–methyl acetate–conc. NH_3, 90:70:40	UV	[69]
Dansyl peptides	ATP-creatine phosphotransferase	methyl acetate–propan-2-ol–conc. NH_3, 9:7:4	UV after being kept in dioxan vapour	[70]
Dansyl peptides	lipoproteins	various (3)	ninhydrin	[71]
Deamino-oxytocin and its analogues		butan-2-ol–25% NH_3–water, 34:3:3 butan-2-ol–90% formic acid–water, 50:9:9	ninhydrin	[72]
Glycopeptides	*E. coli* B. and *B. megaterium*	n-butanol–pyridine–acetic acid-water, 130:20:6:24	Cl_2–tolidine	[73]
Lipopeptides	*Nocardia asteroides*	$CHCl_3$–methanol, 9:1	P_2O_5, 130°	[74]
Lipopeptides	*E. coli*	$CHCl_3$–methanol–water, 68:25:4	radioactivity	[75]
Basic polypeptides	synthesis	butan-2-ol–25%NH_3–water, 34:3:3 $CHCl_3$–methanol, 9:1	ninhydrin–Cl_2–tolidine	[76]
N-Methyloligopeptides	synthesis	$CHCl_3$–methanol, 9:1 $CHCl_3$–acetone, 8:2 or 9:1	Cl_2–tolidine, UV, I_2	[77]
Vasopressin analogues	synthesis	butan-2-ol–acetic acid–water, 4:1:1 propan-2-ol–25% NH_3, 7:3	UV	[78]
Vasopressin and oxytocin (7–9) tripeptide amides	synthesis	butanol–acetic acid–water, 4:1:1	UV	[79]

vestigated the acid and alkaline phosphatases of urine by starch gel thin-layer electrophoresis. The substrate used was phenyl phosphate; paper strips were impregnated with this, and the reaction itself was carried out in TRIS acetate or carbonate buffer on these paper strips.

5.3 THIN-LAYER CHROMATOGRAPHY OF PEPTIDES AND PEPTIDE DERIVATIVES

In practice, the methods used for the TLC of peptides are very similar, or even identical, to those used for the study of amino-acids. Naturally, in the case of peptides containing a large number of members, or of di-, tri-, and oligopeptides provided with protecting groups, the thin-layer chromatographic systems used for the amino-acids must be modified considerably.

5.3.1 Separation of peptides on inorganic layers

Table 5.5 presents some of the possibilities for separating peptides on a silica gel thin-layer.

Thin-layer chromatography may also be utilized effectively for the separation of isomeric dipeptides. The results of Brenner and Pataki [80] on the separation of isomeric dipeptides on a silica gel G thin-layer are given in Table 5.6.

Table 5.6
Separation of isomeric dipeptides on a silica gel G this layer [80]*

Dipeptide pair	$R_f \times 100$	
	S_1	S_2
H . Ala-Gly . OH/H . Gly-Ala . OH	15–16	15–14
H . Gly-Hypro . OH/H . Hypro-Gly . OH	8/13	12–13
H . Gly-Leu . OH/H . Leu-Gly . OH	37–35	27/33
H . Ala-Leu . OH/H . Leu-Ala . OH	45/37	30–31
H . Gly-Phe . OH/H . Phe-Gly . OH	36–38	32/37
H . Gly-Pro . OH/H . Pro-Gly . OH	8–9	8–7
H . Gly-Ser . OH/H . Ser-Gly . OH	12–14	12–14
H . Phe-Ala . OH/H . Ala-Phe . OH	48–45	42–40
H . Gly-Val . OH/H . Val-Gly . OH	32/27	22–24
H . Gly-Leu . NH_2/H. Leu-Gly . NH_2	53/59	25/33

S_1 = butanol–acetic acid–water, 4:1:1
S_2 = propanol–water, 7:3
* Reproduced by kind permission of *Helv. Chim. Acta*

Thin-layer chromatography also finds appreciable use for the separation of diastereomeric peptides. Wieland and Bende [81] successfully separated these on both cellulose and silica gel layers. Some results are given in Table 5.7. It was found by Pravda *et al.* [82] that the differences in the R_f values of the carbobenzoxy derivatives of diastereomeric peptides are smaller than those for the original diastereomers. More recently, Hubert and Dellacherie

Table 5.7
Separation of some diastereomeric peptides on cellulose and silica gel thin layers [81]

Peptide	$R_f \times 100$	
	MN cellulose powder 300 S_1	Kieselgel G S_2
L-Met-L-Ala	47	41
D-Met-L-Ala	34	37
L-Ala-L-Phe	48	38
D-Ala-L-Phe	33	28
L-Ala-L-Tyr	55	44
D-Ala-L-Tyr	39	35
L-Val-L-Tyr	70	53
D-Val-L-Tyr	52	42

S_1 = pyridine–water, 79:18 w/w
S_2 = methanol–water, 99:1

[83] succeeded in separating diastereomeric tripeptides on a silica gel F_{254} thin-layer plate.

The starting materials, the intermediates and the end-products in peptide synthesis can be examined rapidly by TLC. This is reflected by the abundant literature [84]. The carbobenzoxypeptide esters migrate in the vicinity of the front, the carbobenzoxypeptides fall behind them, and the amino-acids and peptides are left in the neighbourhood of the start (Table 5.8).

The (*N*-2-*p*-biphenylyl-2-propoxycarbonyl) amino-acids have been employed with success for solid-phase peptide synthesis. Wang and Merrifield [85] solved the separation of these by thin-layer chromatography.

In certain special cases, aluminium oxide has also proved a suitable support for the separation of peptides [86]. Riniker and Schwyzer [87] easily separated some isomeric analogues of hypertensin on an aluminium oxide layer.

Table 5.8
Separation of amino-acids, peptides and carbobenzoxy derivatives of amino-acids, peptides and peptide esters on a silica gel G thin layer [84]*

Compound	$R_f \times 100$		
	S_1	S_2	S_3
Z-Gly-OH	81	76	64
Z-Digly-OH	66	65	56
Z-Trigly-OH	57	56	50
Z-Tetragly-OH	51	43	45
Z-Gly-Gly-OEt	76	67	69
Z-Gly-Gly-Gly-OEt	76	67	69
Z-DL-Ala-Gly-OEt	81	75	77
Z-DL-Ala-OH	77	77	61
Z-DL-Diala-OH	72	72	59
Z-Gly-DL-Ala-OH	68	68	56
Z-DL-Ala-Gly-OH	68	68	57
Z-Gly-DL-Ala-Gly-OH	61	59	54
Z-DL-Phe-Gly-OEt	86	83	76
Z-Gly-DL-Phe-OEt	84	79	76
Z-Gly-Gly-DL-Phe-OEt	81	75	76
Z-Gly-DL-Phe-Gly-OEt	83	78	74
Z-DL-Ala-DL-Phe-OH	78	78	61
Z-Gly-Phe-OH	72	76	65
Z-Gly-L-Ile-OH	75	73	67
Z-Gly-L-Leu-OH	74	74	65
Z-Gly-Gly-L-Leu-OH	71	69	65
Z-Gly-L-Gly-OH	72	71	62
Z-DL-Phe	74	76	75
H-Gly-Gly-OH	17	25	15
H-DL-Ala-Gly-OH	21	30	22
H-Gly-L-Leu-OH	43	44	41
Gly	22	29	21
DL-Ala	27	34	27
DL-Phe	45	42	46
HCl-DL-Phe-OEt	59	52	65
HCl-Gly-OEt	43	42	53

Z = carbobenzoxy
S_1 = n-butanol–acetone–acetic acid–NH_3 (conc. NH_3–water, 1:4)–water, 9:3:2:2:4
S_2 = n-butanol–acetic acid–NH_3 (as for S_1), 11:6:3
S_3 = n-butanol–acetic acid–water–pyridine, 15:3:12:10

* Reproduced from *J. Chromatog.* **7**, 405 (1962), with the kind permission of Elsevier Scientific Publishing Company.

Table 5.9
Chromatographic study of various peptides on a cellulose-based thin layer

Peptide type	Origin	Solvent mixture	Detection	Ref.
ACTH	natural, synthetic	n-butanol–pyridine–acetic acid–water, 42:24:4:30	Reindel-Hoppe reagent	[88]
Dipeptide	synthesis	n-butanol–formic acid–water, 70:15:15	ninhydrin–collidine	[89]
Oligopeptide	synthesis	propan-2-ol–butanone–1 *M* HCl, 12:3:5	ninhydrin	[90]
Tryptic	"coat" protein	n-butanol–acetic acid–pyridine–water, 15:3:12:12	Cd–ninhydrin, autoradiography	[91]
Tryptic	insulin	ammonia–propanol, 4:6	ninhydrin	[92]
Tryptic	human papilloma virus	n-butanol–acetic-acid–pyridine–water, 15:3:12:12	autoradiography	[93]
Acidic (HCl)	collagen	(1) propan-2-ol–butanone–1*M* HCl, 60:15:25; (2) 2-methyl-2-butanol–butanone–acetone–methanol–conc. NH_3, 5:4:7:1:4	Cd–ninhydrin, *o*-phthaldialdehyde	[94]
Tryptic	retinol-binding protein	(1) butanol–acetic acid–water, 4:1:2; (2) butanol–acetic acid–pyridine–water, 5:1:4:4	Cd–ninhydrin, phenanthrenequinone	[95]
Trifluoroacetate	bradykinin, HO-proline analogues	n-butanol–acetic acid–water, 4:1:5 (upper phase)	ninhydrin	[96]
Oligopeptides	proteins	(1) isoamyl alcohol–ethanol–acetic acid–pyridine–water, 175:50:13:175:150; (2) propan-2-ol–ethanol–NH_4OH, 20:20:15	ninhydrin–ethanol–acetic acid–2,6-lutidine, 1 g:100 ml:210 ml: 29 ml	[97]
Cystine peptides	synthesis	butanol–acetic acid–water, 71:7:22	ninhydrin	[98]
Copper–peptide complexes	urine	50 m*M* borate buffer	ninhydrin	[99]

5.3.2 Separation of peptides on organic layers

Table 5.9 lists some peptide separations on cellulose layers.

It was established by Heathcote *et al.* [100] that some peptides may very easily be confused with amino-acids if they are identified merely on the basis of the ion-exchange chromatography retention time. Only the joint application of TLC and ion-exchange column chromatography offers some defence against such mistakes.

From a comparison of the behaviour of 17 leucine and isoleucine peptides in TLC and ion-exchange chromatography, it was found that the colour densities of the leucine and isoleucine peptides do not differ significantly from

Table 5.10

Chromatographic behaviour of dipeptides on a cellulose thin layer [100]*

Peptide	$R_f \times 100$			Colour yield, $10^{-4} \times (mm^2/\mu mole)$**		Area ratio (490/405)
	S_1	S_2	S_3	405 nm	490 nm	
Ile-Ala	90	68	56	3.5	7.8	2.2
Ile-Gly	84	53	46	3.7	5.2	1.4
Ile-Glu	94	13	13	10.1	23.0	2.3
Ile-Leu	100	88	86	4.6	14.6	3.2
Ile-Lys	58	50	41	8.6	22.5	2.6
Ile-Met	94	79	80	3.7	8.4	2.3
Ile-Phe	100	81	89	4.5	13.6	3.0
Ile-Pro	89	60	63	2.9	3.5	1.2
Ile-Ser	87	53	38	5.8	13.7	2.3
Ile-Trp	100	83	83	2.6	10.8	4.1
Ile-Val	100	80	83	5.8	14.5	2.5
Leu-Ala	97	58	—	2.9	5.1	1.8
Leu-Gly	82	52	—	2.9	5.6	1.9
Leu-Leu	99	66	83	3.8	8.5	2.2
Leu-Met	100	71	75	9.6	15.1	1.6
Leu-Phe	99	73	79	6.9	10.5	1.5
Leu-Ser	88	61	41	4.5	9.8	2.2
Leu-Trp	100	73	77	7.7	13.0	1.7
Leu-Tyr	97	67	66	8.9	10.8	1.2
Leu-Val	100	77	—	3.1	5.0	1.6

S_1 = propan-2-ol–butanone–1 *M* HCl, 60:15:25

S_2 = 2-methylbutan-2-ol–butanone–acetone–methanol–water–conc. NH_3, 50:20:10:5:15:5

S_3 = n-butanol–butanone–water–conc. NH_3, 80:5:17:3.

* Reproduced from *J. Chromatog.* **79,** 187 (1973), with the permission of Elsevier Scientific Publishing Company

** Calculated area under the densitometric curve

one another on a thin-layer chromatogram, whereas on an ion-exchange column the isoleucine peptides display lower colour values. The determination of the peptides on a thin-layer (e.g. with a densitometer) may yield reliable data (Table 5.10).

In special cases ion-exchange cellulose derivatives have also proved to be convenient supports for the separation of peptides.

It has been proved that in the chromatography of stereoisomeric forms on a cellulose thin-layer, the D-compounds always exhibit higher R_f values.

5.3.3 Electrochromatography of peptides

Like the amino-acids and proteins, the peptides can advantageously be subjected to electrophoresis on various sorbents. Particularly good results may be obtained with a combination of TLC and electrophoresis. Either may be performed in the first direction, and the other in the second. This combination may be termed thin-layer electrochromatography. It is excellent for peptide mapping. Burns and Turner [91] found MN-Polygram Cell 300 precoated plates, prewashed first with 1% acetic acid, then with buffer, and suitably dried, to be best for this purpose. Electrophoresis was used in the first direction (glacial acetic acid–98% formic acid–water, 17:5:280, pH = 2), and the following solvent mixtures proved best for chromatography in the second direction: n-butanol–acetic acid–pyridine–water, 5:1:4:4; butan-2-ol–pyridine–water, 7:7:6; isoamyl alcohol–pyridine–water, 7:7:6; n-butanol–acetic acid–water, 5:1:4 (upper phase).

It may be seen from Fig. 5.5 that this method is well suited for the study of the tryptic peptides of labelled proteins.

Table 5.11 gives the more important recent results from electrochromatography of the peptides.

5.3.4 Reactions for detecting peptides and peptide derivatives

The relatively specific chromatographic reactions of the peptides are of great importance in the differentiation of natural and artificial amino-acid–peptide mixtures, in following synthesis of peptides, and in the structural examination of proteins through identification of the products of the various protein-decomposition reactions.

The various ninhydrin reagents and other universal amino-acid reagents used in TLC for detection of the amino-acids, can also be employed for the detection of peptides, but the sensitivities are in general lower; indeed, the existence of peptides giving ninhydrin-negative reactions must also be reckoned with.

Table 5.11
Electrochromatography of peptides of various origins

Peptides	Support	Running mixture	Buffer	Detection	Ref.
Tryptic peptides of human papilloma virus	cellulose	n-butanol–acetic acid–pyridine–water, 15:3:12:12	aqueous buffer, pH = 3.5	autoradiography	[93]
Glycopeptides of glycoprotein	Selecta-1500 (silica gel)	n-butanol–pyridine–acetic acid–water, 68:10:14:25	acetic acid–pyridine–water, 10:1:8	ninhydrin; orcinol–sulphuric acid	[101]
Glutamate dehydrogenase	silica gel + starch	pyridine–butanol–acetic acid–water, 40:68:14:25	water–pyridine–acetic acid, 89:10:1, 50 V/cm	autoradiography	[102]
Peptides of ribosomal proteins of *E. coli*	polyacrylamide gel (8% acrylamide)		Kaltschmidt buffer without urea	0.2% Amido Black in 7.5% acetic acid	[103]
Rat apoferritin	cellulose	n-butanol–acetic acid–water, 4:1:5	pyridine–acetic acid–water, 10:0.25:150	ninhydrin	[104]
Tryptic peptides of heart tissue aldolase	cellulose	n-butanol–pyridine–acetic acid–water, 150:100:30:120	pyridine–acetic acid–water, 100:30:3000, pH = 5.5, 300 V	autoradiography	[105]
Tryptic peptides of polyoma virus	cellulose	n-butanol–acetic acid–water–pyridine, 150:30:120:120	5% acetic acid, 0.5 pyridine, 25 V/cm, pH = 3.5	ninhydrin	[106]
Actinomycin hydrolysate	cellulose	butanol–water–acetic acid, 5:4:1	formic acid–acetic acid, pH = 6.0, 10 mA/45	Cu–ninhydrin, fluorescamine	[107]
Various peptides	silica gel	butanol–acetic acid–water–pyridine, 15:3:12:10	pH = 6.5 buffer, pH = 3.5 buffer	Cd–ninhydrin	[108]
Ribonuclease S-protein	cellulose	butanol–acetic acid–pyridine–water, 30:6:20:24	1.25 *M* pyridine acetate buffer, pH = 6.45, 400 V	Cd–ninhydrin	[109]

Of the reactions suitable for the detection of peptides, the chlorine–tolidine and the morin reactions are fairly widely used.

The *chlorine–tolidine reaction.* The thin-layer chromatogram is moistened over a boiling water-bath, and placed for 10–15 min in a chamber containing gaseous chlorine. After completion of the chlorination, the chromatogram is aerated for a few minutes, and then sprayed with an aqueous acetone solution of *o*-tolidine containing sodium tungstate and potassium iodide. The amino-acids and peptides give green-tinged blue spots, which are stable [110].

Alternatively, the layer is moistened and chlorinated as above, then aerated and sprayed with an aqueous solution of *o*-tolidine containing acetic acid and potassium iodide [111].

The *morin reaction.* The dried layers are sprayed with a 0.05% solution of morin (3,5,7,2′,4′-pentahydroxyflavone) in methanol, and heated for 2 min at 100 °C. In ultraviolet light the *N*-protected amino-acid and peptide derivatives on the silica gel layer give yellowish-green fluorescence on a green fluorescent background, or dark absorption spots. The limit of detection is about 2 μg/spot [112].

The *iodine–starch reaction* detects peptides and amino-acids. The layer is placed in a strong iodine atmosphere for 5 min. The excess of iodine is removed by aeration, and the layer is then sprayed with 1% aqueous starch solution. The amino-acids and peptides show up in the form of blue spots. An excess of iodine makes the background too blue, and such an excess must therefore be avoided [113].

The past 10 years have seen spread of the use of the extremely sensitive fluorescamine and *o*-phthaldialdehyde reactions for the detection of compounds containing α-amino groups.

The *fluorescamine reaction* may be employed with a thin layer in the following way. The layer is treated with a 10% solution of triethylamine in methylene chloride, dried for a few seconds, sprayed with a 0.05% fluorescamine solution, dried again, and then resprayed with triethylamine solution. The limit of detection of the proteins under a 350 nm ultraviolet lamp is 50 pmole [114, 115].

o-*Phthaldialdehyde reaction.* After development, the layer is sprayed with an acetone solution of 0.1% phthaldialdehyde and 0.1% 2-mercaptoethanol, and 5 min later with a 1% solution of triethylamine in acetone. After 10 min the plates are examined under a 350 nm ultraviolet lamp. The limit of detection is 50–100 pmole [116].

Schiltz *et al.* [117] compared the classical ninhydrin reaction with the modern fluorescamine and *o*-phthaldialdehyde reactions for the detection of amino-acids and peptides, and further studied the recoveries of the individual peptides from a layer fingerprint. Cellulose and silica gel thin-layer were used in their investigations. The solvent mixture in the thin-layer electrophoresis was pyridine–acetic acid–water–acetone (2:4:79:15) and for the thin-layer chromatography it was butanol–pyridine–acetic acid–water, 15:10:3:12. Being ultrasensitive reagents, fluorescamine and *o*-phthaldialdehyde permit the detection of even subnanomole quantities of proteins.

Pataki [118] recommends use of 2,4-dinitrofluorobenzene for the decomposition-free detection of the amino-acids; this method is naturally also suitable for the detection (in the native state) of the peptides separated on a thin layer. The peptides detected may be eluted from the layer and subsequently rechromatographed, and identified by ultraviolet or infrared spectrophotometry, for example.

The specific or relatively specific reactions of certain protein amino-acids are retained even when peptide bonding is involved, and this means that these reactions can well be utilized for the specific detection of some amino-acid-containing peptides. In our experience, the well-known Sakaguchi reaction of arginine can be advantageously employed for the detection of arginine peptides among the tryptic peptides of arginine-rich histones. Table 5.12 contains selective peptide-detection results obtained by use of the detection reactions for other amino-acids, similarly to those for the arginine peptides.

Some authors [129] have observed a specific reaction between dicarbonyl compounds and certain groups of peptides containing a low number of members.

5.4 THIN-LAYER CHROMATOGRAPHY OF AMINO-ACIDS AND RELATED COMPOUNDS

5.4.1 Preparation of the test materials

The proteins and peptides containing the amino-acids to be determined must be hydrolysed before thin-layer chromatography (see Chapter 4 for details).

Besides the free amino-acids, body fluids and organ extracts generally contain peptides, proteins, carbohydrates, urea, salts and lipids. The removal of these requires special operations.

Removal of macromolecules. Various precipitants are employed for this purpose. Oepen and Oepen [130] compared a number of deproteinating methods, and found that in certain cases a considerable loss of amino-acid must be taken into account.

Removal of urea. The best procedure for urine samples is the addition of a trace amount of urease [131].

Table 5.12
Detection of peptides on a thin layer by relatively specific amino-acid reactions

Amino-acid	Layer	Development mixture	Detection	Ref.
Arginine	silica gel, cellulose	n-butanol–pyridine–0.2 *M* acetic acid, 10:7:24; n-butanol–ethanol–pyridine–0.2 *M* acetic acid,16:1:2:28	Sakaguchi	[110]
	Whatman No. 3MM	butanol–pyridine–acetic acid–water, 15:10:3:12	1,2-cyclohexanedione	[120]
Glycine	silica gel	$CHCl_3$–ethanol, 9:1	Bromophenol Blue	[121]
Histidine	silica gel + 0.8% NaCl	ethyl acetate–acetic acid–water, 3:3:2; n-butanol–pyridine–water, 1:1:1	*o*-phthalaldehyde, UV	[123]
Lysine	silica gel, Al_2O_3	benzene–ethyl acetate, 1:1	I_2, conc. H_2SO_4, ninhydrin	[123]
		$CHCl_3$–ethyl acetate–acetic acid, 6:3:1		
		benzene–ether, 1:1 or 2:1		
Ornithine	silica gel	$CHCl_3$–methanol–water, 65:25:4	I_2, ninhydrin	[124]
	silica gel + $AgNO_3$	solvent mixtures	Schiff reagent, Dittmer–Lester reagent,	
Proline	silica gel	$CHCl_3$–methanol, 97:5 + 3 solvent mixtures	ninhydrin, I_2–KI–tolidine	[125]
Serine	silica gel	Seiler method	UV	[126]
Tryptophan	cellulose	n-butanol–acetic acid–water, 4:1:1	ninhydrin, Ehrlich reagent	[127]
	silica gel	benzene–pyridine–acetic acid, 16:1:4	Ehrlich reagent	[128]
		$CHCl_3$–tert.-amyl alcohol–acetic acid,70:70:3		

Removal of salts. Salts are conveniently removed by passage of the sample through a cation-exchange resin column (e.g. Dowex 50 × 8) [132].

Enrichment of amino-acids in urine. Aliquots (10 ml) of the urine are lyophilized, and then leached with 1 ml of methanol–1 *M* hydrochloric acid (4:1). The mixture is centrifuged, and 20 μl of the supernatant liquid are applied to the thin layer [133].

Enrichment of N*-methylated amino-acids from biological fluids and protein hydrolysates.* An aliquot of the fluid or hydrolysate is adjusted to be 0.1 *M* in hydrochloric acid, and is then treated with an equal volume of 2% aqueous solution of Reinecke's salt. The solution is left in a refrigerator overnight, and the precipitate is filtered off and dissolved in acetone. This solution is centrifuged, and the supernatant liquid is taken. An equal volume of water is added, and the mixture is extracted several times with ether. The lower layer containing water, acetone and ether is evaporated to dryness, and the residue is dissolved in aqueous 10% propan-2-ol solution. This solution is used for TLC and other tests [134].

5.4.2 Separation on inorganic layers

Inorganic sorbents may be used to great advantage for TLC of amino-acids. Silica gel is mainly used. The most well-known and widely-used products are perhaps Merck silica gel G, and silica gel 60 but the products of other firms can satisfactorily be used for the investigation of amino-acids. It is customary to use a 0.25 mm thick layer, in activated form, but for preparative purposes 1–2 mm thick layers are best.

Table 5.13 lists the most commonly used of the large number of solvent mixtures for separation of amino-acids on silica gel.

Table 5.13
Solvent mixtures for chromatography on a silica gel thin layer

Solvent mixtures	Ratios	Ref.
Propanol–water	7:3	[135]
Butanol–acetic acid–water	4:1:1	
Phenol–water	75:25 (w/w)	
Butanone–pyridine–water–acetic acid	70:15:15:2	[136]
Chloroform–methanol–17% NH_3	2:2:1	
Acetone–urea–water	60:0.5:40	[137]
Butanol–butanone–17% NH_3–water	5:3:1:1	
Butanol–butanone–water	2:2:1	

These solvent mixtures are generally used in combination for two-dimensional chromatography, as none of them on its own is able to resolve all the amino-acids, even of protein hydrolysates. At best, 14 or 15 amino-acids may be resolved, e.g. with the n-butanol–acetic acid–water (4:1:5, upper phase) mixture. With more complex amino-acid mixtures, however, only

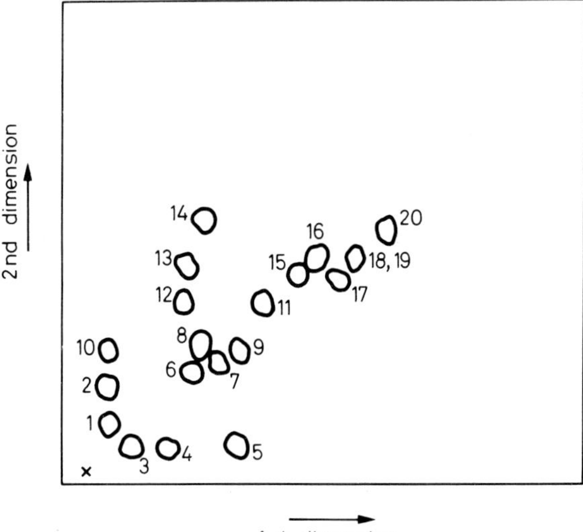

Fig. 5.6. Two-dimensional separation of authentic amino-acids on a silica gel G thin layer [138]. 1st dimension: n-butanol–acetic acid–water, 4:1:1; 2nd dimension: phenol–water, 75:5 w/w. *1*, Lys; *2*, Arg; *3*, $CysSO_3H$; *4*, Asp; *5*, Glu; *6*, Ser; *7*, Tau; *8*, Gly; *9*, Thr; *10*, His; *11*, Ala; *12*, Gln; *13*, Hyp; *14*, Pro; *15*, Val; *16*, Met; *17*, Tyr; *18*, *19*, Leu + Ile; *20*, Phe. (Reproduced from *J. Chromatog.*, **17**, 580 (1965), with the permission of Elsevier Scientific Publishing Company.)

informatory data may be obtained from chromatography with a single solvent.

The following are the solvent pairs most frequently used for two-dimensional examinations:

first direction	second direction
n-butanol-acetic acid–water (4:1:5), upper phase	phenol–water (7:3, w/w)
chloroform–methanol–17% ammonia (2:2:1)	phenol–water (3:1, w/w)
n-butanol–acetic acid–water (4:1:5), upper phase	chloroform–methanol–17% ammonia (2:2:1)

Precoated plates are nowadays widely known, and an increasing number of workers use them for the investigation of amino-acids. However, the results do not always agree with those obtained on the classical silica gel layers, and it is necessary to make certain modifications in the running mixtures.

Figures 5.6 and 5.7 show two-dimensional separations of authentic amino-acids on silica gel layers.

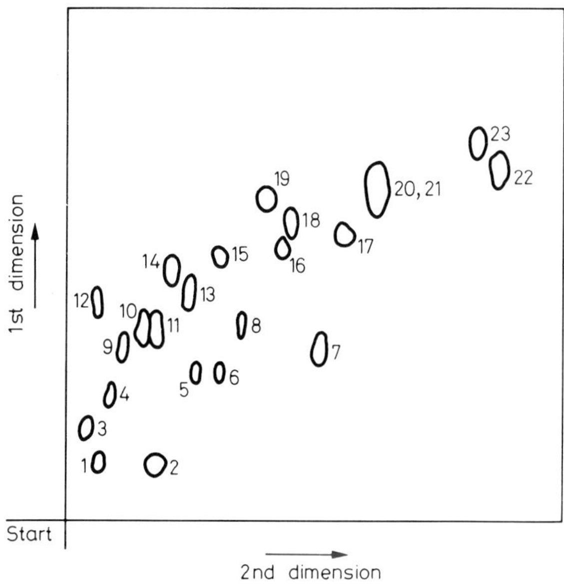

Fig. 5.7. Two-dimensional separation of amino-acids and related compounds on a silica gel G thin layer [139]. 1st dimension: chloroform–methanol–17% NH_3, 2:2:1; 2nd dimension: phenol–water, 75:25 w/w. *1*, Lys; *2*, Arg; *3*, His; *4*, Asp; *5*, β-Ala; *6*, γ-Abut; *7*, Pro; *8*, Hyp; *9*, Ser; *10*, Gly; *11*, Glu; *12*, $CysSO_3H$; *13*, Ala; *14*, Thr; *15*, His; *16*, $MetO_2$; *17*, Tyr; *18*, Val; *19*, Met; *20*, *21*, Leu + Ile; *22*, Trp; *23*, Phe. (Reproduced by permission from *Scand. J. Clin. Lab. Invest.* **16,** 149 (1964)

In Table 5.14 may be seen the R_f values (× 100) of 30 protein amino-acids (including the *N*-methylated basic amino-acids). It is noteworthy that distilled water has proved particularly suitable for the separation of the *N*-methylated bases. Some two-dimensional separations of these amino-acids are presented in Fig. 5.8.

The TLC of a large number of amino-acid derivatives, in addition to that of protein amino-acids, was investigated by Tyihák and Vágújfalvi [141, 141a] on a silica gel G layer. The results are given in Table 5.15.

Table 5.14

$R_f \times 100$ values of 30 protein amino-acids on a silica gel thin layer [133]

Amino-acid	Silica gel G (Woelm)			Silica gel G (Merck)
	S_1	S_2	S_3	S_1
	$R_f \times 100$			
Glycine	88.5	36	46	84
Leucine	70	64	74	66
Isoleucine	69	63	72	65
Valine	77	59	66	73
Threonine	89	45	57	83
Asparagine	89	33	45	83
Aspartic acid	94	10	21	84
Glutamine	93	16	28	86
Methionine	76	63	68	70
HO-proline	85	30	46	78
Proline	64	33	55	60
Cysteine	70(88)	36(27)	51	77
Cystine	88	26	46	76
Serine	92	27	51	77
Alanine	88	40	54	84
Phenylalanine	70	67	65	66
Histidine	64	47	57	63
1-Methylhistidine	47	58	56	45
3-Methylhistidine	32	54	56	31
Lysine	71	9	23	69
N^{ε}-monomethyl-L-lysine	51	7	23	48
N^{ε}-dimethyl-L-lysine	21	5	53	24
N^{ε}-trimethyl-L-lysine	15	1	10	17
Arginine	63	7	16	63
N^{G}-monomethylarginine	44	12	25	46
N^{G},N^{G}-dimethylarginine	36	22	36	35
N^{G},N'^{G}-dimethylarginine	25	19	32	24
Tryptophan	74	64	69	71
Tyrosine	87	61	59	86

Solvent:

S_1 = distilled water

S_2 = chloroform–methanol–25% NH_3 solution, 4:4:1

S_3 = propan-1-ol–25% NH_3 solution, 70:30

Table 5.16 presents comparative data on the Dragendorff and ninhydrin sensitivities of the *N*-methylated basic amino-acids and other amino-acid derivatives [141, 141a].

For study of amino-acids, silica gel has the disadvantages that the layer is not sufficiently stable, and that the background very readily changes in colour. Cellulose layers generally do not suffer from these disadvantages.

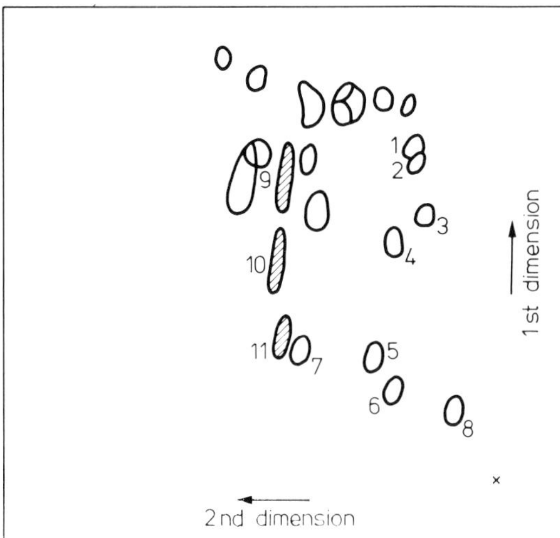

Fig. 5.8. Two-dimensional separation of basic protein amino-acids on a silica gel G (Woelm) thin layer [140]. 1st dimension: distilled water; 2nd dimension: chloroform–methanol–25% NH_3, 4:4:1. *1*, Lys; *2*, Arg; *3*, MML; *4*, MMA; *5*, DMA; *6*, DMA'; *7*, DML; *8*, TML; *9*, His; *10*, 1-MeHis; *11*, 3-MeHis (By the author.)

Other inorganic supports have been employed. Mottier [142] investigated the sodium salts of amino-acids on an aluminium oxide layer. Shasha and Whister [143] used celite as sorbent, and propan-2-ol–water (9:1) as development eluent. Others have used mixtures of silica gel and kieselguhr, the best ratio being 4:11 [144].

Among biological applications, study of the amino-acids of urine is of especial importance. Thin-layer chromatography on silica gel is widely used to diagnose aminoaciduria in children [145].

Figure 5.9 depicts a two-dimensional thin-layer chromatogram of desalinated urine. In general, 10 μl of urine is sufficient for examination.

A silica gel G layer has also been used for investigation of the serum of patients suffering from tumours [147].

Table 5.15

Colour reactions and $R_f \times 100$ values of amino-acids and amino-acid derivatives [141, 141a]*

Amino-acid	Colour reaction		$R_f \times 100$			
	Ninhydrin	Dragendorff + diluted H_2SO_4	KG	MN-300 S_1	KG S_2	KG S_3
Aliphatic monoamino-acids						
Glycine	+	–	47	33	84	46
Sarcosine	+	–	40	38		
L(+)-alanine	+	–	51	51		
β-alanine	+	–	41	31	84	54
L-serine	+	–	42	32	77	51
DL-phosphoserine	+	–		0		
DL-α-amino-n-butyric acid	+	–	82	51		
α-amino-isobutyric acid	+	–		57		
L-threonine	+	–	57	47	83	57
DL-β-aminoisobutyric acid	+	–	44	39		
γ-aminobutyric acid	+	–	42	35		
L(+)-norvaline	+	–				
L(+)-valine	+	–	68	76	73	66
L(+)-norleucine	+	–		85		
L(–)-leucine	+	–	72	84	66	74
L(+)-isoleucine	+	–	70	85	65	72
L-aspartic acid	+		20	6	84	21
L-asparagine hydrate	+	–	43	13	83	45
L-glutamic acid	+	–	35	8	86	28
L-N-methylglutamic acid	+	–		10		
L-glutamine	+	–	50	38	80	49
Aliphatic diamino-acids						
β,γ-diaminopropionic acid	+	–	16	11		
L-α,β-diaminobutyric acid	+	–	5	9		
L(+)-ornithine	+	–	15	45		
Citrullin	+	–	48	37		
L(+)-lysine	+	–	16	48	69	23
DL(+)-allo-δ-hydroxylysine	+	–	5	12		
L-N^{α}-methyllysine	+	–	13	45		
DL-N^{ε}-methyllysine	+	+	12	69	48	23
DL-N^{ε}-dimethyllysine	+	+	32	89	24	53
DL-N^{ε}-trimethyllysine	+	+	7	49	17	10
DL-α,α-diaminopimelic acid	+	–		4		
S-containing mono- and diamino-acids						
L(+)-cysteine	+	–		13	77	51
L-cysteic acid	+	–	24	7		
DL-mesohomocysteine	+	–	17	5		

Table 5.15 (cont'd)
Colour reactions and $R_f \times 100$ values of amino-acids and amino-acid derivatives [141, 141a]*

Amino-acid	Colour reaction		$R_f \times 100$			
	Ninhydrin	Dragendorff + diluted H_2SO_4	KG	MN-300 S_1	KG S_2	KG S_3
L-methionine	+	–	69	73	70	68
DL-ethionine	+	–	77	50		
L(+)-mesolanthionine	+	–	82			
L(–)-cystine	+	–	2	10	76	46
DL-allocystathionine	+	–	4	2		
L-djenkolic acid	+	–	25	5		
Aromatic monoamino-acids						
DL-phenylglycine	+	–	70	38		
DL-phenylalanine	+	–	72	83	66	65
L-*N*-methyl-phenylalanine	+	–	96	83		
DL-*o*-tyrosine	+	–	76	34		
m-tyrosine	+	–	64	20		
L-tyrosine	+	–	63	53	86	59
L-*N*-methyltyrosine	+	–		50		
DL-3,4-dihydroxyphenylalanine	+	–	4	10		
β-phenylserine	+	–	63	27		
N-hetero amino- and imino-acids						
L-2-azotidinecarboxylic acid	+	–	30	30		
L(–)-proline	+	–	43	64	60	55
L-4-hydroxyproline	+	–	49	43	78	46
DL-pipecolic acid	+	–	33	49		
L-tryptophan	+	–	69	64	72	69
DL-5-methyltryptophan	+	–	90	54		
DL-6-methyltryptophan	+	–	90	52		
L-histidine	+	+	56	54	63	57
L-1-methylhistidine	+	+	80	66	45	56
L-3-methylhistidine	+	+	66	70	31	56
L-2-thiohistidine	+	+	73	13		
Ergothioneine	–	+		46		
L-carnosine	+	+	26	15		
Guanidine derivatives of α-amino-acids						
Glycocyanin	–	–				
Creatine	–	–				
L-arginine	+	+	2	22	63	16
N^G-monomethyl-L-arginine	+	–	12		46	25
N^G,N^G-dimethyl-L-arginine	+	+	22		35	36
N^G,N'^G-dimethyl-L-arginine	+	+	19		24	32
N^G,N^G,N^G-trimethyl-L-arginine	+	+	18		18	20

Table 5.15 (cont'd)
Colour reactions and $R_f \times 100$ values of amino-acids and amino-acid derivatives [141, 141a]*

Amino-acid	Colour reaction		$R_f \times 100$			
	Ninhydrin	Dragendorff + diluted H_2SO_4	KG	MN-300 S_1	KG S_2	KG S_3
L-N^{α}-methylarginine	+	+	16	34	46	25
L-homoarginine	+	+	14	19	35	36
L-canavaline sulphate	+	−	36	9	24	32
Creatinine	−	+	92	72		

KG = Kieselgel G;
MN-300 = MN cellulose powder 300;
Solvent media: S_1 = $CHCl_3$–methanol–25% NH_3, 4:4:1;
S_2 = distilled water;
S_3 = propan-1-ol–25% NH_3, 7:3
* Reproduced from *J. Chromatog.*, **49**, 343 (1970), with the permission of Elsevier Scientific Publishing Company

A method has been developed by Okumura *et al.* [174, 175] for the preparation of sintered layers. A suspension is prepared from a mixture of an inorganic support (silica gel or aluminium oxide) and glass powder of appropriate particle size (200 mesh) and purity, and is spread as a thin layer on a soda-glass plate. The layer is then sintered at a temperature between 450 and 700 °C. The fine glass powder particles melt, and a layer is obtained which is mechanically extremely stable, heat- and acid-resistant, and capable of regeneration.

Okumura *et al.* [174–176] used such sintered thin-layers effectively for the investigation of widely varied groups of compounds. The results obtained for amino-acids in the traditional manner and on a sintered layer do not differ substantially, but the sintered layers have somewhat smaller capacity.

5.4.3 Separation of amino-acids on cellulose layers

In the early days of TLC the examination of amino-acids on a cellulose layer was regarded in practice merely as an improved variant of amino-acid investigation by paper chromatography. Currently, however, MN cellulose 300 thin-layers, for example, are among the most widely and effectively used supports in the TLC of amino-acids.

Table 5. 16
Limits of detection (μg) of amino-acids and amino-acid derivatives on silica gel and cellulose thin layers [141, 141a]*

Amino-acid	Ninhydrin		Dragendorff + dilute H_2SO_4	
	After one-dimensional running	After two-dimensional running	MN-300	Silica gel
β-alanine	0.009	0.05		
L-alanine	0.01	0.06		
L-arginine	0.01	0.06	(1.5)	(1.0)
L-aspartic acid	0.1	0.4		
L-cysteic acid	0.01	0.1		
L-glutamic acid	0.04	0.2		
L-glycine	0.001	0.006		
L-histidine	0.05	0.5	(1.5)	(0.8)
L-hydroxyproline	0.05	0.1		
L-leucine	0.01	0.2		
L-lysine	0.005	0.03	(1.5)	(1.0)
L-methionine	0.01	0.4		
L-phenylalanine	0.05	0.2		
L-proline	0.1	0.5		
L-serine	0.008	0.1		
L-threonine	0.05	0.1		
L-tryptophan	0.05	0.5		
L-tyrosine	0.03	0.1		
L-valine	0.01	0.2		
L-N^{α}-methyl-lysine		0.04	(1.0)	(0.7)
DL-N^{ε}-methyl-lysine		0.05	(0.8)	(0.7)
DL-N^{ε}-dimethyl-lysine		0.09	(0.5)	(0.4)
DL-N^{ε}-trimethyl-lysine		0.5	(0.2)	(0.2)
L-1-methylhistidine		0.6	(0.7)	(0.6)
L-3-methylhistidine		0.6	(0.3)	(0.4)
L-2-thiohistidine		0.1	(0.3)	(0.3)

* Reproduced from *J. Chromatog.*, **49**, 343 (1970), with the permission of Elsevier Scientific Publishing Company

The layer is made by 'turbo-mixing' 15 g of MN cellulose 300 for 10 min with 90 ml of distilled water, then spreading it in a 0.25 mm thick layer, and leaving it to dry overnight.

Other firms have also marketed cellulose powders for TLC and ready-made cellulose layers are increasingly used, the best-known probably being

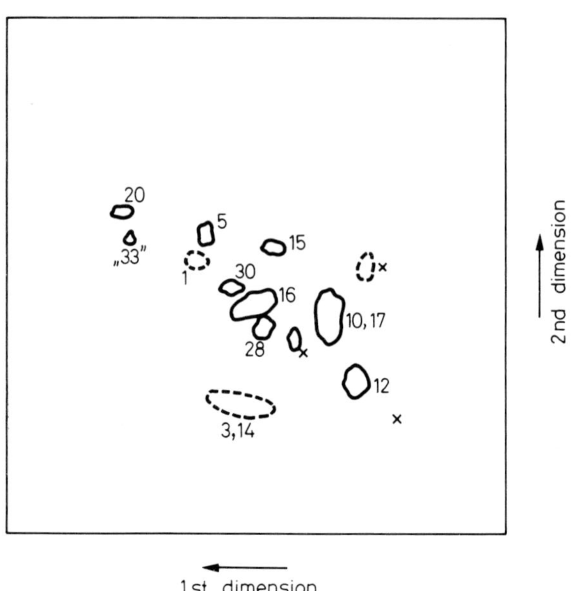

Fig. 5.9. Two-dimensional examination of amino-acids of desalinated human urine on a silica gel G thin layer [146]. 1st dimension: n-butanol–acetic acid–water, 4 : 1 : 1; 2nd dimension: phenol–water, 3 : 1. Sample applied: 20 μl of salt-free urine. *3* + *14*: Asp + Glu; *12*, $(Cys)_2$; *10* + *17*; Arg + His; *28*, Ser; *16*, Gly; *30*, Thr; *15*, Gln; *1*, Ala; *5*, Abu(?); – *33* +, Val; *20*, Leu + Ile. (By kind permission of Elsevier/North-Holland Biomedical Press.)

Table 5.17

Solvent mixtures for chromatography on a cellulose thin layer

Solvent mixtures	Ratios	Ref.
Butanol–acetic acid–water	4:1:5	[148]
Methanol–water–pyridine	20:5:1	
Propan-1-ol–8.8% NH_3	4:1	
Chloroform–methanol–17% NH_3	20:20:9	[149]
Butanol–acetone–diethylamine–water	10:10:2:5	
Phenol–water	75:25 w/w	
Propan-2-ol–butanone–1 *M*HCl	60:15:25	[150]
2-Methylpropan-2-ol–butanone–acetone–methanol–water–conc. NH_3	40:20:20:1:14:5	

Table 5.18
Study of amino-acids and related compounds on a cellulose thin layer [150]*

Amino-acid	$R_f \times 100$		Amino-acid	$R_f \times 100$	
	S_1	S_2		S_1	S_2
Alanine	57	23	Methionine	78	51
Arginine	19	6	γ-Amino-n-butyric acid	51	23
Aspartic acid	48	1	Hydroxylysine	10	17
Glutamic acid	56	1	Norleucine	92	73
Serine	39	27	2-Amino-octanoic acid	93	81
Glycine	37	16	Argininosuccinic acid	31	0
Threonine	51	61	Cadaverine	14	70
Valine	79	44	Putrescine	16	60
Isoleucine	90	63	Histamine	15	85
Leucine	90	69	Kynurenine	38	67
Histidine	11	26	Homoarginine	23	12
Lysine	16	17	Ethionine	83	66
Phenylalanine	82	67	1-Methylhistidine	14	18
Tyrosine	72	35	3-Methylhistidine	12	14
Tryptophan	70	54	Sarcosine	48	24
Proline	58	30	Homoserine	45	27
Hydroxyproline	48	17	3,4-DOPA	43	6
Cysteine	12	5	Ornithine	11	15
Cystine	6	3	Pipekolic acid	66	36
Cysteic acid	53	8	Citrulline	34	12
β-Alanine	46	19	Penicillamine	34	15
α-Aminoadipic acid	62	3	Djenkolic acid	9	5
β-Aminoisobutyric acid	60	26	2,6-Diaminopimelic acid	16	1
Asparagine	21	14	Formiminoglycine	45	16
Glutamine	31	13	2,4-Diaminobutyric acid	10	25
Ethanolamine	39	81	Glucosamine	20	38
Phosphoethanolamine	43	0	Epinephrine	51	52
Taurine	41	33			

Solvent mixtures: S_1 = propan-2-ol–butanone–1 *M* HCl, 60:15:25;
S_2 = 2-methylpropan-2-ol–butanone–acetone–methanol–water–conc. NH_3, 40:20:20:1:14:5

* Reproduced from *J. Chromatog.*, **41**, 380 (1969), with the permission of Elsevier Scientific Publishing Company

the Macherey–Nagel products, such as MN Polygram Cell 300, which contains MN cellulose 300 in appropriately bound form.

Cellulose powders in general contain impurities soluble in water or organic solvents, and it is therefore advisable to wash even MN cellulose 300 several times (e.g. with $0.1M$ acetic acid, methanol and acetone), and redry it before use.

Cellulose layers have several advantages: they are stable, inert towards most development mixtures, can be used with various specific reagents and give reproducible data. They are thus particularly suitable for quantitative evaluation (densitometry). Cellulose layers have the drawbacks that corrosive reagents cannot be used, and that the sensitivities of the detection reactions of certain amino-acids are lower than on silica gel layers.

Of the large number of solvent mixtures that have been described so far, those that have proved best in practice are listed in Table 5.17. However, for good separations they must be used in pairs for two-dimensional chromatography.

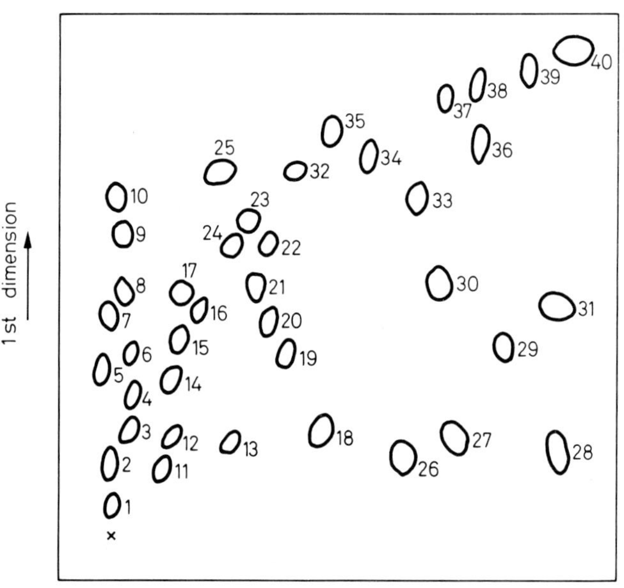

Fig. 5.10. Two-dimensional separation of amino-acids on a Cellulose MN 300 thin layer [150]. 1st dimension: propan-2-ol–butanone–1 M HCl, 60:15:15; 2nd dimension: 2-methylpropan-2-ol–butanone–acetone–methanol–water–conc. NH_3 40:20:20:1:14:5. *1*, Cys; *2*, $(Cys)_2$; *3*, Arg; *4*, Homo-Arg; *5*, Arg-succinic acid; *6*, $CysSO_3H$; *7*, Phosphoethanolamine; *8*, Asp; *9*, Glu; *10*, α-aminoadipic acid; *11*, Orn; *12*, Cys; *13*, His; *14*, Gln; *15*, Gly; *16*, β-Ala; *17*, Hyp; *18*, Glucosamine; *19*, Taurine; *20*, Ser; *21*, $Met(O_2)$; *22*, Pro; *23*, β-aminoisobutyric acid; *24*, Ala; *25*, ε-aminocaproic acid; *26*, Putrescine; *27*, Cadaverine; *28*, Histamine; *29*, Kynurenine; *30*, Thr; *31*, Ethanolamine, *32*, Tyr; *33*, Trp; *34*, Met; *35*, Val; *36*, Phe; *37*, Ile; *38*, Leu; *39*, Norleu; *40*, α-amino-octanoic acid. (Reproduced from *J. Chromatog.*, **41,** 380 (1969), with the permission of Elsevier Scientific Publishing Company.)

Heathcote *et al.* made an especially systematic investigation of the possibilities of amino-acid separation on cellulose thin-layers [94, 151, 152]. Some R_f values are given in Table 5.18 [150].

Figure 5.10 illustrates the two-dimensional separation of authentic amino-acids on a cellulose thin-layer.

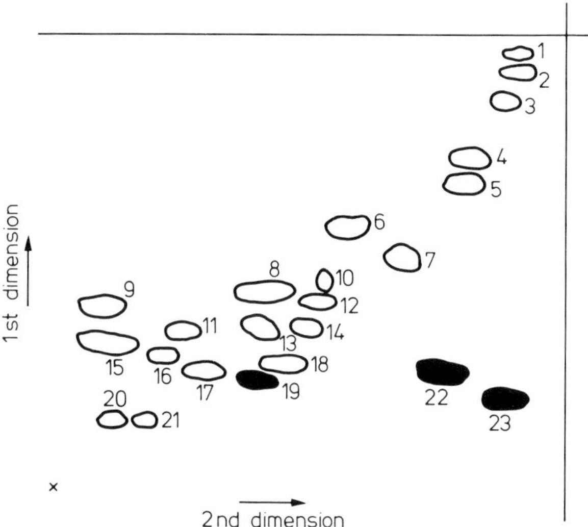

Fig. 5.11. Two-dimensional separation of protein amino-acids on a Cellulose MN 300 thin layer [141]. 1st dimension: n-butanol–acetic acid–water, 4 : 1 : 5 (rechromatography); 2nd dimension: chloroform–methanol–25% NH_3, 4 : 4 : 1. Reagent: ninhydrin, or ethyl acetate + Dragendorff reagent and sensitization with 0.01 *N* sulphuric acid. *1*, Glu; *2*, Asp; *3*, Arg; *4*, Gly; *5*, His; *6*, Lys; *7*, Ser; *8*, Thr; *9*, Hyp; *10*, Ala; *11*, Tyr; *12*, Trp; *13*, Pro; *14*, Met; *15*, Val; *16*, Phe; *17*, Ile; *18*, Leu; *19*, TML; *20*, MML; *21*, DML; *22*, $(Cys)_2$; *23*, Cys. (Reproduced from *J. Chromatog.*, **49**, 343 (1970), with the permission of Elsevier Scientific Publishing Company.)

It may be seen in Fig. 5.11 that, in addition to the classical protein amino-acids, the *N*-methylated lysines can also be well separated; by treatment with Dragendorff reagent in ethyl acetate and sensitization with 0.1 *N* sulphuric acid, they give a specific red colour.

Figure 5.12 presents the separation of 8 ‘minor amino-acids’ on a cellulose layer [141]. Since these amino-acids give characteristic reactions with the Dragendorff reagent on a cellulose layer, this method may be one means of identifying them in protein hydrolysates, and in analysis or synthesis.

The studies by Haworth and Heathcote [150] confirm that the selection of solvent mixtures is of great importance in determining the number of amino-acids that may be separated. Heathcote and Al-Alawi [94] have described an improved technique for the analysis of the amino-acids and peptides from partially hydrolysed proteins, making use of the Cd-ninhydrin reagent and the Zimmermann orthophthaldialdehyde reaction.

Cellulose is often combined with a silica gel support; better and more beautiful spots may be obtained on such a mixed layer than on a layer of

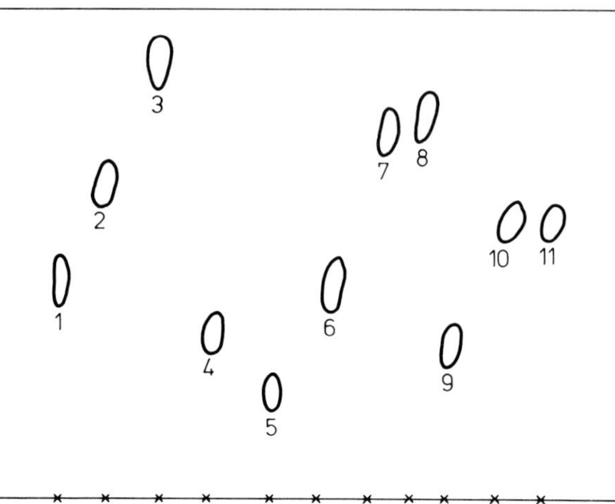

Fig. 5.12 Thin-layer chromatographic separation of basic amino-acids on a Cellulose MN 300 thin layer [141]. Solvent mixture: chloroform–methanol–25% NH_3, 4:4:1. Developing reagent: 0.5 g of ninhydrin in 100 ml of acetone, 90 °C 10 min. *1*, Lys; *2*, MML; *3*, DML; *4*, TML; *5*, Arg; *6*, MMA; *7*, DMA; *8*, DMA′; *9*, His; *10*, 1-MeHis; *11*, 3-MeHis. (Reproduced from *J. Chromatog.*, **49**, 343 (1970), with the permission of Elsevier Scientific Publishing Company.)

cellulose alone, and the ninhydrin-sensitivity is increased by the presence of silica gel [153].

With the exceptions of alanine, glycine and leucine, Horton *et al.* [154] found the ninhydrin-sensitivities of the amino-acids on a microcrystalline cellulose (Avicel) layer to be higher than on the classical silica gel layers.

Figure 5.13 shows a separation suitable for the determination of the amino-acids in a casein hydrolysate [155].

The serum amino-acids have also been studied by two-dimensional separation on a layer of MN cellulose powder 300 [156].

For thin-layer chromatography of guanidino compounds, Völkl and Berlet [157] found a cellulose layer to be the best, with the following solvent mixtures: propan-2-ol–acetic acid–water, 4:3:1; propan-2-ol–DMF–water, 2:1:1.

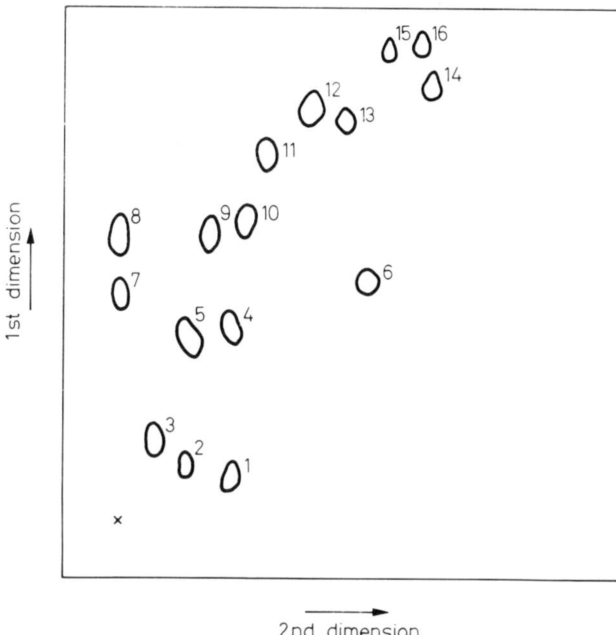

Fig. 5.13. Separation of amino-acids of casein hydrolysate on a cellulose thin layer for quantitative determination [155]. 1st dimension: propan-2-ol–butanone–1*M* HCl, 12:3:5; 2nd dimension: 2-methylbutanol–butanone–acetone–methanol–water–conc. NH_3, 10:4:2:1:3:1. *1*, His; *2*, Lys; *3*, Arg; *4*, Ser; *5*, Gly; *6*, Thr; *7*, Asp; *8*, Glu; *9*, Ala; *10*, Pro; *11*, Tyr; *12*, Val; *13*, Met; *14*, Phe; *15*, Ile; *16*, Leu. (Reproduced from *Biochem. J.*, **114** 667 (1969), with the permission of Cambridge University Press.)

5.4.4 Separation of amino-acids on ion-exchange layers

Besides the inorganic and organic supports presented above, sorbents with ion-exchange properties are also used as the stationary phase for the TLC separation of amino-acids. Either an ion-exchange cellulose can be used, or a resin fixed in the layer.

De la Llosa *et al.* [158] reported R_f values for 20 amino-acids on an MN DEAE-cellulose 300 thin-layer, with various solvent mixtures, each containing an organic solvent. For separation of the main protein amino-acids, Verceanst *et al.* [159] also used mixtures containing an organic solvent [n-butanol–acetic acid–water (4:1:5), upper phase; pyridine–water, (4:1)] in one- and two-dimensional chromatography on a layer prepared from Whatman DEAE cellulose. Levis *et al.* [160] successfully separated iodoamino-acids on a layer prepared from a 3:2-mixture of silica gel and DEAE cellulose.

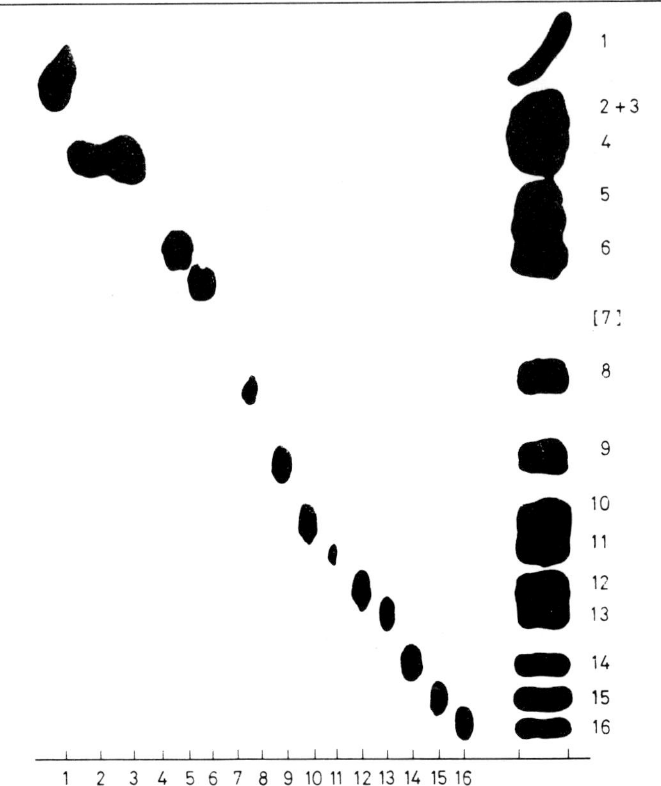

Fig. 5.14. One-dimensional separation of amino-acids on a Fixion 50 × 8 thin layer [167]. Elution buffer: sodium citrate, pH = 3.3, 0.4 *M* Na^+, 0.4 *M* citrate, 50 °C. *1*, Asp; *2*, Thr; *3*, Ser; *4*, Glu; *5*, Gly; *6*, Ala; *7*, Pro; *8*, Val; *9*, Met; *10*, Ile; *11*, Leu; *12*, Tyr; *13*, Phe; *14*, His; *15*, Lys; *16*, Arg. (Reproduced from *Acta Biochim. Biophys. Acad. Sci. Hung.*, **5**, 435 (1970)).

Gozzi *et al.* [161] studied the ion-exchange TLC behaviour of 31 amino-acids on a layer prepared from alginic acid.

Kiyasu [162] attempted to layer mixtures of Dowex I or Dowex II resin and silica gel in a ratio of 1 : 2 on a glass plate, but did not obtain a stable layer.

Ogiya *et al.* [163] examined the amino-acids on a layer prepared from powdered resin, but were not able to standardize the conditions unambiguously.

To examine the amino-acids in urine, Kraffczyk and Helger [164] made use of a double layer. They prepared a slurry of 45 g of Merck cellulose SF and 5 g of Merck Ion-Exchanger I (a strong acid cation-exchanger) by turbo-mixing for 15 min in 0.05% aqueous carboxymethylcellulose, and spread a 0.25 mm thick layer of this in a 2 cm band on a 20 × 20 cm glass plate. The remainder of the glass plate was coated with a similar thickness of a cellulose SF suspension. After drying at 80–90 °C, the layers were used to examine the urine amino-acids. The salts in the urine (which was added dropwise to the thin band) were bound by the ion-exchanger.

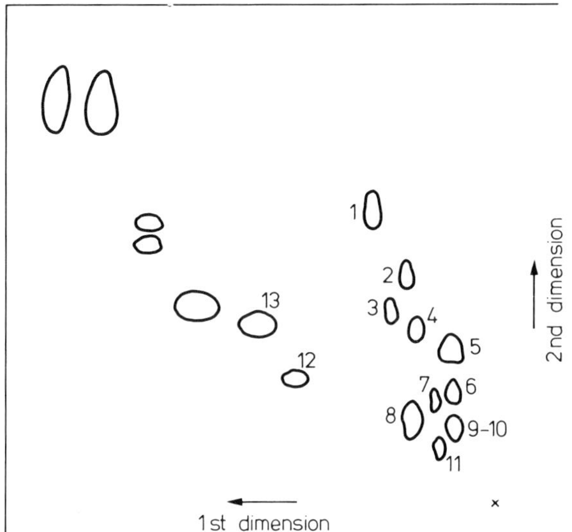

Fig. 5.15. Two-dimensional separation of methylated basic amino-acids and aromatic amino-acids on a Fixion 50 × 8 thin layer [173]. 1st dimension: sodium citrate, pH = 5.28, 0.35 *M* Na^+; 2nd dimension: 0.5 *M* trisodium citrate + 1 *M* NaCl. *1*, Lys; *2*, His; *3*, MML; *4*, 1-MeHis; *5*, Arg; *6*, MMA; *7*, 3-MeHis; *8*, DML; *9*, *10*, DMA + DMA′; *11*, TML; *12*, Phe; *13*, Tyr. (Reproduced from *J. Chromatog.*, **102**, 257 (1974), with the permission of Elsevier Scientific Publishing Company.)

A mixed layer of cellulose and the ion-exchanger Amberlite CG-120 was effectively used in a similar way by Copley and Truter [165], who used development with water in one dimension.

A significant development in ion-exchange TLC, with particular reference to amino-acids, resulted from the method published first by Dévényi and Zoltán [166], in which they succeeded in fixing an ion-exchange resin with high resolving power on a plastic foil. The commercially-available Fixion 50 × 8 (Chinoin) and the identical Polygram Ionex-25SA-Na (Macherey–Nagel) contain the Na^+ form of a strong acid cation-exchange resin of the Dowex 50 × 8 type.

Figure 5.14 illustrates the examination of authentic amino-acids on a Fixion 50 × 8 plate [167].

Figure 5.15 shows the two-dimensional separation of methylated basic amino-acids and aromatic amino-acids on a Fixion 50 × 8 layer.

Fixion 50 × 8 plates have been used effectively for detecting tryptophan in peptide hydrolysates [169], to examine free and bound amino-acids in plant seeds [170], and to evaluate the urine of various types of acidaemic patients [171].

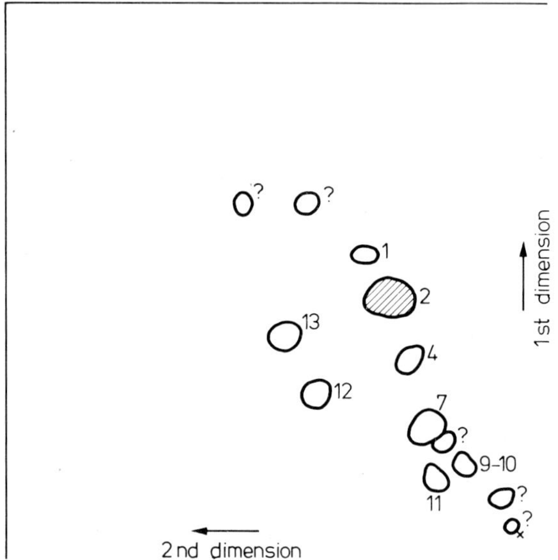

Fig. 5.16. Two-dimensional examination of healthy human urine on a Fixion 50 × 8 thin layer [133]. Conditions and amino-acid notations as in Fig. 5.15. (Reproduced from *Kísérl. Orvostud.* **27,** 532 (1975).)

Another Fixion preparation, the Fixion 2 × 8 anion-exchange plate, is suitable with a pyridine–acetic acid buffer (pH 3.81) for the simple examination of cysteic acid [172].

Figures 5.16 and 5.17 depict two-dimensional thin-layer chromatograms of the urine of healthy individuals and cancer patients, respectively, and indicate that the method is very suitable for the production of 'thin-layer fingerprints'.

5.4.5 Electrochromatography of amino-acids

As early as 1946, Consden *et al.* [177] used a silica gel layer for the electrophoretic examination of amino-acids. Nowadays, both inorganic and organic supports are employed for the electrophoresis of amino-acids; either

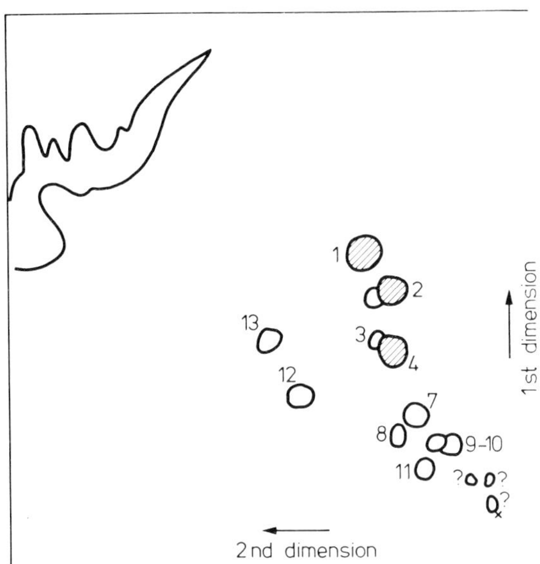

Fig. 5.17. Two-dimensional examination of human urine from a cancer patient, on a Fixion 50 × 8 thin layer [133]. Conditions and amino-acid notations as in Fig. 5.15. (Reproduced from *Kísérl. Orvostud.*, **27**, 532 (1975).)

one- or two-dimensional electrophoresis may be used, but a combination of TLC and electrophoresis (fingerprint method) gives the most reliable results. In general, the sequence of the two methods is immaterial, but if the electrophoresis involves the use of a salt-containing buffer, it is preferable to perform the TLC first, as this procedure is disturbed by a salt-containing layer (this does not apply if an ion-exchange layer is employed).

Table 5.19 gives the conditions used is some electrochromatographic examination.

5.4.6 Detection of amino-acids on thin-layer chromatograms

Although the ninhydrin reaction may be very sensitive, it depends on many factors. On occasion, this fact may be utilized for the relatively specific identification of the amino-acids (e.g. with polychromatic reagents) but after alkaline development, for example, the sensitivity of the ninhydrin reaction can only be kept at the desired level by addition of acetic acid to make the pH not more than 5. Constancy of the colour formed may be attained by the addition of complex-forming cations (Cu, Cd, Ca), and specific colours may be produced by the addition of bases (collidine, benzylamine). A very large number of ninhydrin reagent compositions is currently known.

Table 5.19
Electrochromatography of amino-acids

Electrophoresis			Chromatography	Ref.
Buffer	Support	Conditions	Solvent medium	
2 *M* acetic acid–0.6 *M* formic acid, 1:1; pH = 2	Kieselgel impregnated with citrate buffer	460 V; 12.6 mA; 160% humidity; 1 h	butanol–acetic acid–water, 4:1:1	[169]
0.7% formic acid	MN cellulose powder 300	400 V; 15 min	butanol–formic acid–water, 4:1:2 propanol–pyridine–water, 5:1:2	[170]
Citric acid–phosphate buffer, pH = 4.6	polyamide	500 V; 20 mA		[171]
50 m *M* borate buffer, pH = 9.2	cellulose	40 V/cm; 1 h		[172]
pH = 1.9	cellulose	50–60 V/cm; 0.4–1 mA/cm; 11–30 min	butanol–acetone–acetic acid–water, 7:7:2:1	[173]

Ninhydrin reagents used for silica gel layers. In the procedure of Brenner *et al.* [183], the well-dried, solvent-free layer is sprayed with a solution of 0.3 g of ninhydrin in 100 ml of n-butanol and 3 ml of acetic acid, and then heated for 30 min at 60 °C or for 10 min at 110 °C.

Opienska-Blauth *et al.* [184] first mark the spots formed by the action of the ninhydrin reagent at room temperature, and heat the plates to 60 °C. Many amino-acids can be identified even in the cold.

The method of Barrolier *et al.* [185–187] is particularly successful for determination of amino-acids. The layer is thoroughly impregnated with a solution of 100 mg of cadmium acetate and 1 g of ninhydrin in 10 ml of water, 5 ml of acetic acid and 100 ml of acetone. About 50 ml of this reagent are necessary for a 20 × 20 cm plate. Colour development is performed in the dark at room temperature (24 h).

Reagents for cellulose-based layers. In the method of von Arx and Neher [149], the solvent-free cellulose layer is sprayed with a solution of 1 g of ninhydrin in 700 ml of absolute ethanol, 29 ml of 2,4,6-collidine and 210 ml of acetic acid, and then dried for 20 min at 90 °C.

Reagents for layers containing ion-exchange resin. Dévényi *et al.* [167, 188] use an acetone solution containing 1% ninhydrin and 10% collidine. Development is for 24 h at room temperature, or 10 min at 70 °C.

Polychromatic reagents. A classical polychromatic ninhydrin reagent was devised by Moffat and Lyttle [189]. Solution 1 is 50 ml of 0.2% ninhydrin solution in absolute ethanol + 10 ml of acetic acid + 2 ml of 2,4,6-collidine; solution 2 is 1% $Cu(NO_3)_2 \cdot 3\,H_2O$ in absolute ethanol. Before use, the two solutions are mixed in a ratio of 50:3. This reagent was modified by Krauss and Reinbothe [190], who replaced the ethanol by methanol, which gives better wetting properties. The ninhydrin–cadmium chloride reagent can be employed on both MN cellulose 300 and silica gel layers [191].

Certain authors have achieved polychromatic amino-acid detection by the joint application of ninhydrin and primary, secondary or tertiary amines. In the procedure described by Circo and Freeman [192] and modified by Krauss and Reinbothe [190] the layer is first uniformly sprayed with diethylamine, and then dried for 3 min at 110 °C; after cooling, the layer is sprayed with 0.2% methanolic ninhydrin reagent, and heated for 10 min at 110 °C; the spots of the amino-acids appear on a pale blue background. The reagent described by Kolor and Roberts [193] as modified by Krauss and Reinbothe [190] is 0.27 g of ninhydrin + 0.13 g of isatin + 2 ml of triethylamine + 100 ml of methanol. The layer is sprayed with this, and heated for 10 min at 90 °C. The amino-acids show up on a yellow background.

Table 5.20
Specific reactions for detection of amino-acids

Amino-acid	Reagent	Ref.
Arg	8-hydroxyquinoline	[194]
Arg	α-naphthol, urea, Br_2	[195]
Asp	ninhydrin, borate soln., HCl	[196]
Cys, $(Cys)_2$, Met	sodium nitroprusside, NaCN	[197]
Cys, $(Cys)_2$, Met	NaN_3, iodine	[199]
Gly	*o*-phthalaldehyde, KOH	[200]
His	sulphanilic acid	[201, 202]
His	Echtblausalz	[203]
Hypro	diethylamine, ninhydrin	[204]
Lys, Orn	vanillin, KOH	[205]
Lys, Orn	$K_2S_2O_5$, furfurol	[206]
Arg(Me), His(Me), Lys(Me)	BiI_3	[141]
Pro	diethylamine, ninhydrin	[204]
Ser, Thr, Tyr	sodium metaperiodate, Nessler reagent	[207]
Trp	*p*-dimethylaminobenzaldehyde	[208]
Trp	orthophosphoric acid	[209]

Jellinek [22] used polychromatic amino-acid detection on silica gel bound with agar-agar.

In another, non-specific, amino-acid detection reaction, Pataki [118] sprays the layer with 2,4-dinitrofluorobenzene; this detects the amino-acids without decomposing them. Amino-acids detected in this way may be eluted from the layer, and further identified in other TLC systems.

The comparatively specific detection reactions for amino-acids are given in Table 5.20.

5.4.7 Semi-quantitative and quantitative determination of the amino-acids

The amino-acids on thin-layer chromatograms produced in various ways may be determined quantitatively either directly on the layer or after elution.

The simplest means of direct amino-acid determination is semi-quantitative evaluation by measurement of the size and intensity of the spots, after chromatography of various known amounts of authentic compounds on the same layer, followed by detection. Purdy and Truter [210] report that there is a linear correlation between the logarithm of the quantity of material applied and the square root of the spot size. This algebraic method has the advantage that practically no apparatus is needed.

Table 5.21
Thin-layer chromatographic and ion-exchange column chromatographic determination of amino-acid composition of casein hydrolysate [213]*

Amino-acid	g/100 g of casein		
	Layer (cellulose)	Ion-exchange (Technicon)	Literature data
Alanine	3.0	3.3	3.2
Arginine	3.8	3.9	4.1
Aspartic acid	7.6	7.3	7.1
Glutamic acid	22.0	20.8	22.4
Glycine	2.4	2.2	2.0
Histidine	3.1	3.2	3.1
Isoleucine	5.6	5.0	6.1
Leucine	8.8	8.2	9.2
Lysine	7.8	8.1	8.2
Methionine	2.9	3.1	2.8
Phenylalanine	4.8	5.0	5.0
Proline	10.1	10.5	10.6
Serine	6.4	6.3	6.3
Threonine	5.5	6.9	4.9
Tyrosine	5.8	4.9	6.3
Valine	6.5	6.3	7.2
Cystine	0.4	1.4	0.4

* Reproduced from *J. Chromatog.*, **43**, 84 (1969), with the permission of Elsevier Scientific Publishing Company

More reliable results may be obtained, however, if the amino-acids separated on the layer are evaluated quantitatively with a densitometer. Hara *et al.* [211] constructed a densitometer with which a two-dimensional layer chromatogram can be evaluated. A particularly useful densitometer is marketed by Joyce and Loebl [212]. In the quantitative determination of 22 protein amino-acids with this instrument, Heathcote and Haworth [213] obtained values agreeing well with the data found with an automatic amino-acid analyser, and other values in good agreement with the literature data [214], e.g. in the case of a casein hydrolysate (Table 5.21).

Frodyma and Frei [215] used a Beckman DK spectrophotometer for emission measurements, and found that the amino-acids may be determined with an error of 5–12%. Pataki [216] measured the λ_{max} values of various amino-acids on a silica gel layer with a Zeiss Chromatogram-Spectrophotometer, and constructed appropriate calibration curves.

Seiler and Wiechmann [217] and Pataki [216] described the direct fluorimetric evaluation of thin-layer chromatograms.

After use of the appropriate colour reaction (e.g. the ninhydrin reaction), the amino-acids separated on the layer may be eluted and evaluated quantitatively by colorimetry. In the case of the ninhydrin reaction, an error of 2–5% may be expected. Clark [218] described a simple, fast method for the determination of amino-acids, using MN Polygram Cell 300 precoated plates. The spots were cut out and eluted with 2 ml of 50% propanol, and the absorbances were measured with a microspectrophotometer at 570 nm. In one day, 4–6 chromatograms can be processed with this method. Evaluation is possible by use of standard curves.

With 2,4-dinitrofluorobenzene, the free amino-acids can be converted into the corresponding dinitrophenyl (DNP) derivatives; these can be well separated by TLC and determined by colorimetry on the basis of their own colours. The error of the method is 4–5% [219]. Heins and Hauber [220] separated the amino-acids as dinitrophenylamino-acid ^{14}C-methyl esters, on two thin-layer chromatograms. The radioactivity was measured by a liquid-scintillation method, and accurate measurements could be made even in the case of 1.5×10^{-12} mole (specific activity: 34.5 mCi/mmole). The reproducibility of the method is $\pm 6\%$.

Spivak *et al.* [221] constructed an apparatus (incorporating a microscope) for the ultramicro determination (10^{-11}–10^{-10} mole) of dansyl amino-acids. By utilizing ^{14}C-dansyl chloride, Briel *et al.* [222] performed the microanalysis of amino-acids in various biological materials, with two-dimensional chromatography on a 3×3 cm micro-polyamide layer. Varga and Richards [223] developed a direct fluorescence technique for use on a polyamide layer for the detection of 10^{-14} mole of dansyl amino-acid.

5.5 UTILIZATION OF THIN-LAYER CHROMATOGRAPHY FOR THE SEQUENCE ANALYSIS OF PEPTIDES AND PROTEINS

5.5.1 Study of 2,4-dinitrophenyl (DNP) amino-acids

The method of Sanger [224], described in 1945, revealed that 2,4-dinitrofluorobenzene is very suitable for the determination of the *N*-terminal amino-acids of proteins and peptides by production of the corresponding dinitrophenyl (DNP) amino-acid.

Very many methods are now known for the preparation of DNP amino-acids [225–233]. The DNP amino-acids are usually divided into two main groups: those soluble in water, and those soluble in ether.

Table 5.22
Identification of water-soluble DNP-amino-acids on a silica gel thin layer [234]*

Material	$R_f \times 100$	Colour	Colour with ninhydrin
mono-DNP-cystine	29	yellow	brown
DNP-cysteic acid	29	yellow	yellow
α-DNP-arginine	43	yellow	yellow
ε-DNP-lysine	44	yellow	brown
o-DNP-tyrosine	49	colourless	violet
α-DNP-histidine	57	yellow	yellow
di-DNP-histidine	65	yellow	yellow

Solvent mixture: propan-1-ol–34% NH_3, 7:3

* By kind permission of Birkhäuser Verlag.

Table 5.23
Solvent mixtures used for chromatography of ether-soluble DNP-amino-acids on a silica gel G thin layer (from [235])

Solvent mixtures	Ratios
Toluene–pyridine–2-chloroethanol–0.8 *M* NH_3	100:30:60:60
Toluene–pyridine–2-chloroethanol–25% NH_3	50:15:35:7
Chloroform–benzyl alcohol–acetic acid	70:30:3
Chloroform–tert.-amyl alcohol–acetic acid	70:30:3
Benzene–pyridine–acetic acid	80:20:2
Chloroform–methanol–acetic acid	95:5:1
Chloroform–methanol–acetic acid	70:30:5
Chloroform–methanol–acetic acid	98:2:1

Brenner *et al.* [234] separated the water-soluble DNP amino-acids on a silica gel layer with a propanol–ammonia (7:3) mixture. Reactions for the identification of the water-soluble DNP amino-acids are given in Table 5.22.

A large number of development solvents are known for the examination of the ether-soluble DNP amino-acids on a silica gel G layer (Table 5.23). A considerable number of the ether- and water-soluble DNP amino-acids may also be separated by two-dimensional chromatography. The DNP amino-acids may also be separated on cellulose layers, but it is advisable first to impregnate the layer, e.g. with a 0.2 *M* sodium citrate solution in alcohol. After impregnation, the layer is dried rapidly. Munier and Sarrazin [236]

Table 5.24
Solvent mixtures used for chromatography of DNP-amino-acids on a cellulose thin layer [236]

Solvents	Ratios
Toluene–ethylene chlorohydrin–0.8 *M* NH_3–water	150:90:45:90
Saturated $(NH_4)_2SO_4$ soln.–water–sodium dodecyl sulphate	100:700:0.576 g

compiled several very useful solvent mixtures for the separation of the DNP amino-acids (Table 5.24). These solvent mixtures may be utilized particularly effectively in two-dimensional combinations, as may be seen in Fig. 5.18.

Polyamide sorbents have likewise proved to be suitable supports for the separation of the DNP amino-acids [237]. Wang and Weinstein [238] recommend a number of suitable solvent mixtures. Figure 5.19 shows the advantage of using a polyamide layer [237].

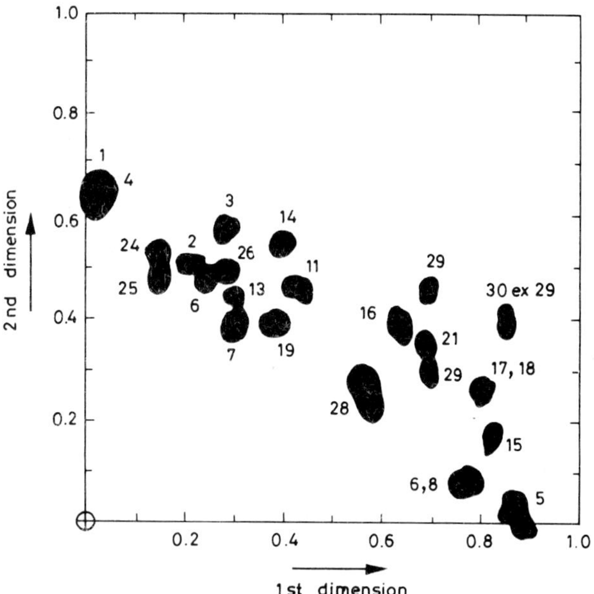

Fig. 5.18. Two-dimensional separation of DNP-amino-acids on a cellulose thin layer [337]. 1st dimension: toluene–ethylene chlorohydrin–pyridine–0.8 *M* NH_3, 150:90:45:90; 2nd dimension: saturated $(NH_4)_2SO_4$–water–sodium dodecyl sulphate, 100 ml:700 ml:0.576 g. *1*, Asp; *2*, Ser; *3*, Thr; *4*, Glu; *5*, Lys; *6*, His; *7*, Gly; *8*, Trp; *11*, Ala; *13*, Ala; *14*, Pro; *15*, Phe; *16*, Val; *17*, Leu; *18*, Ile; *19*, 2,4-dinitroalanine; *21*, Norval; *24*, Asn; *25*, Gln; *26*, $MetO_2$; *27*, Tyr; *28*, Met; *29*, Et; *30*, EtO_2. (Reproduced from *Bull. Soc. Chim. France*, 2959 (1965) by kind permission of publisher.)

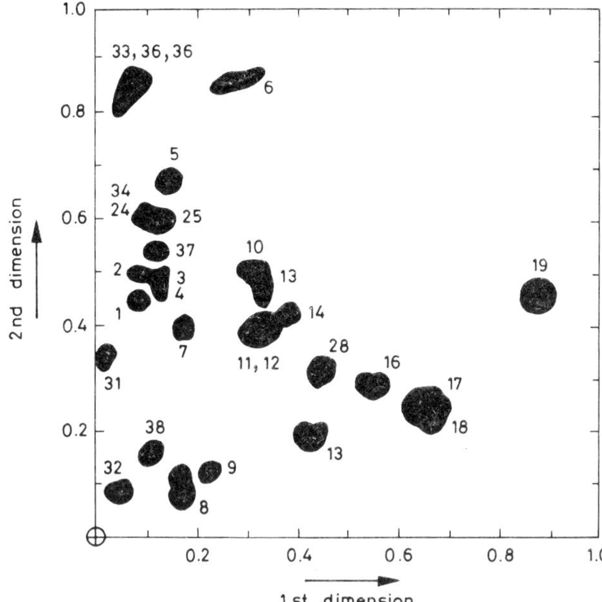

Fig. 5.19. Two-dimensional separation of DNP-amino-acids on a polyamide thin layer [237]. 1st dimension: carbon tetrachloride–acetic acid, 80 : 20; 2nd dimension: 90% formic acid–water, 1 : 1. *31*, $CysSO_3H$; *32*, $(Cys)_2$; *33*, Arg; *34*, MetO; *35*, Orn; *36*, Lys; *37*, Hyp; *38*, Orn; *39*, alloThr; other notations as in Fig. 5.19. (Reproduced from *J. Chromatog.*, **27**, 318 (1967), with the permission of Elsevier Scientific Publishing Company.)

5.5.2 Thin-layer chromatography of dansyl (Dns), disyl (Dis) and bansyl (Bans) amino-acids

1-Dimethylaminonaphthalene-5-sulphonyl chloride (Dns-Cl) is frequently used for the determination of the *N*-terminal amino-acids of peptides and proteins [239, 240]. The dansyl (Dns) amino-acids exhibit especially strong fluorescence, and therefore the sensitivity of this method is 2–3 orders of magnitude better than that of the DNP method.

Table 5.25 shows solvent mixtures employed for the study of the Dns amino-acids on a silica gel G layer.

Polyamide layers are especially useful for the separation of the Dns amino-acids. Table 5.26 lists the R_f values of Dns amino-acids on a commercial polyamide layer [242] and Fig. 5.20 shows a two-dimensional chromatogram of Dns amino-acids on a polyamide layer [242a].

More recently, Seiler and Knödgen [243] separated nanogram amounts of the Dns amino-acids and determined them by two-dimensional chromatography on an HPTLC silica gel layer. The solvent in the first direction was propan-2-ol-acetic acid–water (4 : 3 : 1), and that in the second direction was methyl acetate–propan-2-ol–25% ammonia (9 : 7 : 2). Detection was performed with triethanolamine.

The very favourable results obtained with Dns-Cl provided a stimulus for the preparation of new reagents with similar activities. Ivanov and

Table 5.25
Solvent mixtures for chromatography of Dns-amino-acids on a silica gel G thin layer

Solvents	Ratios	Ref.
Chloroform–benzyl alcohol–acetic acid	100:30:5	
Butanone–triethylamine–dimethylsulphoxide–benzene	4:2:1:3	
Benzene–pyridine–acetic acid	80:20:2	
Chloroform–benzyl alcohol–acetic acid	30:10:1	
Butanone–propionic acid–water	15:5:6	[241]
Butanone–acetic acid	13:5	
Butanone–acetic acid–chloroform	13:5:15	
Butanone–propionic acid	2:1	
Butanone–propionic acid–water	5:5:1	
Hexane–tert.-butanol–acetic acid	3:3:1	[238]

Vladovska-Yukhnowska [244] established that 2-*p*-chlorosulphophenyl-3-phenylindenone (Dis-Cl), which is structurally similar to Dns-Cl, is also very similar as regards chemical reactivity. They also found that if the Dis amino-acids are treated with sodium ethoxide on a silica gel layer, diphenyliso-benzofuran derivatives are formed, which give a yellowish-green fluorescence in ultraviolet light (365 nm). In this way, even 10^{-12} mole of amino-acid may be detected.

Seiler *et al.* [245] recently replaced the dimethylamino group of Dns-Cl by a di-n-butylamino group. With the new reagent, amino-acid derivatives (Bans amino-acids) could be produced which had similarly intensity of fluorescence to the compounds obtained with Dns-Cl, but were less polar, so Bans aspartic acid, Bans glutamic acid and similar amino-acid derivatives could be extracted from the reaction mixture with ethyl acetate. Naturally, the less polar solvent mixtures are suitable for the separation of the Bans amino-acids.

Table 5.26
$R_f \times 100$ values of Dns-amino-acids on a polyamide thin layer [242]

Dns-amino-acids	$R_f \times 100$		
	S_1	S_2	S_3
Dns-L-glutamic acid	21	5	49
Dns-L-threonine	31	9	64
Dns-glycine	34	17	53
Dns-L-cysteine	*	2	2
Dns-L-serine	24	5	64
Dns-L-methionine	43	33	30
Dns-L-phenylalanine	45	36	32
Dns-leucine	61	45	24
Dns-L-isoleucine	64	50	27
Dns-L-tryptophan	17	12	10
Dns-L-arginine	6	2	88
Dns-L-lysine	20	21	11
Dns-L-valine	59	48	39
Dns-L-hydroxyproline	40	14	62
Dns-L-aspartic acid	16	3	51
Dns-L-alanine	46	33	54
Dns-L-histidine	16	3	88
Dns-L-tyrosine	32	45	3
Dns-L-cysteic acid	0	0	19
Dns-L-proline	57	53	41
Dns-NH_2	40	38	54
Dns-OH	0	0	33

S_1 = hexane–tert.-butanol–acetic acid, 3:3:1;
S_2 = benzene–acetic acid, 9:1;
S_3 = 90% formic acid–water, 200:3;
* = tailing

5.5.3 Thin-layer chromatography of phenylthiohydantoin (PTH) amino-acids

The 3-phenyl-2-thiohydantoin (PTH) amino-acids are formed by treatment of proteins or peptides with phenyl isothiocyanate. In this reaction a phenylthiocarbamyl derivative is first produced. On addition of acid, this derivative undergoes rearrangement, and then cleavage and cyclization. The molecular moiety splitting off the peptide or protein is in fact the PTH derivative of the *N*-terminal amino-acid, while the residual part is the original compound, deprived of its end-group, but otherwise intact. Since most PTH

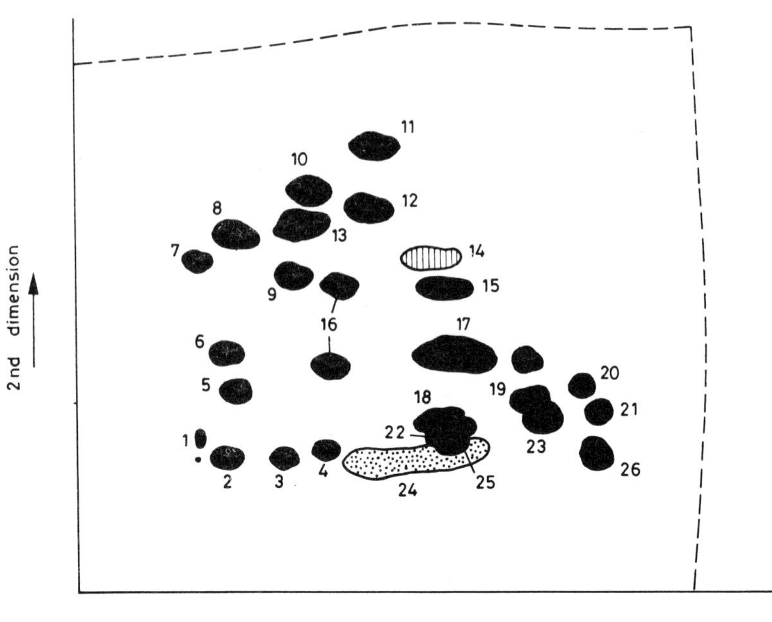

Fig. 5.20. Two-dimensional separation of Dns-amino-acids on a polyamide thin layer [242a]. 1st dimension: water–90% formic acid, 100:0.5; 2nd dimension: benzene–glacial acetic acid, 9:1. *1*, ε-Lys; *2*, Cys; *3*, $CysSO_3H$; *4*, Tyr; *5*, Trp; *6*, Lys; *7*, bisTyr; *8*, His; *9*, Phe; *10*, Ile; *11*, Pro; *12*, Val; *13*, Leu; *14*, $Dns\text{-}NH_2$; *15*, Ala; *16*, Met; *17*, Gly; *18*, Glu; *19*, Thr; *20*, Gln; *21*, Asn; *22*, Asp; *23*, Ser; *24*, Dns-OH; *25*, O-Tyr; *26*. Arg (Reproduced from *Biochim. Biophys. Acta*, **133,** 369 (1967), with the permission of Elsevier/North-Holland Biomedical Press.)

Table 5.27

Solvent mixtures for study of PTH-amino-acids on a Kieselgel G thin layer

Solvents	Ratios	Ref.
Chloroform		
Chloroform–methanol	9:1	
Chloroform–formic acid	100:5	[249]
Chloroform–methanol–formic acid	70:30:2	
Chloroform–propan-2-ol–water	28:1:1	
Chloroform–ethyl acetate	19:1	[250]
Chloroform–ethyl acetate–water	6:3:1	
Heptane–propionic acid–ethylene chloride	58:17:25	[251]
Heptane–n-butanol–75% formic acid	50:30:9	
n-Heptane–n-butanol–acetic acid	40:30:9	[252]

Table 5.28
Specific colour reactions of PTH-amino-acids with ninhydrin–collidine reagent [255]*

PTH-amino-acid	Heating time, min	Colour
Serine	1	red-violet
Glycine	1.5	intense orange
Alanine	1.5	red-violet
Methionine sulphone	2	brown
Cystine	3	intense red
Cysteic acid	3	light red
Asparagine	4	light yellow
Glutamine	4	greenish-yellow
Methionine	4.5	brown
Glutamic acid	4.5	dark brown with blue edge
Histidine	4.5	weak yellow
Aspartic acid	4.5	red
Arginine	4.5	very weak yellow
Tryptophan	5	intense yellow
Tyrosine	5	light yellow
Threonine	5	light brown
Lysine	7	very weak red
Proline	8	very weak red
Phenylalanine	9	very weak yellow
Leucine	15	very weak grey
Isoleucine	15	very weak grey
Valine	–	no colour

* Reproduced from *J. Chromatog.*, **44**, 392 (1969), with the permission of Elsevier Scientific Publishing Company

amino-acids dissolve well in organic solvents, they can be separated from interfering substances by this means.

It follows from the nature of the reaction that the decomposition with phenyl isothiocyanate can be repeated a number of times; that is, the amino-acid sequence of the *N*-terminal peptide may be determined.

An extensive literature is now available on the analytical and preparative applications [246, 247] of the method initially developed by Edman [248].

A compilation of the solvent mixtures useful in the chromatography of the PTH amino-acids on a silica gel G layer is given in Table 5.27.

Wang *et al.* [252] investigated the PTH amino-acids by one- and two-dimensional chromatography on a polyamide layer. In the two-dimensional examination, the best solvent mixture in the first direction proved to be n-heptane–n-butanol–acetic acid (40:30:9) or carbon tetrachloride–acetic acid (9:1), and that in the second direction was formic acid–water (1:1). Recently, Kulbe [253] successfully separated the PTH amino-acids on a polyamide layer. The fast and sensitive separation of the methylthiohydantoin amino-acids on a polyamide layer has also been established [254].

Several reagents have been found for the detection of the PTH amino-acids. The chlorine–tolidine reaction and the butyl hypochlorite–potassium iodide–starch reaction can be employed effectively. Roseau and Pantel [255] observed specific colour reactions between the PTH amino-acids and a ninhydrin–collidine reagent (Table 5.28).

REFERENCES

[1] Izmailov, N. A., Schraiber, M. S.: *Farmacia (Moscow)*, **3,** 1 (1938).
[2] Békésy, N.: *Biochem. J.*, **312,** 100 (1942).
[3] Kirchner, J. G., Miller, J. M., Keller, G. J.: *Anal. Chem.*, **23,** 420 (1951).
[4] Stahl, E., Schröter, G., Kraft, G., Renz, R.: *Pharmazie*, **11,** 633 (1956).
[5] Schlitt, H., Geiss, F.: *J. Chromatog.*, **67,** 261 (1972).
[6] Geiss, F., Schlitt, H.: *Chromatographia*, **1,** 392 (1968).
[7] Niederwieser, A., Honegger, C. C.: *Adv. Chromatog.*, **2,** 123 (1968).
[8] De Zeeuw, R. A.: *Thesis*, State University, Groningen, 1968.
[9] Niederwieser, A., Brenner, M.: *Experientia*, **21,** 50 (1965).
[10] Niederwieser, A., Brenner, M.: *Experientia*, **21,** 105 (1965).
[11] Stahl, E.: *Die Dünnschichtchromatographie*, Springer-Verlag, Berlin, 1962, p. 28.
[12] Brenner, M., Niederwieser, A.: *Experientia*, **17,** 237 (1961).
[13] De Zeeuw, R. A.: *J. Chromatog.*, **32,** 43 (1968).
[14] Tyihák, E., Held, G.: in *Progress in TLC and Related Methods*, Vol. II. Niederwieser, A., Pataki, G. (eds.), Ann Arbor, 1971.
[15] Tyihák, E.: *Magy. Kém. Lapja*, **29,** 195 (1974).
[16] Snyder, L. R., Saunders, D. L.: *J. Chromatog.*, **44,** 1 (1969).
[17] Mangravite, R. V.: *Anal. Chem.*, **40,** 250 (1968).
[18] Kaess, A., Mathis, C.: *Ann. Pharm. Franc.*, **23,** 739 (1967).
[19] Wilk, M., Hoppe, U., Taupp, W., Rochlitz, J.: *J. Chromatog.*, **27,** 311 (1967).
[20] Vágujfalvi, D.: *Planta Med.*, **13,** 79 (1965).
[21] Biagi, G. L., Barbaro, A. M., Gamba, M. F., Guerra, M. C.: *J. Chromatog.*, **41,** 371 (1969).
[22] Jellinek, M.: *J. Chromatog.*, **69,** 402 (1972).
[23] Ho, I. K., Loh, H. H., Way, E. L.: *J. Chromatog.*, **65,** 577 (1972).

[24] Kaiser, R. E.: in *Einführung in die Hochleistungs-Dünnschicht-Chromatographie*, Kaiser, R. E. (ed.): Institute of Chromatography, Bad Dürkheim, 1976.
[25] Zlatkis, A., Kaiser, R. E.: *HPTLC—High-Performance Thin-Layer Chromatography*, Elsevier, Amsterdam, and Institute of Chromatography, Bad Dürkheim, 1977.
[26] Bertsch, W., Hara, S., Kaiser, R. E., Zlatkis, A.: *Instrumental HPTLC*, Hüthig, Heidelberg, 1980.
[27] Jänchen, D.: in *Instrumental HPTLC*, Bertsch, W., Hara, S., Kaiser, R. E., Zlatkis, A. (eds): Hüthig, Heidelberg, 1980, pp. 133–164.
[28] Brinkmann, U. A. Th., De Vries, G.: in *Instrumental HPTLC*, Bertsch, W., Hara, S., Kaiser, R. E., Zlatkis, A. (eds.): Hüthig, Heidelberg, 1980, pp. 39–53.
[29] Halpaap, H., Krebs, K. F.: *J. Chromatog.*, **142,** 823 (1977).
[30] Tyihák, E., Kalász, H., Mincsovics, E., Nagy, J.: *Proc. 17th Hung. Annual Meet. Biochem.*, Kecskemét, 1977. *Chem. Abstr.*, **88**, 15386 (1978).
[31] Tyihák, E., Mincsovics, E., Kalász, H.: *J. Chromatog.*, **174,** 75 (1979).
[32] Mincsovics, E., Tyihák, E., Kalász, H.: *J. Chromatog.*, **191,** 293 (1980).
[33] Kalász, H., Nagy, J., Mincsovics, E., Tyihák, E.: *J. Liquid Chrom.*, **3,** 845 (1980).
[34] Tyihák, E., Mincsovics, E., Kalász, H., Nagy, J.: *J. Chromatog.*, **211,** 45 (1981).
[35] Bauer, K.: *J. Chromatog.*, **32,** 529 (1968).
[36] Güven, K. C., Ozsari, G.: *Eczacilik Bül. Bull. Pharmacy*, **9,** 12 (1967).
[37] Kibardin, S. A., Lasurkina, V. B.: *Biokhimiya*, **30,** 559 (1965).
[38] Gornall, D. A., Kuksis, A.: *Canad. J. Biochem.*, **49,** 44 (1971).
[39] Gornall, D. A., Kuksis, A.: *Canad. J. Biochem.*, **49,** 51 (1971).
[40] Blaton, V., Vandarnine, D., Peters, H.: *Chrom. Symp. VI., Bruxelles*, 1970, p. 254.
[41] Trautmann, K. H.: *Z. Naturforsch.*, **27b,** 263 (1972).
[42] Determann, H.: *Gelchromatographie*. Springer-Verlag, Berlin, 1967.
[43] Wieland, T., Lüben, G., Determann, H.: *Experientia*, **18,** 430 (1962).
[44] Andrews, P.: *Biochem. J.*, **91,** 222 (1964).
[45] Morris, C. J. O.: *J. Chromatog.*, **16,** 167 (1964).
[46] Determann, H., Michel, W.: *J. Chromatog.*, **25,** 303 (1966).
[47] Heinz, F., Prosch, W.: *Anal. Biochem.*, **40,** 327 (1971).
[48] Radola, B. J.: *J. Chromatog.*, **38,** 61 (1968).
[49] Hanson, L. A., Johansson, B. G.: *Clin. Chim. Acta*, **8,** 66 (1963).
[50] Hanson, L. A., Johansson, B. G., Rymo, L.: *Clin. Chim. Acta*, **14,** 391 (1966).

[51] Carnegie, P. R., Pacheco, G.: *Proc. Soc. Exp. Biol. Med.*, **117,** 137 (1964).
[52] Hanson, L. A., Roos, P., Rymo, L.: *Nature*, **212,** 948 (1966).
[53] Agostoni, A., Vergani, C., Lomanto, B.: *J. Lab. Clin. Med.*, **69,** 522 (1967).
[54] Peeters, T. L., Vantrappen, G. R.: *Clin. Chim. Acta*, **47,** 437 (1973).
[55] Maier, C. L. Jr.: *J. Chromatog.*, **32,** 577 (1968).
[56] Adams, J. B., Chulavatnatol, M.: *Biochim. Biophys. Acta*, **146,** 509 (1967).
[57] Cartwright, L. N., Hullin, R. P.: *Biochem. J.*, **101,** 781 (1966).
[58] Nonomura, S., Imose, J., Tatsumi, C.: *Agric. Biol. Chem., Tokyo*, **33,** 1223 (1969).
[59] Copius-Peereboom, J. W., Beckes, H. W.: *J. Chromatog.*, **39,** 339 (1969).
[60] Zaharia, O., Soru, E.: *Eur. J. Biochem.*, **18,** 28 (1971).
[61] Nassi, P., Treves, C., Ramponi, G.: *J. Chromatog.*, **33,** 553 (1968).
[62] Plummer, D. T., Leatwood, P. D.: *Biochem. J.*, **103,** 172 (1967).
[63] Ramsey, H. A.: *Anal. Biochem.*, **5,** 83 (1963).
[64] Maruna, R. F.: *Clin. Chim. Acta*, **25,** 133 (1969).
[65] Paiva, A. C. M., Grandino, A.: *Experientia*, **29,** 154 (1973).
[66] Paiva, A. C. M., Nonailhetas-Miyamoto, M., Paiva, T. B.: *J. Med. Chem.*, **16,** 280 (1973).
[67] Myokei, R., Sakurai, A., Chang, C. F., Kodaira, Y., Takahashi, N., Tamura, S.: *Agric. Biol. Chem., Tokyo*, **33,** 1491 (1969).
[68] Vasileva, G. A., Mironov, A. F., Shtileva, Y. A., Yevstigneyeva, R. P.: *Zh. Obshch. Khim.*, **41,** 2195 (1971).
[69] Babel, I., Stella, R. C., Prado, E. S.: *Biochem. Pharmacol.*, **17,** 2232 (1968).
[70] Atherton, R. S., Thomson, A. R.: *Biochem. J.*, **111,** 797 (1969).
[71] Utermann, G., Wiegandt, H.: *Z. Physiol. Chem.*, **352,** 938 (1971).
[72] Procházka, Z., Jošt, K., Šorm, F.: *Collection Czech. Chem. Commun.*, **37,** 289 (1972).
[73] De Zélée, P., Brikas, E.: *Biochemistry*, **9,** 823 (1970).
[74] Guinand, M., Vacheron, M. J., Michel, G., Das, B. C., Lederer, E.: *Tetrahedron* Suppl., **7,** 271 (1966).
[75] Hantke, K., Braun, V.: *Eur. J. Biochem.*, **34,** 284 (1973).
[76] Bláha, K., Štokrová, Š., Sedláček, B., Šponar, J.: *Collection Czech. Chem. Commun.*, **41,** 2273 (1976).
[77] Stverteczky, J., Bajusz, S.: *Acta Chim. Acad. Sci. Hung.*, **88,** 67 (1976).
[78] Panpsuievich, O. S., Krikis, A. J., Chipens, G. I.: *Zh. Obshch. Khim.*, **42,** 224 (1972).

[79] Brtník, F., Trka, A., Zaoral, M.: *Coll. Czech. Chem. Commun.*, **40**, 179 (1975).
[80] Brenner, M., Pataki, G.: *Helv. Chim. Acta*, **44**, 1420 (1961).
[81] Wieland, T., Bende, H.: *Chem. Ber.*, **98**, 504 (1965).
[82] Pravda, Z., Poduška, K., Bláha, K.: *Collection Czech. Chem. Commun.*, **29**, 2626 (1964).
[83] Hubert, P., Dellacherie, E.: *J. Chromatog.*, **80**, 144 (1973).
[84] Ehrharbt, E., Cramer, F.: *J. Chromatog.*, **7**, 405 (1962).
[85] Wang, S. S., Merrifield, R. B.: *Intern. J. Prot. Chem.*, **1**, 235 (1969).
[86] Obchinnikov, Y. A., Kirynshkin, A. A., Kozhevnikova, I. V.: *Zh. Obshch. Khim.*, **41**, 2085 (1971).
[87] Riniker, B., Schwyzer, R.: *Helv. Chim. Acta*, **44**, 2357 (1964).
[88] Sieber, P., Ritter, W., Riniker, B.: *Helv. Chim. Acta*, **55**, 1243 (1972).
[89] Callahan, P. X., McDonald, J. K., Ellis, S.: *Methods in Enzymology*, **25B**, 282 (1973).
[90] Haworth, C., Oliver, R. W. A.: *J. Chromatog.*, **64**, 305 (1972).
[91] Burns, D. J. W., Turner, N. A.: *J. Chromatog.*, **30**, 469 (1967).
[92] Kemmler, F., Peterson, J. D., Steiner, D. F.: *J. Biol. Chem.*, **246**, 6786 (1971).
[93] Stark, C. R., Crawford, L. V.: *Nature*, **237**, 146 (1972).
[94] Heathcote, Y. G., Al-Alawi, S. Y.: *J. Chromatog.*, **129**, 211 (1976).
[95] Abe, T., Muto, Y., Hosoya, N.: *J. Lipid Res.*, **16**, 200 (1975).
[96] Stewart, J. M.: *J. Med. Chem.*, **17**, 537 (1974).
[97] Wainwright, I. M., Shapshak, P.: *J. Chromatog. Sci.*, **17**, 535 (1980).
[98] Kamber, B., Rittel, W.: *Liebigs Ann. Chem.*, 1928 (1979).
[99] Giliberti, P., Niederwieser, A.: *J. Chromatog.*, **66**, 261 (1972).
[100] Heathcote, J. G., Keogh, B. J., Washington, R. J.: *J. Chromatog.*, **79**, 187 (1973).
[101] Moczár, E.: *J. Chromatog.*, **76**, 417 (1973).
[102] Holbrook, J. J., Jeckel, R.: *Biochem. J.*, **111**, 689 (1969).
[103] Spitnik-Elson, P., Breiman, A.: *Biochim. Biophys. Acta*, **254**, 457 (1971).
[104] Richter, G. W., Moppert, G. A., Lee, J. C. K.: *Comp. Biochem. Physiol.*, **32**, 451 (1970).
[105] Allen, B. L., Gracy, R. W., Harris, B. G.: *Arch. Biochem. Biophys.*, **155**, 325 (1973).
[106] Crawford, L. V., Gesteland, R. F.: *J. Molec. Biol.*, **74**, 627 (1973).
[107] Bogdansky, F. M.: *J. Chromatog. Sci.*, **13**, 567 (1975).
[108] Bates, D. L., Perham, R. N.: *Anal. Biochem.*, **68**, 175 (1975).
[109] Gutte, B.: *J. Biol. Chem.*, **250**, 889 (1975).
[110] Barrolier, J.: *Naturwiss.*, **48**, 404 (1961).

[111] Brenner, M., Niederwieser, A., Pataki, G.: in *Dünnschichtchromatographie*, Stahl, E. (ed.), Springer-Verlag, Berlin, 1962.

[112] Schellenberg, P.: *Angew. Chem.*, **74,** 118 (1962).

[113] Barrett, G. C.: *Nature*, **194,** 1171 (1962).

[114] Mendez, E., Lai, C. Y.: *Anal. Biochem.*, **65,** 81 (1975).

[115] Felix, A. M., Jimenez, M. H.: *J. Chromatog.*, **89,** 361 (1974).

[116] Gunnar, E., Lindeberg, G.: *J. Chromatog.*, **117,** 439 (1976).

[117] Schiltz, E., Schnackerz, K. O., Gracy, R. W.: *Anal. Biochem.*, **79,** 33 (1977).

[118] Pataki, G.: *J. Chromatog.*, **16,** 541 (1964).

[119] Gillesen, D., Studer, R. O., Rudinger, J.: *Experientia*, **29,** 170 (1973).

[120] Patthy, L., Smith, E.: *J. Biol. Chem.*, **250,** 557 (1975).

[121] Černý, A., Kotva, R., Semonský, M.: *Collection Czech. Chem. Commun.*, **37,** 2606 (1972).

[122] Edvinson, L., Hakanson, R., Rönnberg, A. L., Sandler, F.: *J. Chromatog.*, **67,** 81 (1972).

[123] Rostovtseva, L. I., Reshetov, P. D., Khokhlov, A. S.: *Zh. Obshch. Khim.*, **39,** 96 (1969).

[124] Wilkinson, S. G.: *Biochim. Biophys. Acta*, **270,** 1 (1972).

[125] Thompson, R. Ç., Blout, E. R.: *Biochemistry*, **12,** 51 (1973).

[126] Mikeš, O., Turková, J., Toan, N. B., Šorm, F.: *Biochim. Biophys. Acta*, **178,** 112 (1969).

[127] Vasenev, V. I., Kuznetsov, Y. S., Stepanov, V. M.: *Biochemistry (USSR)*, **35,** 690 (1970).

[128] Kravchenko, N. A., Capuk, Y. K.: *Biokhimiya*, **35,** 53 (1970).

[129] Petershofer, G., Rey, O.: *Microchim. Acta*, 381 (1969).

[130] Oepen, M., Oepen, I.: *Klin. Wochenschrift*, **41,** 921 (1963).

[131] Boissonas, R. A., Lo Bianco, S.: *Experientia*, **8,** 425 (1952).

[132] Harris, C. K., Tigane, E., Hanes, C. S.: *Canad. J. Biochem. Physiol.*, **39,** 439 (1961).

[133] Tyihák, E., Patthy, A., Ferenczi, S., Eckhardt, S., Kralovánszky, J., Lapis, K., Szende, B.: *Kísérl. Orvostud.*, **27,** 532 (1975).

[134] Tyihák, E., Patthy, A.: *Acta Agronom. Acad. Sci. Hung.*, **20,** 445 (1973).

[135] Brenner, M., Niederwieser, A.: *Experientia*, **16,** 378 (1960).

[136] Fahmy, A. R., Niederwieser, A., Pataki, G., Brenner, M.: *Helv. Chim. Acta*, **44,** 2022 (1961).

[137] Codern, L., Gadea, J., Gal, E., Montagut, M.: *Afinidad, Barcelona*, **20,** 159 (1963).

[138] Pataki, G.: *J. Chromatog.*, **17,** 580 (1965).

[139] Rokkones, T.: *Scand. J. Clin. Lab. Invest.*, **16,** 149 (1964).

[140] Tyihák, E.: *C. Sc. Thesis*, Budakalász, 1978.

[141] Tyihák, E., Vágujfalvi, D.: *J. Chromatog.*, **49**, 343 (1970).

[141a]Tyihák,E.: (Unpublished data).

[142] Mottier, M.: *Mitt. Lebensmittel. Hyg. Bern*, **49**, 343 (1970).

[143] Shasha, B., Whister, R. L.: *J. Chromatog.*, **14**, 532 (1964).

[144] Bancher, E., Scherz, H., Prey, V.: *Microchim. Acta*, 712 (1963).

[145] Detterbeck, F. Y., Lillevik, H. A.: in *Progress in TLC and Related Methods*, Vol. II, Niederwieser, A., Pataki, G. (eds), Ann Arbor, 1971.

[146] Opienska-Blauth, J., Kraczkowski, H., Bruszkiewicz, H.: in *Thin-Layer Chromatography*, Marini-Bettolo, G.-B. (ed.), Elsevier, Amsterdam, 1964.

[147] von Euler, H., Hasselquist, H., Limnell, I.: *Arkiv Kemi*, **21**, 259 (1963).

[148] Wollenweber, P.: *J. Chromatog.*, **9**, 369 (1962).

[149] von Arx, E., Neher, P.: *J. Chromatog.*, **12**, 329 (1963).

[150] Haworth, C., Heathcote, J. G.: *J. Chromatog.*, **41**, 380 (1969).

[151] Heathcote, J. G., Haworth, C.: *J. Chromatog.*, **43**, 84 (1969).

[152] Heathcote, J. G., Keogh, B. J., Washington, R. J.: *J. Chromatog.*, **79**, 187 (1973).

[153] Turner, N. A., Redgwell, R. J.: *J. Chromatog.*, **21**, 129 (1966).

[154] Horton, D., Tanimura, A., Wolfrom, M. L.: *J. Chromatog.*, **23**, 309 (1966).

[155] Heathcote, J. G., Haworth, C.: *Biochem. J.*, **114**, 667 (1969).

[156] Pataki, G.: *Z. Klin. Chem.*, **2**, 129 (1964).

[157] Völkl, A., Berlet, H. A.: *Clin. Chem. Clin. Biochem.*, **15**, 267 (1967).

[158] de la Llosa, P., Tertin, C., Jutisz, M.: *Experientia*, **20**, 204 (1964).

[159] Verceanst, R., Blaton, V., Peeters, H.: *J. Chromatog.*, **43**, 132 (1969).

[160] Levis, G. M., Kontras, D. A., Vagenakis, A., Messaris, G., Miras, C., Malamos, B.: *Clin. Chim. Acta*, **20**, 127 (1968).

[161] Gozzi, D., Desideri, P. G., Lepri, L., Coas, V.: *J. Chromatog.*, **40**, 138 (1969).

[162] Kiyasu, J. Y.: *Die Dünnschichtchromatographie*, Stahl, E. (ed.), Springer-Verlag, Berlin 1962, p. 453.

[163] Ogiya, S., Baba, S., Tsubaki, S.: *Tohoku Yakka Daigaku Kenkyu Nempo*, **14**, 19 (1967); *Chem. Abstr.*, **71**, 109, 664 (1969).

[164] Kraffczyk, F., Helger, R.: *Z. Anal. Chem.*, **243**, 536 (1968).

[165] Copley, M. N., Truter, E. V.: *J. Chromatog.*, **45**, 480 (1969).

[166] Dévényi, T., Zoltán, S.: *7th Intern. Symp. Chem. Natural Products, Riga*, 1970, p. 52.

[167] Dévényi, T.: *Acta Biochim. Biophys. Acad. Sci. Hung.*, **5**, 435 (1970).

[168] Dévényi, T., Hazai, I., Ferenczi, S., Báti, J.: *Acta Biochim. Biophys. Acad. Sci. Hung.*, **6,** 385 (1971).
[169] Dévényi, T., Báti, J., Fábián, F.: *Acta Biochim. Biophys. Acad. Sci. Hung.*, **6,** 133 (1971).
[170] Ferenczi, S., Báti, J., Dévényi, T.: *Acta Biochim. Biophys. Acad. Sci. Hung.*, **6,** 123 (1971).
[171] Dévényi, T., Báti, J., Kovács, J., Kiss, P.: *Acta Biochim. Biophys. Acad. Sci. Hung.*, **7,** 237 (1972).
[172] Ferenczi, S., Dévényi, T.: *Acta Biochim. Biophys. Acad. Sci. Hung.*, **6,** 389 (1971).
[173] Tyihák, E., Ferenczi, S., Hazai, I., Zoltán, S., Patthy, A.: *J. Chromatog.*, **102,** 257 (1974).
[174] Okumura, T., Kadono, T., Nakatani, M.: *J. Chromatog.*, **74,** 73 (1972).
[175] Okumura, T., Kadono, T.: *Bunseki Kagaku*, **21,** 321 (1972).
[176] Okumura, T., Kadono, T.: *Yakugaku Zasshi*, **92,** 708 (1972).
[177] Consden, R., Gordon, A. H., Martin, A. J. P.: *Biochem. J.*, **40,** 33 (1946).
[178] Honegger, C. G.: *Helv. Chim. Acta*, **44,** 173 (1961).
[179] Nybom, N.: *Physiologia Plantarum*, **17,** 434 (1964).
[180] Nassif-Makki, H., Constabel, F.: *Z. Pflanzenphys.*, **67,** 201 (1972).
[181] Rove, P. B., Lewis, G. P.: *Biochemistry*, **12,** 1962 (1973).
[182] Ersser, R. S., Krywawych, S.: *Med. Lab. Technol.*, **31,** 235 (1974).
[183] Brenner, M., Niederwieser, A., Pataki, G.: in *Die Dünnschichtchromatographie*, Stahl, E. (ed.), Springer-Verlag, Berlin, 1962, p. 405.
[184] Opienska-Blauth, J., Gebala, A.: *Biochem. Clinics, New York*, 205 (1963).
[185] Barrolier, J.: *Naturwiss.*, **48,** 404 (1961).
[186] Barrolier, J., Heilmann, J., Watzke, F.: *Z. Physiol. Chem.*, **309,** 219 (1957).
[187] Barrolier, J., Heilmann, J.: *Z. Physiol. Chem.*, **304,** 21 (1956).
[188] Dévényi, T.: *Ion-exchange Chromatography in Biochemistry*, in *Recent Results in Chemistry* (in Hungarian), Csákvári, B. (ed.), Akadémiai Kiadó, Budapest, 1977.
[189] Moffat, E. D., Lyttle, R. L.: *Anal. Chem.*, **31,** 926 (1959).
[190] Krauss, G.-J., Reinbothe, H.: *Biochem. Physiol. Pflanzen*, **161,** 577 (1970).
[191] Tyihák, E.: *Herba Hung.*, **8,** 113 (1969).
[192] Circo, R., Freeman, B. A.: *Anal. Chem.*, **35,** 262 (1963).
[193] Kolor, M. G., Roberts, H. R.: *Arch. Biochem. Biophys.*, **70,** 620 (1957).
[194] Jepson, J. B., Smith, J.: *Nature*, **172,** 1100 (1953).
[195] Bhattacharya, K. R., Dutta, J., Roy, D. K.: *Arch. Biochem. Biophys.*, **84,** 377 (1959).

[196] Pasieka, A. E., Borowiecki, M. T.: *Biochim. Biophys. Acta*, **111,** 553 (1965)

[197] Toennies, G., Kolb, J. J.: *Anal. Chem.*, **23,** 823 (1951).

[198] Winegard, H. M., Toennies, G., Block, R. J.: *Science*, **108,** 506 (1948).

[199] Awe, W., Reinecke, I., Thum, J.: *Naturwiss.*, **41,** 528 (1954).

[200] Patton, A. R., Foreman, E. M.: *Science*, **109,** 339 (1949).

[201] Frank, H., Petersen, H.: *Z. Physiol. Chem.*, **299,** 1 (1955).

[202] Grimmelt, M. R., Richards, E. L.: *J. Chromatog.*, **20,** 171 (1965).

[203] Jatzkewitz, H., Lenz, U.: *Z. Physiol. Chem.*, **305,** 53 (1956).

[204] Tyihák, E.: (Unpublished data).

[205] Curzon, G., Giltrow, J.: *Nature*, **172,** 356 (1953).

[206] Patthy, M., Patthy, A.: (Private communication).

[207] Consden, R.: *Nature*, **162,** 359 (1948).

[208] Heacock, R. A., Mahon, M. E.: *J. Chromatog.*, **17,** 338 (1965).

[209] Boctor, F. N.: *J. Chromatog.*, **67,** 371 (1972).

[210] Purdy, S. J., Truter, E. V.: *Analyst*, **87,** 802 (1962).

[211] Hara, S., Tanaka, H., Takeuchi, M.: *Chem. Pharm. Bull., Tokyo*, **12,** 626 (1964).

[212] *Instrumental Manual*, Joyce-Loebl and Co., Ltd., Gateshead-on-Tyne, Great Britain.

[213] Heathcote, J. G., Haworth, C.: *J. Chromatog.*, **43,** 84 (1969).

[214] West, E. S., Todd, W. P., Mason, M. S., van Bruggen, J. T.: *Textbook of Biochemistry*, 4th Ed., Macmillan, New York, 1966, p. 1415.

[215] Frodyma, M., Frei, R. W.: *J. Chromatog.*, **17,** 131 (1965).

[216] Pataki, G.: *Chromatographia*, **1,** 492 (1968).

[217] Seiler, N., Wiechmann, H.: *Experientia*, **20,** 559 (1964).

[218] Clark, M. E.: *Analyst*, **93,** 810 (1968).

[219] Brandner, G., Virtanen, A. J.: *Acta Chem. Scand.*, **17,** 2563 (1963).

[220] Heins, K., Hauber, R.: *Z. Physiol. Chem.*, **348,** 357 (1967).

[221] Spivak, V. A., Orlov, V. M., Shcherbukhin, V. V., Varshavsky, Ya. M.: *Anal. Biochem.*, **35,** 227 (1970).

[222] Briel, G., Neuhoff, V., Maier, M.: *Z. Physiol. Chem.*, **353,** 540 (1972).

[223] Varga, J. M., Richards, F. F.: *Anal. Biochem.*, **53,** 397 (1973).

[224] Sanger, F.: *Biochem. J.*, **39,** 507 (1945).

[225] Porter, R. R., Sanger, F.: *Biochem. J.*, **42,** 287 (1948).

[226] Rao, K. R., Sober, H. A.: *J. Am. Chem. Soc.*, **76,** 1328 (1954).

[227] Levy, A. L., Chung, D.: *J. Am. Chem. Soc.*, **77,** 2899 (1955).

[228] Bunnett, J. F., Hermann, D. H.: *Biochemistry*, **9,** 816 (1970).

[229] Biserte, G., Holleman, J. W., Holleman-Dehove, J., Sautiere, P.: *Chromatog. Rev.*, **2,** 59 (1960).

[230] Deyl, Z., Rosmus, J., Bump, S.: *Biochim. Biophys. Acta*, **140,** 515 (1967).
[231] Levy, A. L., Li, C. H.: *J. Biol. Chem.*, **213,** 487 (1955).
[232] Phillips, D. M. P.: *Biochem. J.*, **68,** 35 (1958).
[233] Sanger, F., Thompson, E. O. P.: *Biochem. J.*, **53,** 353 (1953).
[234] Brenner, M., Niederwieser, A., Pataki, G.: *Experientia*, **17,** 145 (1961).
[235] Brenner, M., Niederwieser, A., Pataki, G.: in *Dünnschicht–Chromatographie*, Stahl, E. (ed.), Springer Verlag, Berlin (1967).
[236] Munier, R. L., Sarrazin, G.: *Bull. Soc. Chim. France*, 2959 (1965).
[237] Wang, K.-T., Wang, I. S. Y.: *J. Chromatog.*, **27,** 318 (1967).
[238] Wang, K.-T., Weinstein, B.: in *Progress in Thin-Layer Chromatography*, Vol. III, Niederwieser, A., Pataki, G. (eds), Ann Arbor, 1972, p. 171.
[239] Gray, W. R.: *Methods in Enzymology*, **25,** 121, 333 (1972).
[240] Rosmus, J., Deyl, Z.: *Chromatog. Rev.*, **13,** 163 (1971).
[241] Dawid, I. B., French, T. C., Buchanan, J. M.: *J. Biol. Chem.*, **238,** 2178 (1963).
[242] Wang, K.-T., Wang. I. S.-Y., Yuan, S.-S., Wu, P. S.: *J. Chinese Chem. Soc. Taiwan*, **15,** 59 (1968).
[242a] Woods, K. R., Wang, K.-T.: *Biochim. Biophys. Acta*, **133,** 369 (1967).
[243] Seiler, N., Knödgen, B.: *J. Chromatog.*, **131,** 109 (1977).
[244] Ivanov, Ch. P., Vladovska-Yukhnowska, Y.: *C. R. Acad. Bulg. Sci.*, **20,** 1299 (1967).
[245] Seiler, N., Schmidt-Glenewinkel, T., Schneider, H. H.: *J. Chromatog.*, **84,** 95 (1973).
[246] Sjöquist, J.: *Arkiv Kemi*, **11,** 129 (1957); **14,** 291 (1959).
[247] Needleman, S. B. (ed.): *Protein Sequence Determination*, Springer-Verlag, Berlin, 1975.
[248] Edman, P.: *Acta Chem. Scand.*, **4,** 277 (1950).
[249] Cherbuliez, E., Baehler, B., Rabinowitz, J.: *Helv. Chim. Acta*, **47,** 1350 (1964).
[250] Jeppson, J. O., Sjöquist, J.: *Anal. Biochem.*, **18,** 264 (1967).
[251] Dus, K., De Klerk, H., Sletten, K., Bartsch, R. G.: *Biochim. Biophys. Acta*, **140,** 291 (1967).
[252] Wang, K.-T., Wang, S.-Y., Lin, A.-L., Wang, C.-S.: *J. Chromatog.*, **26,** 323 (1967).
[253] Kulbe, K. D.: *Anal. Biochem.*, **44,** 548 (1971).
[254] Kulbe, K. D., Nogueira-Hattesohl, Y. M.: *Anal. Biochem.*, **72,** 123 (1976).
[255] Roseau, G., Pantel, P.: *J. Chromatog.*, **44,** 392 (1969).

CHAPTER 6

Gel chromatography. Gel filtration and related methods

H. Kalász

6.1 FUNDAMENTALS OF GEL CHROMATOGRAPHY

Gel chromatography is currently one of the most widely used of the separation methods based on differences in molecular dimensions. Its importance is enhanced by the fact that it may be employed for both analytical and preparative purposes, and that separation depending on molecular size is a good supplement to procedures based on the ionic, polar, etc. properties of the molecules.

Because of the inhomogeneity of the pores of gels, gel chromatography not only differentiates molecules into those smaller and larger than the pores of the gel, but also permits the separation of the various molecular species with a relatively broad range of molecular size (selective permeation). Molecules which penetrate into the interior of the gel pores, and are thereby retarded, need a larger volume of solvent for elution than the molecules which are larger than the pores of the gel, and therefore excluded from it. When a mixture of substances is separated, penetration into the pores of the gel particles can be achieved only by those molecules which are smaller than the pore size of the gel; accordingly, the components of the mixture are distributed between the stationary and the mobile phases on the basis of the sizes of their molecules (Fig. 6.1).

Gel chromatography may be considered as occupying a middle position among separation methods as regards both its loading capacity and its separation power. For purification purposes, gel chromatography is applied (mainly with traditional methods) to substances that have already been purified to a certain extent: thus, the gel chromatography of a sample of biological or industrial origin is preceded by centrifugation, precipitation with protamine sulphate or sodium sulphate, ultrafiltration or ion-exchange chromatography. On the other hand, components separated from mixtures by gel chromatography are generally further purified by electrophoresis, isoelectric focusing, ion-exchange gel chromatography or affinity chromatography, etc.

In general, the selective permeation range of the gel is utilized in separations [1]. However, it is nowadays generally considered that all

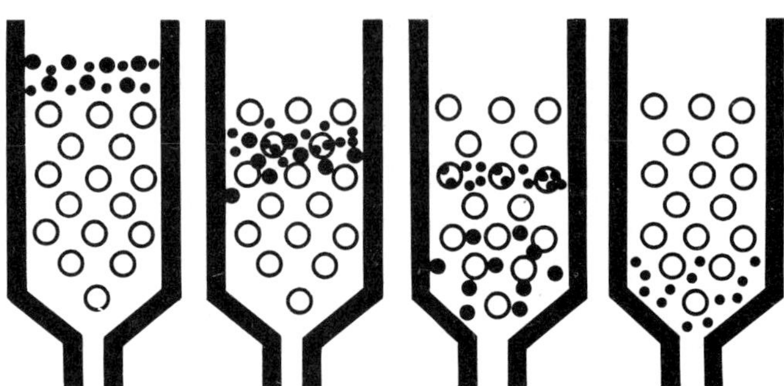

Fig. 6.1. The process of gel chromatography. In the course of the separation the small molecules penetrate into the interior of the gel particles, thereby undergoing a delay, while the large molecules are excluded from the interior of the gel particles and are eluted without a delay. (By kind permission of Prof. H. Determann and Springer-Verlag.)

fractions with bulkier molecules than those of the component to be obtained should be removed in the purification step(s) preceding the gel chromatography, and that in the gel chromatography itself the component in question should be the first to be eluted, in the void volume V_0. In this way the chromatogram will not be affected by the tailing of any earlier-eluted component. Ultrafiltration is particularly suitable for such a prepurification [2].

The advantages of gel chromatography include the following.

(a) The order of magnitude of the molecular weight is enough to allow selection of the optimum gel type.

(b) Parallel experiments with appropriate standards provide a means of calculating the molecular weight.

(c) The conditions such as the pH, ionic strength, and temperature of the eluent can be varied widely, the only critical factor being the stability of the substances to be purified.

(d) There is no mutual interference between the retarded species and excluded species.

(e) The optimum chromatographic parameters such as column dimensions, fraction volumes, and sample volume can be calculated beforehand.

(f) The simple procedures require only minimal instrumentation, apparatus and material.

(g) Modern equipment permits continuous monitoring, and certain techniques (e.g. recycling) enhance the separation considerably.

(h) The gel columns are practically completely self-regenerating, i.e. on completion of a separation (when all components have been eluted) the process may be recommenced. In this respect, gel chromatography differs fundamentally not only from thin-layer chromatography, where the layers may be used only once, but also from ion-exchange or adsorption chromatography.

6.1.1 Gels, gel types and their characteristics

In general, the gels used must satisfy the following requirements.

(a) The matrix of the gel must be inert. The substances to be separated must not interact with it, and there should be no irreversible binding or denaturation of solute molecules on the gel.

(b) The gel-solute interactions must not cause any irreversible changes in the secondary or tertiary structures of labile macromolecules. Although there is a comparatively high danger of inactivation in the case of biologically important proteins and nucleic acids, gel chromatography is one of the few methods in which the danger of denaturation is minimal, and the yield (measured in biological units) is practically quantitative.

(c) Ion-exchange groups must be either absent or present to only a slight extent on substances used for gel chromatographic purposes. (This requirement does not apply to ion-exchange gels.)

(d) Any interactions between the gel and the solute may be at most weak and reversible; however, the separation of small molecules is frequently facilitated by hydrophobic interactions, and thus in certain cases the gel–solute interactions must be altered in a particular way, generally so that there is a specific increase or decrease. It is necessary to eliminate the gel–solute interactions mainly in the determination of molecular weight and molecular size; measurement anomalies caused by aggregation or adsorption are usually eliminated by urea or guanidine hydrochloride in a concentration of 6 *M*.

(e) The gel must be stable both mechanically and chemically and must yield the same separation for a long time under the conditions generally used, i.e. the spatial network of the gel, its original pore sizes, and the volumes within and between the gel particles must remain unchanged.

(f) A fundamental requirement is stability in aqueous solution between pH 2 and 10. Methanol and ethanol must cause no permanent change in the gel.

(g) The distribution of particle size of the gel must be carefully controlled. In recent years, practically only polymeric bead-form gels have been produced commercially; these ensure an enhanced separation. With other forms of gel, there is considerable inhomogeneity in the channels between the particles, and the effect of this inhomogeneity is further increased by the heterogenity of the particle size distribution.

6.1.2 Gels used in chromatographic separations

The gels that may be obtained for gel chromatographic operations satisfy virtually all demands. Of the commercially-available gels, Sephadex, Sephacryl S and Sepharose [3–5], Bio-Gel P and Bio-Gel A [6] and Ultragel AcA [7] are the most widely used. Their approximate exclusion limits are listed in Table 6.1.

For molecular weights up to several thousand, Sephadex G [3, 8] and Bio-Gel P [9] are most commonly used. These two gel types are fundamentally different, since the Sephadex G gels (the number after the G indicates ten times the water regain of the gel) and the Bio-Gel P gels (the number after the P is 1/1000 of the exclusion limit) have different matrixes: Sephadex gels are built up from chains of glucose units, cross-linked by glycerol ether bridges, and Bio-Gel P gels are polyacrylamide. In spite of this basic difference, the individual members display strikingly similar swelling capacities and separation ranges; thus, a parallel is observed between the parameters of the Sephadex G-15 and Bio-Gel P-2 gels, the Sephadex G-75 and Bio-Gel P-30 gels, etc.

The chromatographic properties of agar gels have been described in detail by Araki [10], Russel *et al.* [11] and Polson [12]. Although the gels have a large pore size (the fractionation range is for molecular weights ranging from several hundred thousand to a few million), their mechanical stability is fairly good. That of Ultragel AcA gels is similarly good [13].

Sephadex LH-20 (a hydroxypropylated derivative of Sephadex G-25 [14]) is capable of swelling not only in water, but also in organic solvents. It has found widespread application in analytical and preparative separation, especially of amino-acids (e.g. iodinated tyrosine derivatives), peptides and substituted peptides. The recently-introduced spherical modification of Sephadex LH-20 (Sephasorb HP Ultrafine) provides hope for the extension of high-performance gel chromatography in the investigation of iodinated amino-acids and peptides [15].

Table 6.1
Approximate exclusion limits of gels used for gel chromatography, referred to globular proteins

Type of gel	Approximate exclusion limit for globular proteins	Type of gel	Approximate exclusion limit for globular proteins
Sephadex G-10	7×10^2	Sepharose 6B	4×10^6
Sephadex G-15	1.5×10^3	Sepharose 4B	2×10^7
Sephadex G-25	5×10^3	Sepharose 2B	4×10^7
Sephadex G-50	3×10^4		
Sephadex G-75	8×10^4	Bio-Gel A-0.5 M	5×10^5
Sephadex G-100	1.5×10^5	Bio-Gel A-1.5 M	1.5×10^6
Sephadex G-150	3×10^5	Bio-Gel A-5 M	5×10^6
Sephadex G-200	5×10^5	Bio-Gel A-15 M	1.5×10^7
		Bio-Gel A-50 M	5×10^7
Bio-Gel P-2	1.8×10^3	Bio-Gel A-150 M	1.5×10^8
Bio-Gel P-4	4×10^3		
Bio-Gel P-6	6×10^3	Sephacryl S-200	2.5×10^5
Bio-Gel P-10	2×10^4	Sephacryl S-300	1.5×10^6
Bio-Gel P-30	4×10^4	Ultragel AcA 54	7×10^4
Bio-Gel P-60	6×10^4	Ultragel AcA 44	1.3×10^5
Bio-Gel P-100	1×10^5	Ultragel AcA 34	4×10^5
Bio-Gel P-150	1.5×10^5	Ultragel AcA 22	1×10^6
Bio-Gel P-200	2×10^5		
Bio-Gel P-300	4×10^5	Sephadex LH-20	4×10^3
		Sephadex LH-60	1×10^4
		Sephasorb HP	4×10^3

6.1.3 Equipment for gel chromatography using column technique

A considerable proportion of gel chromatography operations may be carried out even under simple laboratory conditions. Some straightforward gel chromatographic set-ups may be seen in Fig. 6.2.

The elution flow rate can be regulated by using a constant hydrostatic head (cf. Fig. 6.2), or a peristaltic pump.

The separations may be monitored continuously with a suitable detector or measurements may be made on individual fractions. A detailed account of such methods is given in Kirkland's and Morris's books [16, 17].

In preparative work, however, the properties of the substance to be purified are often not known, and only the biological effect can be monitored. Sometimes the biological activity is parallel by a characteristic physico-chemical property (e.g. ultraviolet absorption), but the connection between the two and the substances in question has not been confirmed. In such a case the physico-chemical characteristic may be followed, but if the parameters of

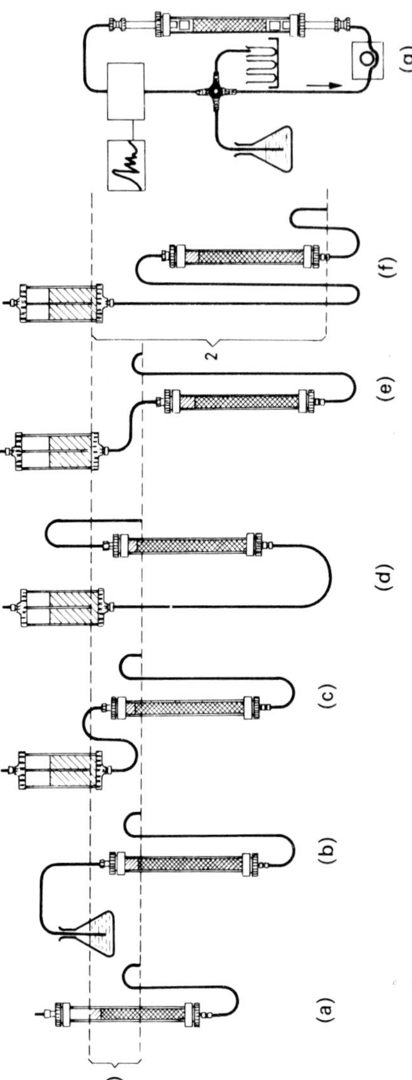

Fig. 6.2. Gel chromatography set-ups. The flow-rate may be regulated by hydrostatic pressure, (a)–(f), or with a peristaltic pump, (g); the magnitude of the hydrostatic pressure employed is indicated by the difference in level (h)

the method are changed or a new method is employed, it is necessary to check that the characteristic followed and the biological activity do in fact match. The detection methods usually monitor both the component to be purified and the impurity eluted closest to it, allowing both location of the substance and a check on the progress of purification. Some equipment provides dual wavelength measurement, and some instruments record the change in refractive index.

6.1.4 Concepts and calculations of gel chromatography

Parameters of the gel column

The total volume of the gel bed, V_t, is the volume of column filled with the appropriately packed gel, and has three components: the volume of the gel itself (V_m), the volume within the gel (V_i), and the volume external to the gel (V_0):

$$V_t = V_m + V_i + V_0$$

If a molecule is smaller than the pores of the gel used, it will be eluted at an effluent volume equal to $V_i + V_0$; if the molecule is larger than the pores of the gel, it will be eluted at effluent volume equal to V_0.

Several means are available for determination of these chromatographic parameters.

V_t, may be calculated from the dimensions of the column.

V_0 is the void volume (interstitial volume). It is generally measured by finding the effluent volume corresponding to elution of a substance that is large enough to be completely excluded from the gel. The molecular weight of the Blue Dextran molecule is 2×10^6; it is bright blue in solution, with appreciable absorption at 254 and 280 nm. It is used in general for determination of the void volumes of Sephadex and Bio-Gel P gels.

When Blue Dextran is used, however, some is adsorbed at the top of the gel bed, and this bound material is desorbed in the next experiment, so the fractions and substances appearing at effluent volume V_0 in a separation following an experiment with Blue Dextran will be coloured blue.

V_∞ is the total liquid volume of the gel bed, i.e. the sum of V_0 and V_i. With gels of large pore size, where the volume of the gel itself is negligible, V_∞ is nearly equal to the total volume of the gel bed, but with gels of small pore size the difference is significant.

V_∞ may be obtained by calculation:

$$V_\infty = V_t - V_m = V_0 + V_i$$

Physically, V_∞ means the volume of eluent needed to elute a substance not adsorbed on or interacting with the gel. Marsden [18] recommends the use of DHO, D_2O, THO or T_2O for measurement of V_∞, but detection of the isotopes is difficult. Acetone or potassium chloride [19, 20] is more suitable for the measurement of V_∞. Neither interacts with the gels, and the measured values are independent of the temperature, ionic strength and pH, except when a change in these parameters leads to variation in V_∞.

Other characteristic parameters are the flow rate or elution rate, generally given in units of ml/h, ml/min or $ml \cdot h^{-1} \cdot cm^{-2}$, and the velocity of the elution front, in units of cm/h or cm/min.

On a given column, behaviour of a given compound is characterized by the elution time, t_e, or the elution volume, V_e. If the width of the peak is also given (in the same units), the number of theoretical plates (see p. 280), a value characteristic of the chromatographic column, can be calculated.

Relative elution parameters provide a means of describing the behaviour of a particular compound on a given gel, irrespective of the condition employed. The relative elution time or volume is generally used:

$$V_{rel} = \frac{V_e}{V_0} \quad \text{and} \quad t_{rel} = \frac{t_e}{t_0}$$

and these quantities are naturally equal for constant flow rate. The retention coefficient, R, is given by the reciprocals of the relative retention parameters:

$$R = \frac{V_0}{V_e} = \frac{t_0}{t_e}$$

Gel chromatography may be regarded as a special kind of liquid–liquid partition chromatography, and the elution volume can be calculated from the distribution coefficient K_d, and K_d from the elution data:

$$V_e = V_0 - K_d V_i; \quad K_d = \frac{V_e - V_0}{V_i} = \frac{V_e - V_0}{V_\infty - V_0}$$

the last equation being the most often used for determination of K_d. Since a single liquid phase of given nature is involved, the partition mechanism being steric hindrance, the maximum value of K_d is 1.00. V_∞ is measured with acetone.

Laurent and Killander [21] propose the calculation of K_{av} instead of K_d:

$$K_{av} = \frac{V_e - V_0}{V_t - V_0} = \frac{V_e - V_0}{V_i + V_m}$$

thereby eliminating the difficulties arising from the determination of V_i.

The use of K_{av} undoubtedly has a certain advantage, as only V_e, V_0 and V_t need to be determined. V_t is calculated from the dimensions of the column, V_0 is measured with Blue Dextran, and V_e by means of detection of the substance under examination. However, the following must be noted.

(*i*) Only K_d has the physical meaning for which these calculations were introduced. K_{av} does not. K_{av} is not a distribution coefficient, as the substance clearly does not penetrate into the volume of the gel matrix itself.

(*ii*) K_{av} is not sensitive to compacting of the gel in the same way as K_d is. This is one reason why its use is preferred. At the same time, it is not suitable for checking on the compacting of the gel bed.

The experimental data are the most important of the numerical values given as gel chromatographic characteristics; from them the retention values and the distribution data (R, K_{av}, etc.) can be calculated [5, 22, 23]. Thus, it is of fundamental importance to give the experimental data in description of the separation. A typical calculation is set out below.

For a gel with small pore size, the total volume of the column is 122.5 ml. Blue Dextran is eluted at 36 ml, acetone at 88.5 ml, and the sample at 60 ml. What are the values of K_d and K_{av}?

From the given data, we may calculate V_i, V_m, K_d and K_{av}:

$$V_i = V_\infty - V_0 = 88.5 - 36 = 52.5 \text{ ml}$$

$$V_m = V_t - V_\infty = 122.5 - 88.5 = 34 \text{ ml}$$

$$K_d = \frac{V_e - V_0}{V_\infty - V_0} = \frac{60 - 36}{188.5 - 36} = 0.457$$

$$K_{av} = \frac{V_e - V_0}{V_t - V_0} = \frac{60 - 36}{122.5 - 36} = 0.257$$

If K_d is calculated for acetone, naturally a value of 1.00 is obtained, whereas K_{av} will be less than 1.00:

$$K_d = \frac{88.5 - 36}{88.5 - 36} = 1.00; \quad K_{av} = \frac{88.5 - 36}{122.5 - 36} = 0.607$$

Gel chromatographic values have been compared by Fischer [23]. Naturally, comparison refers only to one particular gel type, and even then only to its "normal" packing state, and thus the chromatographic parameters and the calculable data are compared on the basis of a "normally packed" column.

As already mentioned, the theoretical value of K_d cannot exceed 1.00. If the experimentally found K_d value is higher than 1.00, gel–solute interactions must be involved, such as adsorption of aromatic compounds or enhanced retardation due to hydroxyl or amino groups. On the other hand retardation can be decreased by repulsion between anions and carboxyl groups (ion-exclusion). These effects are known as secondary gel chromatographic phenomena and contribute to the elution parameters governed by the size of molecule. With large molecules the effects are barely observed, but for small molecules the elution curve is shifted into the range $K_d > 1$. The range of gel chromatography may be extended from $0 \leq K_d \leq 1$ even to $0 \leq K_d \leq 10$. This increases the peak capacity of gel chromatography, and permits the separation of molecules which are of nearly the same size, but differ as regards their hydroxyl, amino or carboxyl groups, or the number of aromatic rings.

The symmetrical elution curves obtained on Bio-Gel P-2 approximate to a Gaussian distribution function; they are characterized by the peak height, A, the peak width, 4σ, and the position of the maximum of the elution peak (x_0). The formula used in the approximation [24] is

$$y = A \exp\left[-\frac{(x - x_0)^2}{2\sigma^2}\right]$$

The effectiveness of the column and of the separation is determined by the number of theoretical plates and the peak separation.

Number of theoretical plates

This concept is general in both gas chromatography and liquid chromatography, and is mainly characteristic of the column. The number (N) can be calculated on the basis of Fig. 6.3, by dividing the 'distance' of the peak

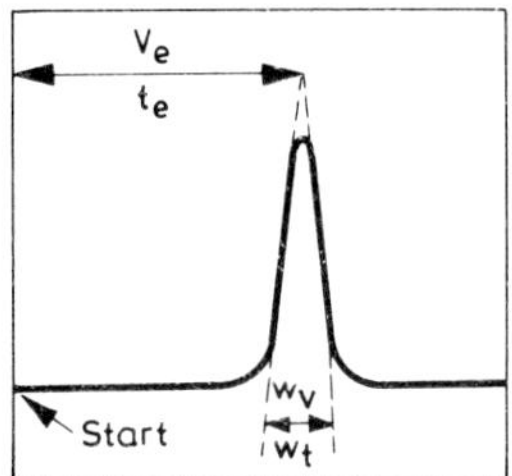

Fig. 6.3. Calculation of number of theoretical plates. Data: elution volume (V_e), elution time (t_e), peak width (w_t or w_V)

of the curve from the start (t_e or V_e) by the standard deviation of the curve (σ_t or σ_V), and squaring the result. The calculation may be performed with either time or volume units. If the peak width measured at the baseline (w_t or w_V) is taken instead of the standard deviation, the value of the square must be multiplied by 16:

$$N = \left(\frac{t_e}{\sigma_t}\right)^2 = \left(\frac{V_e}{\sigma_V}\right)^2 = 16\left(\frac{t_e}{w_t}\right)^2 = 16\left(\frac{V_e}{w_V}\right)^2$$

The height equivalent to a theoretical plate *(HETP, H)* is given by [22, 25]:

$$H = \frac{L}{N}$$

where L is the length of the column.

Many experimental and theoretical studies have been devoted to improving the HETP, and they have revealed that:

(*i*) the HETP decreases with decreasing sample size;
(*ii*) increasing the elution rate increases the HETP;
(*iii*) increasing the temperature reduces the HETP;
(*iv*) the use of a bed made of uniformly sized particles decreases the HETP.

Calculation of peak separation

Peak separation or resolution, R_s, is a parameter characterizing the degree of chromatographic separation of two peaks. In accordance with Fig. 6.4, it is calculated from

$$R_s = \frac{2(V_2 - V_1)}{w_1 + w_2} = \frac{2(t_2 - t_1)}{w_1 + w_2}$$

The value of R_s provides clear-cut information on the separation achieved, but the formula above is not suitable for utilizing the separation attained on a

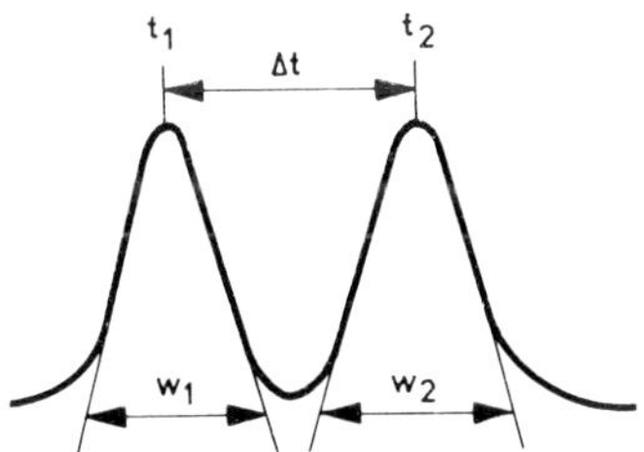

Fig. 6.4. Calculation of peak separation or resolution from difference of elution times ($t_2 - t_1 = \Delta t$) and from peak widths (w_1 and w_2)

certain column to calculate the dimensions (or number of theoretical plates) needed for a column to yield a desired degree of separation. Another formula [26], is needed for this, which contains a further concept, the relative peak distance, m:

$$m = \frac{t_2 - t_1}{t_1}$$

The relation between m and R_s is

$$R_s = \frac{mN^{1/2}}{4}$$

If the relative peak distance of chromatographic curves representing two substances is 0.4, then a column with only about 100 theoretical plates will give $R_s = 1$. If $m = 0.01$, then 16 000 theoretical plates are needed for $R_s = 1$. If $N = 1600$, then $R_s = 1$ for $m = 0.1$. Some authors give the extent of mutual contamination to be found in the various peaks. Thus, for values of R_s of 1.5, 1.25, 1.00, 0.75 and 0.575, the extent of contamination is 0.13%, 0.62%, 2.27%, 6.68% and 12.51% respectively.

In general, a separation is regarded as satisfactory if the R_s is 1.00 or higher.

The peak capacity defined by Giddings [27] is the maximum number of components giving a peak separation of 1.00 in the chromatographic system. However, in Gidding's calculations the maximum value of V_e was restricted to 2.3 V_0. This does not remain valid in the case of oligopeptides and amino-acids since the gel–solute interactions shift the maximum of V_e to well above this value. As $V_e = 2.3\ V_0$ would mean $K_d = 1$, it is worthwhile giving the increased peak capacities for various values of K_d larger than 1. (Table 6.2, based on our own data [28].)

Table 6.2
Peak capacity values calculated by Giddings [27] and Kalász *et al.* [28]

Number of theoretical plates, N, of column	Peak capacity [27], assuming $0 \le K_d \le 1$	Peak capacities [28], assuming			
		$0 \le K_d \le 2$	$0 \le K_d \le 3$	$0 \le K_d \le 4$	$0 \le K_d \le 9$
100	3	5	7	9	14
400	5	9	11	14	24
1000	7	12	16	19	33
2500	11	19	25	31	52
10000	21	35	48	58	99

The fundamental operations of gel chromatography include collection of the effluent in fractions. It is generally sufficient if a chromatographic peak is collected in four fractions, each corresponding to one-quarter of the peak width measured at the baseline (i.e. the standard deviation). In that case the number of eluate fractions from application of the sample until the maximum of the chromatographic peak appears is equal to N [29] (Fig. 6.5).

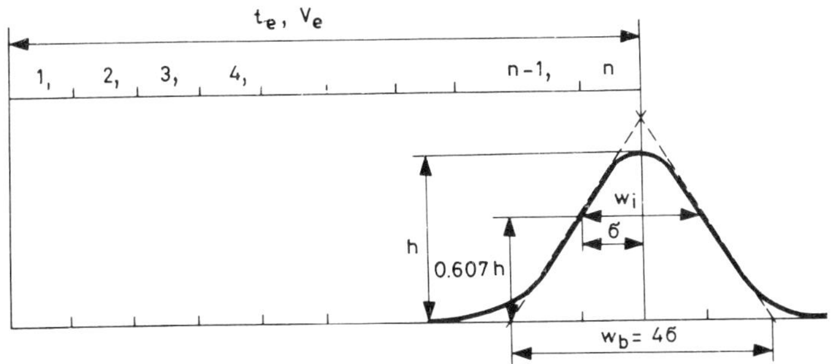

Fig. 6.5. Optimum fractionation. If each fraction represents a standard deviation, i.e. the elution peak is collected in four parts, the number of fractions up to the maximum of the elution peak is V_e/σ_V or t_e/σ_t

Mathematically, the number of fractions to be collected up to the peak maximum, f_r, is given by

$$f_r = \frac{t_e}{\sigma_t} = \frac{V_e}{\sigma_V}$$

but by definition

$$N = \left(\frac{t_e}{\sigma_t}\right)^2 = \left(\frac{V_e}{\sigma_V}\right)^2$$

so

$$f_r = \sqrt{N}$$

Naturally, numerical analysis of the optimum number of fractions and the corresponding peak separation cannot replace the analytical control of the homogeneity of the purified fractions (by chemical or chromatographic analysis).

Determination of sample volume

Fischer [23] states that in analytical separations the volume of the sample may be 1–2% of the total volume of the column, and in preparative separations (in the case of group separations) it should be at most 25–30% of the column volume.

In particular, if the sample is to be limited to less than the separation volume, V_{sep}, defined as the difference in the effluent volume at which the peaks of substances 1 and 2 appear in the eluate, then,

$$V_{sample} < V_{sep} = V_{e_2} - V_{e_1} = (K_{d_2} - K_{d_1})V_i$$

It is clear that the volume of sample to be separated can be increased by:

(*i*) obtaining the largest possible difference in K_d values by selection of the gel [5];

(*ii*) increasing the value of V_i as much as possible by selection of the gel with larger pore size [5, 23].

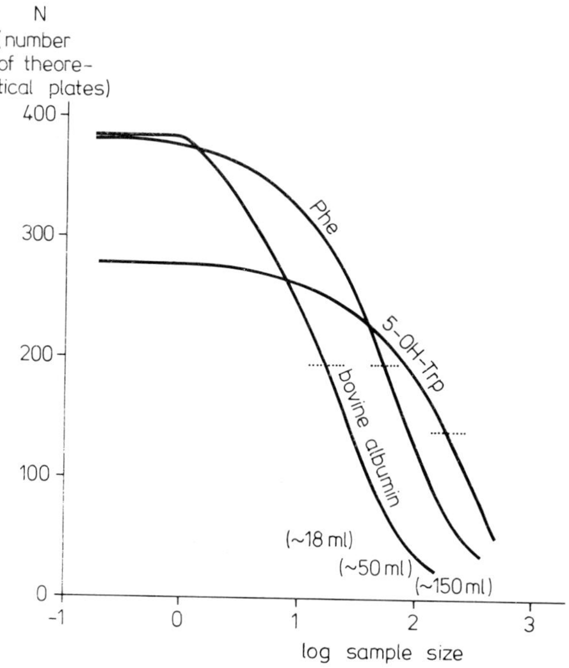

Fig. 6.6. Maximization of sample volume by experimental means. If samples with the same material content, but with different volumes, are separated by gel-chromatography, the plots of number of theoretical plates vs. sample volume are like those in the figure

There are two practical possibilities for deciding on the maximum sample volume. A plot of the number of theoretical plates vs. log of the number of ml of sample gives a sigmoid curve, and the maximum sample volume is taken either as the value of the inflection point of the curve or at the end of the initial horizontal on the graph (Fig. 6.6). The ordinate may also be expressed as a percentage of the maximum number of theoretical plates [29].

The other method is also graphical. If aliquots of the sample are applied, separated from each other with the appropriate amount of eluent, then an elution curve is obtained which is similar to that for a large volume of sample, as in Fig. 6.7.

Effects of sample composition and concentration

In general, the separation is not influenced by the composition of the sample. If the sample has a different pH, or salt concentration or composition from that of the eluent, the medium is exchanged in the first part of the gel bed; although the danger of precipitation may exist (this is foreseeable, however), the chromatographic picture does not usually change. An exception is when there is an extremely large difference between the viscosities of the sample and the eluent.

In principle, it should not cause a problem if the concentration of the sample applied is very high, but in many cases the elution peaks are then broadened or distorted. It is possible to compensate for this and improve the final picture by means of recycling. Figure 6.8 shows that the deterioration in the peak separation caused by applying a tenfold quantity of sample is practically totally compensated for in the second cycle; hence, one recycling stage means a fivefold saving in time [28].

Exploratory group separations and fractionations are performed on small-scale gel columns. There are two ways to increase the efficiency of the separations: either the dimensions of the column (especially the length) are increased, or recycling is employed.

Recycling was first used in gel chromatography by Porath and Bennich to purify human coeruloplasmin. A considerable part of their apparatus is a simple gel chromatographic set-up [30], i.e. eluent reservoir, peristaltic pump, gel column, detector, fraction collector, but flow-switches are inserted between the detector and the fraction collector; by this means, substances eluted from the column may be recycled into the column. The method has since been used widely, in the most varied fields of protein separations. Such a set-up is presented in Fig. 6.9.

Bombaugh [31] derived correlations between the number of cycles (n) and the peak-width (w) or the peak separation, R_0

$$w_n = w_0 n^{1/2} \quad \text{and} \quad R_n = R_0 n^{1/2}$$

where subscripts o and n refer to the first and the nth cycle. The formulae suggest that the resolution increases by a factor of n as the number of cycles increases by a factor of n^2. This is naturally not the case, since the peaks spread more at each cycle and eventually occupy the whole of the column. Also, after a time a more rapidly moving substance will catch up with a more slowly moving one; thus, up to a certain number of cycles the separation will improve but from that number on will deteriorate. The separation is optimum when the peak of the slower-moving substance in the nth cycle is equidistant from the peaks of faster-moving substance in the nth and $(n+1)$th cycles. This is the optimum number of cycles [32], n_{opt}, and can be calculated from the elution times of the individual peaks [26]:

$$n_{opt} = \frac{t_1}{2(t_2 - t_1)} = \frac{1}{2m}$$

where m is the relative peak distance. The separation in the nth cycle, i.e. the maximum separation that can be attained on the column, can also be calculated:

$$R_{opt} = \frac{1}{4\sqrt{2}} m^{1/2} N^{1/2} = 0.177 \sqrt{m} \sqrt{N}$$

If R_{opt} is not sufficiently high, it can be increased by

(*i*) better selection of the gel type;

(*ii*) for small molecules, increasing the value of m by more appropriate choice of the conditions (temperature, pH, ionic strength, etc.);

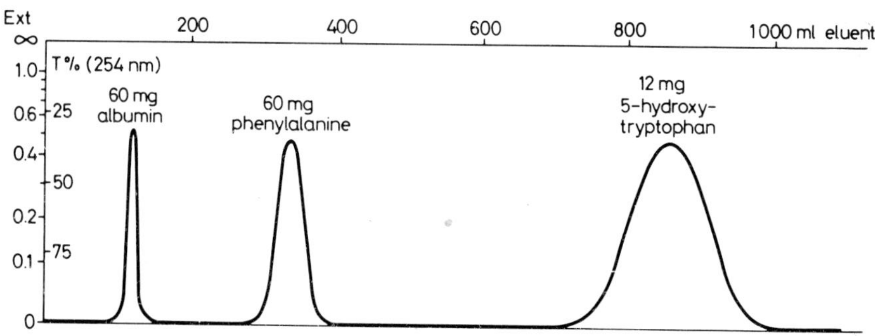

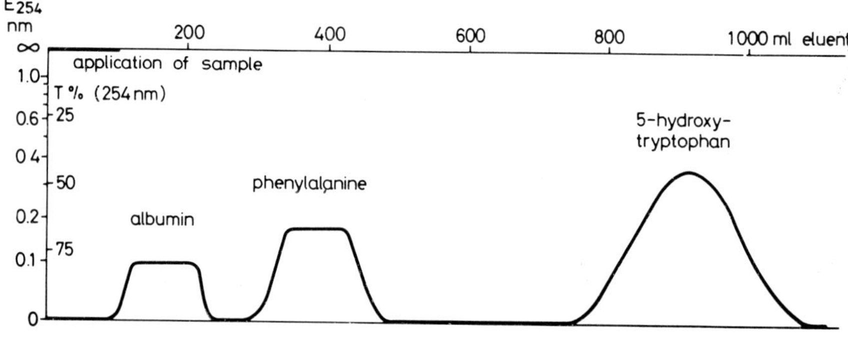

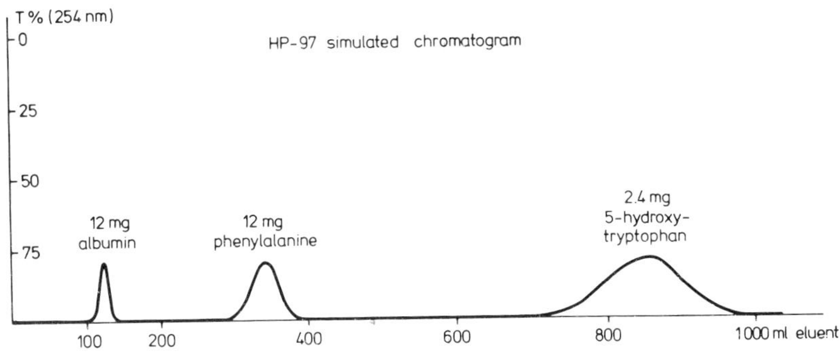

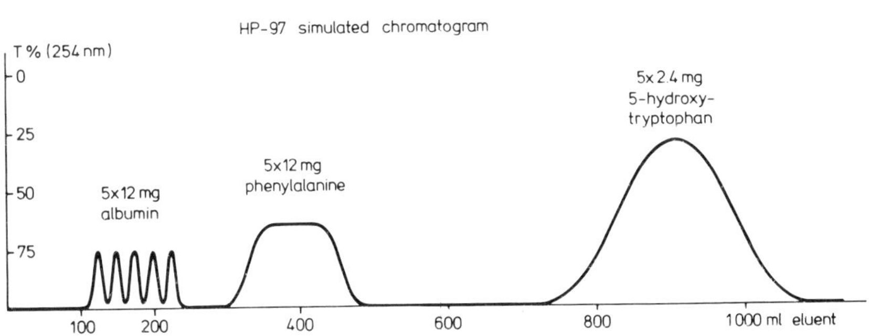

Fig. 6.7. Identification of limiting sample volume by graphical means. (a) 60 mg of albumin, 60 mg of phenylalanine and 12 mg of 5-hydroxytryptophan in a sample of 4 ml; (b) 60 mg of albumin, 60 mg of phenylalanine and 12 mg of 5-hydroxytryptophan in a sample of 100 ml; (c) 12 mg of albumin, 12 mg of phenylalanine and 2.4 mg of 5-hydroxytryptophan in a sample of 0.8 ml; (d) separation of quantities as in (c), but with five samples applied one after the other, each separated by 24 ml of eluent

(*iii*) increasing N by increasing the dimensions of the column, using finer gel particles, selecting a lower flow-rate, decreasing the dead-space in the separation system, and possible by refilling the column;

(*iv*) using a two-column recycling system such as that in Fig. 6.10. The separation that can be achieved with such a two-column set-up is presented in Fig. 6.11 [33].

The peak capacity, defined above, appears as a peak-cycle capacity in recycling. It is the number of peaks, k, that can still be separated with a peak resolution $R_s = 1$ after n cycles on a column with N theoretical plates:

$$n = \frac{N}{16k^2} \quad \text{or} \quad k = \frac{\sqrt{N}}{4\sqrt{n}}$$

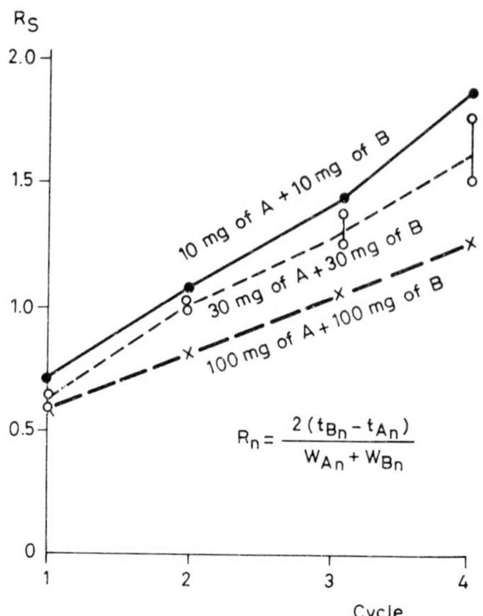

Fig. 6.8. Focusing of peak separation by recycling [28]. The peak separation decreases if the material content of the sample increases. This is demonstrated by the three recycling experiments: the peak separation vs. cycle number for 100 mg each of substances A and B are shown by the bottom curve (– – –), for 30 mg of each by the middle curve (------), and for 10 mg of each by the top curve (——). (By kind permission of the publisher.)

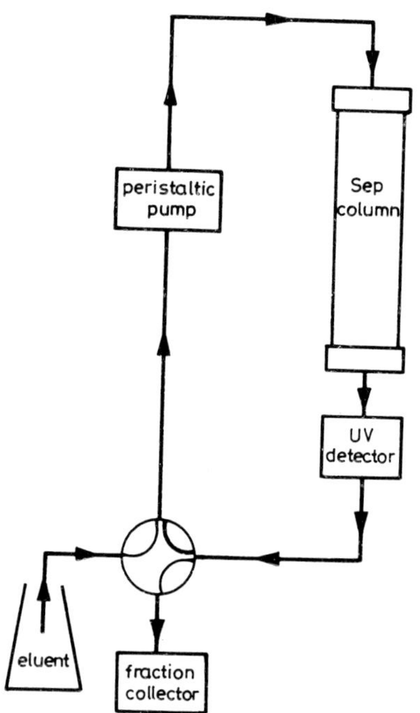

Fig. 6.9. Simple one-column recycling set-up

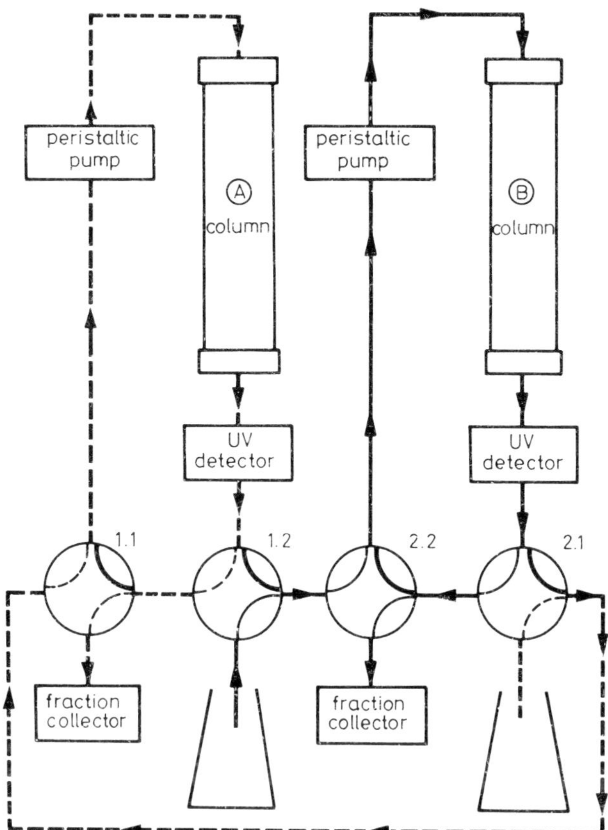

Fig. 6.10. Two-column gel chromatography set-up for isolation of two components to be purified by recycling gel chromatography [33]. The gel-chromatographic separation proceeds in columns A and B, with recycling in the individual columns by means of the flow switches 1.1 and 2.2 in the position (——). Flow switches 1.2 and 2.1 are used to switch over from column A to column B, and from column B to column A, respectively. (By kind permission of LKB Produkter AB.)

The peak capacity, defined above, appears as a peak-cycle capacity in recycling. It is the number of peaks, k, that can still be separated with a peak resolution $R_s = 1$ after n cycles on a column with N theoretical plates:

$$n = \frac{N}{16k^2} \quad \text{or} \quad k = \frac{\sqrt{N}}{4\sqrt{n}}$$

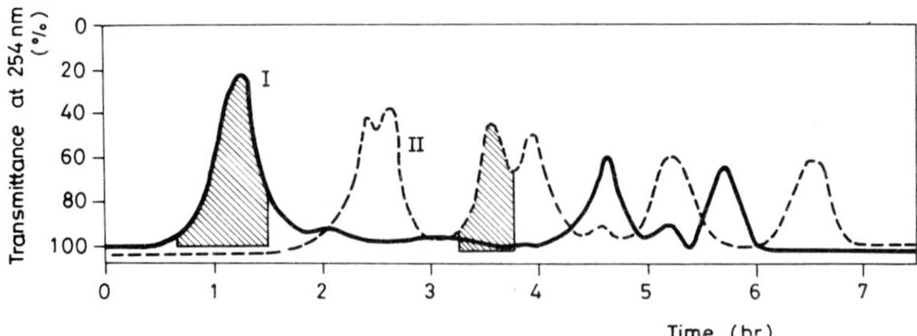

Fig. 6.11. Purification of cellulin-A by recycling gel-chromatography [33]. Recycling was performed with the apparatus set-up shown in Fig. 6.10 [33]. The solution of the crude product displaying a specific biological effect was transferred to column A, and after a single gel-chromatography run the main component (first shaded part) was transferred to column B, where it was cycled twice. After the second cycle, the first component of the partially-separated pair of substances was returned to column A, while the second was left in column B, and recycling was performed in both columns. The continuous line (———) and the dashed line (-------) show the signals obtained on the recorders of columns A and B, respectively. The detector was a Uvicord II, with reading at 254 nm. (By kind permission of LKB Produkter AB.)

6.2 PRELIMINARIES AND METHODOLOGY IN GEL CHROMATOGRAPHIC EXPERIMENTS

6.2.1 Preparation of the gel

Both Sephadex G and Bio-Gel P are distributed commercially as dry powders. Before the gel is packed into a column, it must be swollen with a solution corresponding to the eluent. For practical reasons, the gel is added to constantly stirred eluent solution.

Sepharose B, Bio-Gel A, Sephacryl S and Ultragel AcA are commercially available in swollen form, and do not need pretreatment.

The swollen gels may also be stored in the wet state; they can be protected against microbial decomposition with a 0.05–0.1% solution of sodium azide.

If they are not expected to be used again within several months, Sephadex G and Bio-Gel P may be dried. The gel is washed with water, with aqueous ethanol and finally with ethanol, and dried in vacuum at 60–80 °C.

6.2.2 Packing the column

The packing of the gel bed is the most important preliminary step in gel chromatography. Since a gel column is normally used 10–50 times without being repacked, any resulting errors will appear that number of times, spoiling the chromatographic results and limiting the separation.

The first operation is to fill the bottom of the carefully cleaned tube, which contains a bed support (glass wool, polyamide cloth, porous plastic, etc.), normally by upward flow of buffer, until about one-tenth of the tube is filled. The dead-volume under the support can be filled with glass beads to minimize dead-volume and prevent mixing of the eluted zones. Bubbles of air adhering to the tube and bed support are removed by continuous shaking as the liquid flows in.

Next, the column is filled with 1 : 1 ratio gel–eluent suspension and the tube is kept closed for about 10 min to allow the coarser gel particles to sediment out, thereby improving the seal provided by the bed support. The sedimentation is then continued by passage of eluent through the column. With Sephadex G-10–G-50 and Bio-Gel P-2–P-30 gels, the flow rate under gravity is adequate, but with gels of higher number the appropriate pressure must be applied (see Fig. 6.2).

6.2.3 Avoidance of contamination of the gel; purification of contaminated gel

To avoid contamination of the gel, any insoluble matter in the sample should be removed by centrifugation before the gel chromatography.

In properly performed gel chromatography, the vast majority of the separated substances are completely eluted from the gel column, i.e. gel chromatography is a self-regenerating process, in contrast to most adsorption, ion-exchange and partition chromatographic procedures.

In experiments involving distilled water, retardation and zone spreading may occur, especially by binding of basic substances. These may be desorbed from the gel with 1–2% sodium chloride or calcium chloride solution.

Certain lipoproteins may be precipitated on the gel. Margolis and Langdon [34] report that these can be removed with 0.5 *M* sodium chloride +0.1 *M* EDTA +1.0 *M* potassium phosphate.

6.3 MAIN APPLICATIONS OF GEL CHROMATOGRAPHY

Gel chromatographic separations are carried out for both analytical and preparative purposes.

Analytical investigations are performed to establish the sizes of proteins and peptide molecules, to examine the nature of the binding and of the complex in the case of the coupling of molecules to proteins or peptides, and for qualitative and quantitative studies on complex mixtures of proteins, peptides and amino-acids.

Preparative separations include group separation and preparative fractionation, and especially fractionation of peptides and amino-acids.

There are two main reasons for differentiating between the characteristics of preparative and analytical separations.

(a) The sensitivity of the detectors currently available is such that in general a larger amount of material (10–1000 mg of protein) is necessary for monitoring of a preparative separation.

(b) In an analytical separation the components produced are pure and are very rarely discarded, since it is both practical and customary to employ other methods in conjunction with analytical gel chromatography. An example is presented below in connection with the study of sera of various origins, the separation serving both preparative and analytical purposes simultaneously.

6.3.1 Gel chromatographic measurement of molecular weight

Both the planar technique (thin-layer gel chromatography) and the column method are employed. The high-pressure column procedure for determination of molecular weight is detailed later.

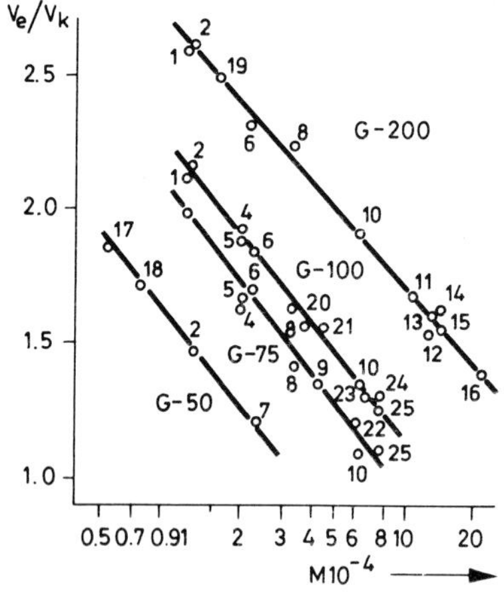

Fig. 6.12. Gel-chromatography molecular weight measurement. Elution parameter *vs.* molecular weight for various Sephadex gels [48a]. 1, Kallikrein inhibitor; 2, trypsin inhibitor; 3, cytochrome, C; 4, ribonuclease A; 5, methaemoglobin; 6, myoglobin; 7, trypsin inhibitor (soya); 8, chymotrypsin; 9, trypsin; 10, chymotrypsinogen A; 11, pepsin; 12, peroxidase-1; 13, egg-white albumin; 14, HO-steroid-dehydrogenase; 15, phosphoglycero-mutase; 16, serum albumin, monomer (bovine); 17, malic acid dehydrogenase; 18, enolase; 19, creatine-phosphokinase; 20, phosphoglycerinaldehyde-dehydrogenase; 21, serum albumin, dimer (bovine); 22, γ-globulin (human); 23, aldolase (fungal); 24, alcohol-dehydrogenase; 25, catalase. (By kind permission of the authors and of the copyright owner.)

The use of gel chromatography for the measurement of molecular size has the following advantages.

(a) With suitable types of gels, the method can be used for a wide range of molecular weight, as indicated in Fig. 6.12.

(b) If the substance under examination possesses some specific property (enzymatic activity, metal ion content, specific biological activity) and a sensitive method is available for the detection of this, then it may be possible to calculate the molecular weight of even a very small amount of the substance in the presence of a large excess of contaminant, without any special preliminary purification.

(c) Molecular weights and molecular weight distributions may also be obtained.

(d) Both column and thin-layer techniques can be used.

(e) The results are not affected by the temperature.

(f) The apparatus is relatively cheap, and the procedure rapid and simple.

The following should be considered in order to avoid errors.

(a) The lower limit of reliable molecular weight measurement is 1000–3000. For molecular weights below 1000, the elution characteristics are determined by the gel–solute interactions [28]. The addition of guanidine hydrochloride or urea merely decreases, and does not eliminate, these phenomena.

(b) For molecular weights below 10^4, it is recommended that the measurements be checked by ultrafiltration, etc. It is also worthwhile stating the temperature at which the molecular weight was determined.

(c) Only molecules of the same type can be compared. In the preparation of the calibration curve, the standard compounds used must be of the same type as the sample substances. If globular proteins are to be studied, peptides of a fibrillar nature and glucose polymers should not be employed for the calibration. The various types of protein may also differ in behaviour: the glycoproteins or the lipoproteins may behave differently from the proteins consisting only of amino-acids.

(d) A concentration of urea or guanidine hydrochloride between 3 and 6 *M* in the eluent generally eliminates the association of proteins and reduces the gel-solute interactions.

(e) In the measurement, attention must be paid to the possibility of compaction of the gel column. Only an appropriately 'pre-eluted' column should be utilized.

(f) In the calculations and evaluation of the measurements it must be taken into account that most molecular weights in the literature are based on measurement of the Stokes radius of the molecule. This does not cause any problem as a rule, as the Stokes radius is generally proportional to the molecular weight; only rarely is there a substantial difference. Siegel and Monty [35] describe anomalous behaviour in the cases of fibrinogen and ferritin, where the gel chromatographic behaviour of the molecules is characteristic only of the Stokes radius, and does not give the molecular weight.

Table 6.3
Equations used in gel-chromatography measurement of molecular weights

Compound	Eluent	Equation	Ref.
Proteins	aqueous	$K_d^{1/3} = k_1 - k_2 M^{1/2}$ (E)	[38]
Proteins	aqueous	$V_e/V_0 = k \log M$ (E)	[39]
Proteins	aqueous	$V_e = k \log M$ (E)	[40]
Proteins	aqueous	$K_d = f(M^{1/3}/k)$ (T)	[41]
Proteins	aqueous	$(V_e/V_0)^{1/3} = k_1 - k_2 M^{1/3}$ (T)	[42]
Oligopeptides	phenol-acetic acid-water	$K_d^{1/3} = k_1 - k_2 M^{1/2}$ (E)	[43]

(E) = empirical equation
(T) = theoretically-derived equation
k, k_1, k_2 = constants
M = molecular weight

The haemoglobin family can be regarded as a general exception. The gel chromatographic behaviour of human haemoglobin, for example, suggests a molecular weight not of 6.45×10^4 but of $(1.5–1.8) \times 10^4$ [23, 36].

Equations of use in gel chromatographic measurement of the molecular weights of proteins are given in Table 6.3. Some of the equations were compared by Siegel and Monty [35, 44], while Anderson and Stoddard [45] compared the results of Laurent and Killander [46] and of Ackers and Thompson [47]. Molecular weight measurements made in different laboratories on the basis of plots of V_e/V_0 vs. log of the molecular weight (see Fig. 6.12) were compared by Determann and Michel [48].

A method is given by Determann and Michel [48] which allows determination of the molecular weight of the substance in question by numerical calculation from the equation for a straight line, without calibration measurement, if one of the gels mentioned is used and V_e and V_0 are known.

Selectivity curves for Sephacryl S-200 Superfine can be seen in Fig. 6.13.

Molecular weight calibration of Bio-Gel A-5m is illustrated in Fig. 6.14. The eluent was 6 *M* guanidine hydrochloride and 0.1 *M* mercaptoethanol [49].

Pusztai and Watt [50] carried out experiments on Bio-Gel P-100, the eluent being a 1 : 1 : 1 mixture of phenol, acetic acid and water. If the range of molecular weight measurement is taken into consideration ($10^3 - 6.5 \times 10^4$), the individual points fit the straight line fairly well.

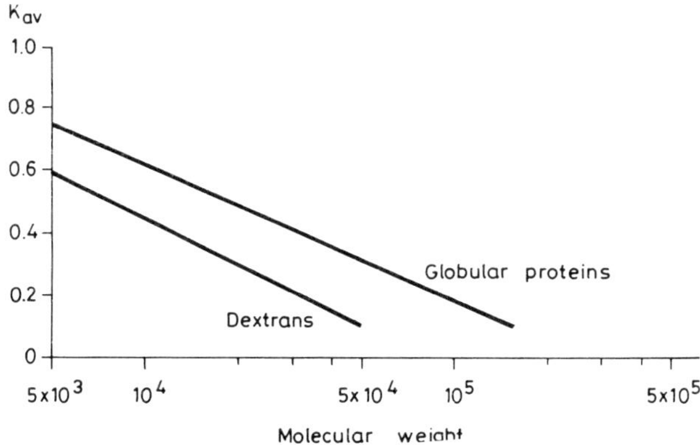

Fig. 6.13. Selectivity curves for Sephacryl S-200 Superfine (reproduced with kind permission of Pharmacia Fine Chemicals, Uppsala, Sweden)

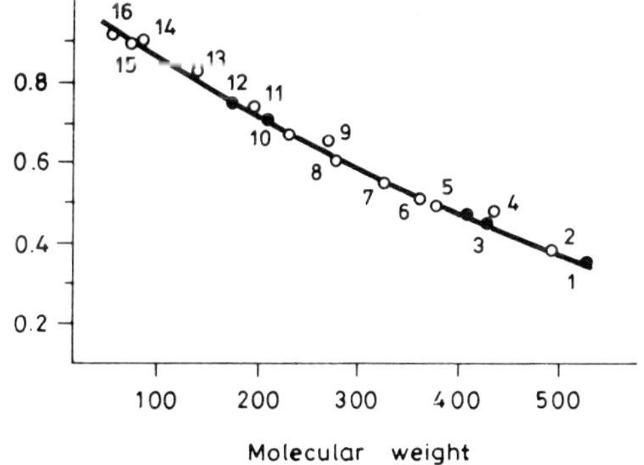

Fig. 6.14. Curve for establishment of molecular weight on Bio-Gel A-5m [49]. 1, Transferrin; 2, bovine serum albumin; 3, Taka-amylase; 4, γ-globulin (heavy); 5, aldolase; 6. MDH: 7, chymotrypsinogen; 8, γ-globulin (light); 9, β-lactoglobulin; 10, haemoglobin; 11, lysozyme; 12, cytochrome, C; 13, cytochrome C I; 14, insulin; 15, cytochrome C II; 16, cytochrome C III. (By kind permission of the author and of The American Society of Biological Chemists, Inc.)

Fish *et al.* [49] plotted both the V_e/V_o and the K_d values against the log of molecular weight, but also evaluated the possibilities of utilizing $K_d^{1/3}$. They obtained a sigmoid curve; in the linear part of this calibration curve, in the molecular weight range ($1 \times 10^4 - 4 \times 10^4$) the molecular weight could be determined from the chromatographic data within an error of 7%. Their experiments were performed on Bio-Gel P in 6 *M* guanidine hydrochloride, 0.1 *M* mercaptoethanol.

In many cases the measurement of molecular weight by gel chromatography reveals not only the pure protein monomer, but also its dimeric, trimeric, tetrameric forms. For instance, Pedersen [51] demonstrated the dimer, trimer and tetramer in addition to albumin itself, while Siegel and Monty [44] detected associated forms of urease. The quantitative investigation of the reversible association of proteins is also possible by gel chromatography.

6.3.2 Gel chromatographic study of association of protein molecules

It is characteristic of the gel chromatographic study of the association of macromolecules that the equilibrium can be investigated between wide limits of pH, protein concentration, electrolyte concentration and temperature. A further advantage is that the substances obtained in an analytical determination can be isolated.

A study on reversible enzyme aggregations is presented by Kakiuchi *et al.* [52]. In a medium containing zinc, the equilibrium between the zinc-free amylase monomer and the zinc-containing amylase dimer develops in accordance with the zinc and amylase concentrations. If the dimer peak fraction is rechromatographed in a medium that is zinc-free but contains EDTA, a proportion of the dimer molecule dissociates and two peaks are again obtained.

The method of Sheperd and Petersen [53] uses mathematical analysis to determine not only the molecular weight of monomer, but also its association equilibrium constant, when only a minimal amount of monomer is detected in the presence of the associated form, even as a tail of the latter.

Examples of investigation of the mode of binding are given by authors dealing with the linkage between haemoglobin and haptoglobin. This question is of importance not only from the aspect of gel chromatography but also with regard to physiology and diagnostics, since haptoglobin, one of the α_2-globins of the plasma, participates in the regulation of the free haemoglobin level by means of the physiological binding of haemoglobin. On a Sephadex G-100 column, free haemoglobin is eluted much later than albumin, which has a similar molecular weight, whereas the corresponding complex, with a molecular weight of 1.7×10^5 is eluted much earlier. Hence, the complex can be separated fairly easily from haemoglobin. The determination was originally described by Lionetti *et al.* [54], but the problem was first discussed in depth by Ratcliff and Hardwicke [55]. The procedure was standardized by Hodgson and Sewell [56], whose method is suitable for the rapid determination of haemoglobin and the complex.

Separation of the haemoglobin-haptoglobin complex from haptoglobin is a more involved task. It was achieved by Killander *et al.* [57] by recycling gel chromatography.

The binding formed between insulin and proteins has been studied by Manypol and Spitzy [58]. Insulin treated with radioactive iodine was incubated with plasma, and then chromatographed on a Sephadex G-75 column. The radioactivity was measured in the fraction eluted at V_0. In the method developed by Rivera *et al.* [59], not only the insulin-binding capacities of the proteins, but also the total protein binding were investigated. Examination of complexes of insulin antibodies and insulin was described by Toro-Goyco *et al.* [60] and by Ceska [61], and examination of insulin-binding proteins by Gjedde [62] and by Pearson and Martin [63]. Insulin–protein complexes [64] and insulin receptors [65] have also been studied by gel chromatography. Dixon *et al.* [66] reported on insulin–serum protein investigations in a case of diabetes.

Examination of the binding between small molecules and serum proteins is widespread; the Hummel-Dryer method [67] may be used to measure the equilibrium constant of such complexes. The gel column is brought into equilibrium with the eluent to be used in the measurement, which thus has the appropriate pH and necessary ion concentration. One of the reactants (A) is contained in the eluent, while the other (B) is included in the sample (the pH and ion concentration of which are naturally the same as those of the eluent, and the concentrations of A and B are equivalent). As the zone of B passes down the column, it reacts with A to form the complex AB. If the zones of B and AB proceed at different rates, B always encounters fresh A, and continues to react until the equilibrium state has been established. In this state there is no longer any change in the amounts of A, B and AB. After elution these components may be determined separately.

Figure 6.15 illustrates an investigation of the cytidylic acid binding of ribonuclease. A Sephadex G-25 column was used with 9×10^{-5} *M* cytidylic acid in 0.1*M* acetate buffer as eluent, i.e. cytidylic acid was component A. Component B, ribonuclease, was added together with the sample. The position of the maximum indicates elution of the complex. The method is also applicable to competition studies, e.g. between acetyltryptophan and acetyltryptophanamide [69].

Other work has concerned the ability of proteins to bind drugs, drug metabolites, metal ions and dyestuffs. As the pH, the ionic strength and the temperature conditions may be varied over wide ranges, it is also possible to measure the constants. Spitzy *et al.* [70] detected binding between ionic

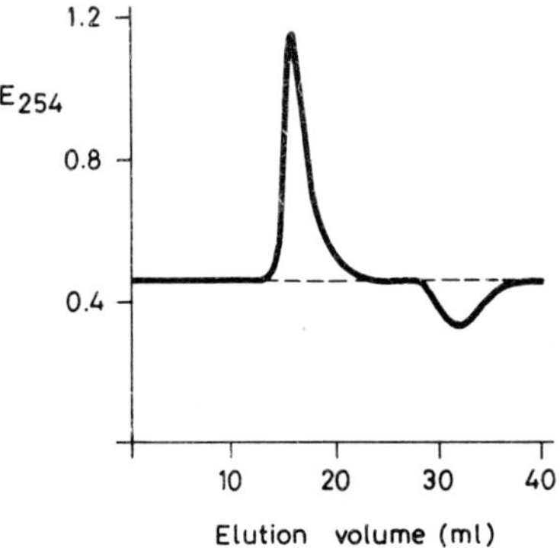

Fig. 6.15. Study of the cytidylic acid binding of ribonuclease [67]. (By kind permission of the author and of the Elsevier/North-Holland Biochemical Press.)

iodine and protein in blood plasma. Lissitzky *et al.* [71] performed similar work, and determined not only the bound but also the free iodine. Shapiro and Rabinowitz [72] and Ceska *et al.* [73] carried out experiments with radioactive iodine, thereby facilitating the determination considerably.

6.3.3 Preparative group separations

Two fundamental aspects must be taken into consideration in the selection of the gel. Large molecules must be excluded, and at the same time, because of the scale of working, and the comparatively high flow rate, mechanical rigidity of the gel is also essential. Hence Sephadex G-10, G-15, G-25 and G-50 and Bio-Gel P-2 P-4, P-6 and P-10 are generally used for group separations. According to the recommendations of the manufacturers, Sephadex G-25 and Bio-Gel P-6 are particularly suitable for large scale group separations.

The volume of sample may be at most 25–30% of the volume of the gel bed, i.e. the column must have a volume 3–4 times that of the sample.

The rate of elution is high; in general 10–20 cm/h. This means that for a column 20–50 cm long the desalting takes 3 h, and the column is regenerated within 6 h. The gel must be swollen and brought into equilibrium with the medium in which the substance is to be isolated (distilled water, a solution of volatile buffer, etc.).

Samuelsson *et al.* [74] have published a method for the production of protein-enriched milk on an industrial scale. They used three Sephadex G-25 coarse gel columns, at 5–10 °C. In this way, practically continuous operation can be achieved, i.e. the material to be purified is applied in turn to one or other of the columns, and the salt-, fat-, and sugar-free milk proteins are eluted continuously. Protein-rich milk may also be produced by evaporation, but the evaporation also increases the quantities of lactose, fat and salts, whereas in the gel chromatographic procedure the small molecules are removed from the milk in parallel with concentration of the proteins.

In other processes of industrial importance, pyrogenic substances are removed from serum [75] or from penicillin [76]. In the case of the latter the protein is added to the crude preparation of penicillin; allergens are obtained, which are separated from the penicillin by gel chromatography.

The desalting is most generally done following the ion-exchange or affinity chromatographic purification of peptides and proteins. Elution from the ion-exchange or affinity column is performed with a fairly high concentration of small molecules, and the removal of these is necessary as the final step of the preparation. This can be done by dialysis or ultrafiltration, but gel chromatography is also frequently employed.

Group separation is generally applied for the separation of chemically modified proteins and peptides from the free reactant. A short column is often used to remove fluorescent isothiocyanates, small aromatic molecules being strongly retarded on the gel [77]. Dyestuffs remaining (adhering) on the gel are removed from the column by means of their reaction with albumin. The high yield, low dilution and fast separation attainable with gel chromatography are particularly advantageous in radioactive labelling. Wood *et al.* [78] and Numa *et al.* [79] studied the removal of the radioactive reagent from proteins. In their micropreparative procedures, Hunter and Greenwood [80] and Greenwood *et al.* [81] incubated 3–4 μg of growth hormone with radioactive iodine and with chloramine-T for the necessary time. The excess of chloramine-T was then destroyed by the addition of sodium metabisulphite, and the labelled product was separated from the unreacted iodine and from the small molecules formed, on a 1 × 10 cm Sephadex G-50 column.

In certain cases the group separation serves both preparative and analytical purposes. Thus, Englander [82] investigated the exchange of tritium for hydrogen atoms of ribonuclease. The incubated enzyme was separated from the unbound tritium on a 3 × 6 cm column of Sephadex G-25;

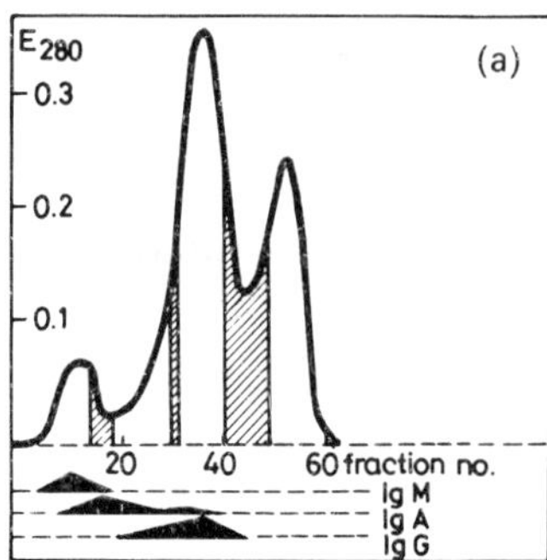
E280
0.3
0.2
0.1
(a)
20
40
60 fraction no.
Ig M
Ig A
Ig G

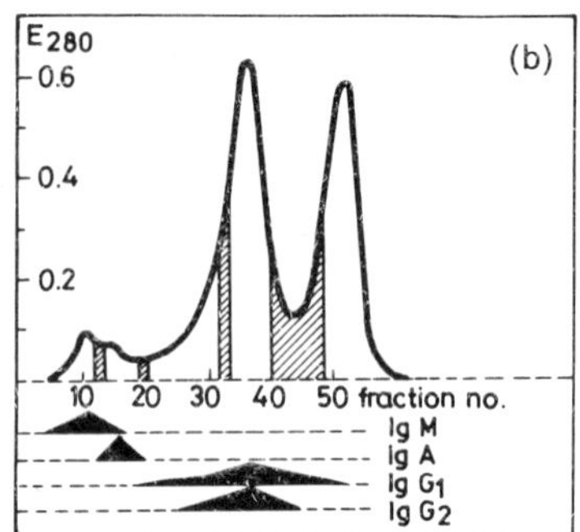
E280
0.6
0.4
0.2
(b)
10 20 30 40 50 fraction no.
Ig M
Ig A
Ig G1
Ig G2

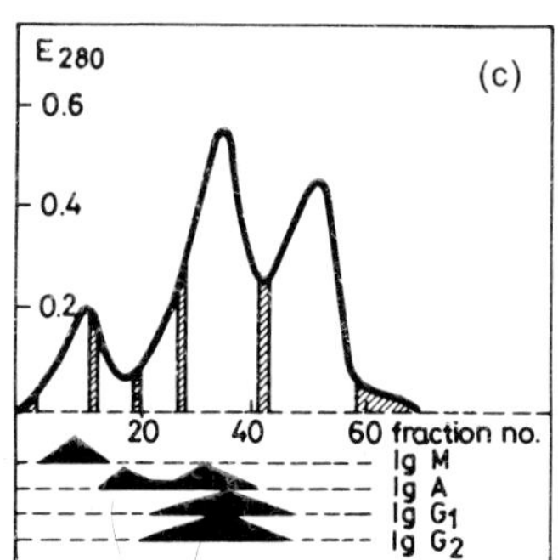
E280
0.6
0.4
0.2
(c)
20
40
60 fraction no.
Ig M
Ig A
Ig G1
Ig G2

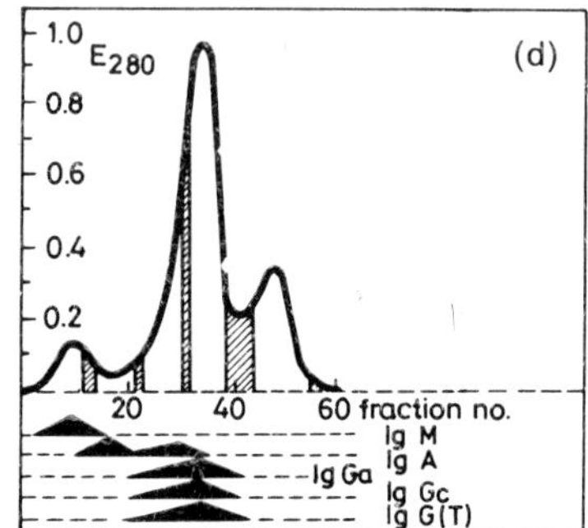
1.0
E280
0.8
0.6
0.4
0.2
(d)
20
40
60 fraction no.
Ig M
Ig A
Ig Ga
Ig Gc
Ig G(T)

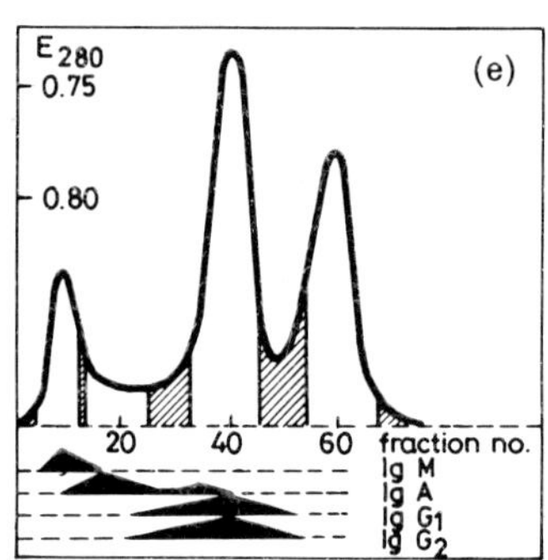
E280
0.75
0.80
(e)
20
40
60
fraction no.
Ig M
Ig A
Ig G1
Ig G2

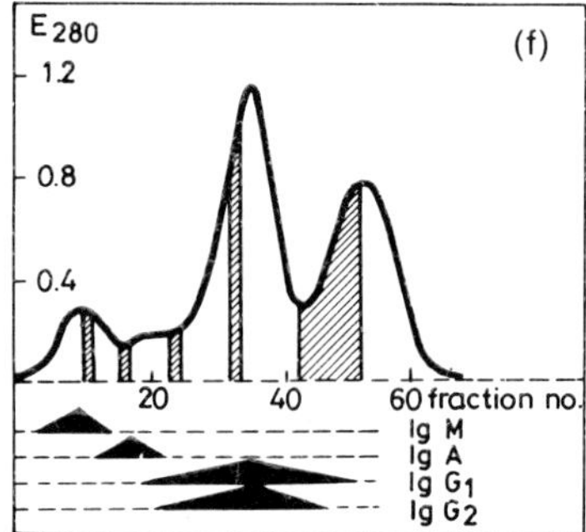
E280
1.2
0.8
0.4
(f)
20
40
60 fraction no.
Ig M
Ig A
Ig G1
Ig G2

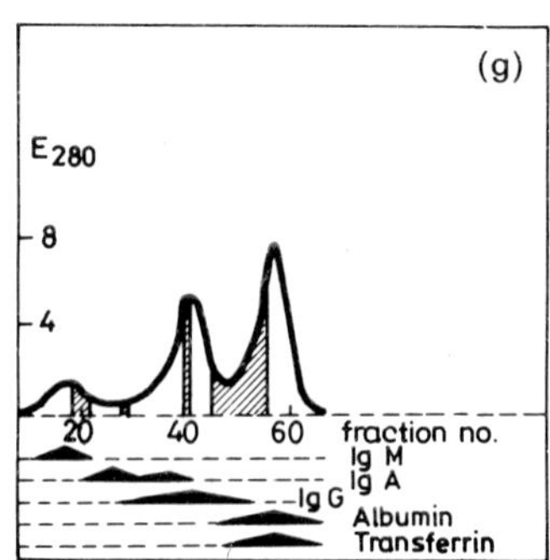
(g)
E280
8
4
20
40
60
fraction no.
Ig M
Ig A
Ig G
Albumin
Transferrin

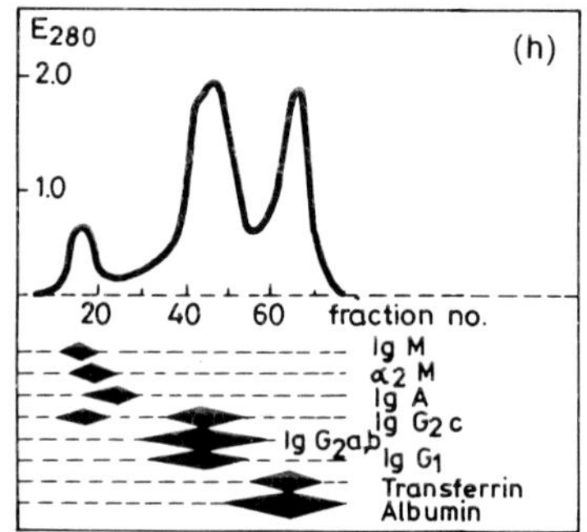
E280
2.0
1.0
(h)
20
40
60
fraction no.
Ig M
α2 M
Ig A
Ig G2c
Ig G2a,b
Ig G1
Transferrin
Albumin

elution was carried out up to half the column height. This was followed by intervals of various durations, which resulted in the splitting off of the bound tritium. The first tritium peak eluted after the ribonuclease, therefore, is the split-off tritium that has undergone exchange with hydrogen during the interval in the elution. With this rapid method, valuable conclusions were drawn on the hydrogen–tritium exchange, its kinetics, and on the structure of the protein helix.

6.3.4 Preparative fractionation of proteins

We speak of preparative fractionation when a mixture, generally containing many components, is to be separated into its constituents by gel chromatography, or when a given constituent of a multicomponent mixture is to be separated in more or less pure form. One of the main fields of application of gel chromatography is the preparative fractionation of proteins, polypeptides or oligopeptides. In addition to those mentioned in the introduction to this chapter, the main advantages of gel chromatography are that the individual fractions from preparative fractionation can be used directly for analysis, and that the pH, the ion concentration, etc. remain constant during the separation.

The review by Vaerman [83] presents gel chromatographic comparisons of cat, sheep, hedgehog, horse, goat, cattle, pig and rabbit sera, and immunoelectrophoretic comparisons of the individual labelled fractions (Fig. 6.16), together with a comparative examination of the immunoglobulins from samples of sheep serum, sheep colostrum and sheep milk (Fig. 6.17).

A new and interesting type of gel chromatography materials is the family of Ultragel AcA gels. These gels contain a three-dimensional network matrix of polyacrylamide with agarose. Following the description by Boschetti *et al.* [7], it became widely known that Ultragel AcA displays an enhanced resistance to the deforming effect of high-rate elution. A number of authors have used Ultragel AcA for separation of human serum proteins. Serum fractions were separated on Ultragel AcA 34 by Ryley and Matthews [84], who observed that the elution profile is influenced by the temperature of fractionation. A considerable difference is to be seen in the experiments carried out at 3 and 24 °C (Fig. 6.18). Separation at 24 °C is more clear-cut,

←

Fig. 6.16. Separation and study of immunoglobulins of serum of various animal species on Sephadex G-200. The individual immunoglobulins were located by immunoelectrophoresis [83]. (a) Cat serum; (b) sheep serum; (c) hedgehog serum; (d) horse serum; (e) goat serum; (f) bovine serum; (g) pig serum; (h) rabbit serum. (By kind permission of the author and of the Imprimerie Sintal.)

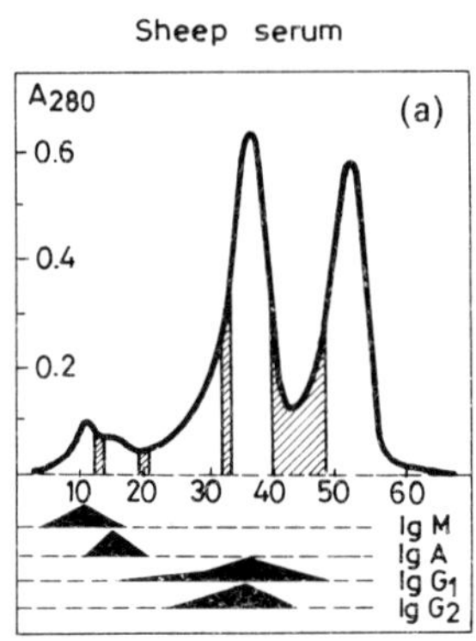

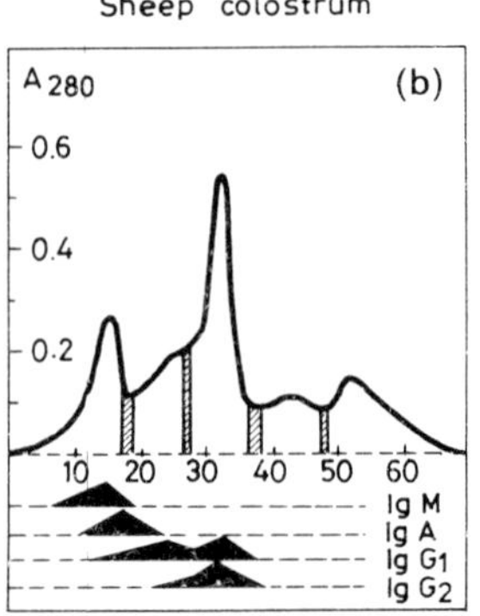

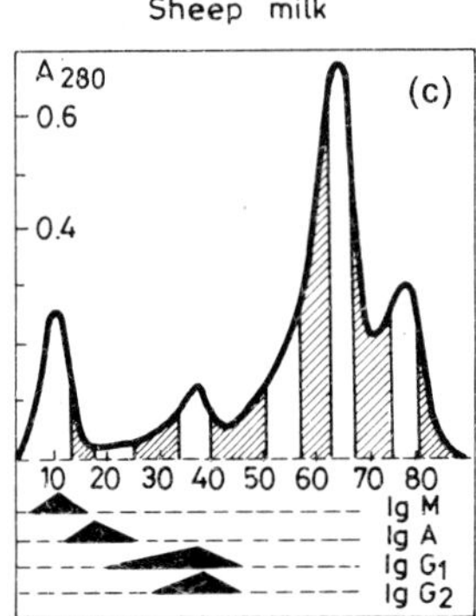

Fig. 6.17. Study of immunoglobulins of sheep. The separations were performed by gel-chromatography on Sephadex G-200. Immunoelectrophoresis was used for identification of the individual immunoglobulins [83]. (By kind permission of the author and of the Imprimerie Sintal.)

but further elevation of the temperature is not permitted by the stability conditions.

Human IgG fragments (fragments F_c and F_{ab}) cleaved with plasmin were chromatographed and rechromatographed on Ultragel AcA 44 by Skvaril and Theilkaes [85] (Fig. 6.19). The purification was followed with agar-agar immunoelectrophoresis.

Blumenthal and Smith [86] made use of gels to carry out the sixth step in the purification of NADP-dependent glutamate-dehydrogenase (Fig. 6.20).

For the purification of aldolase obtained from *Micrococcus aerogenes*, Lebherz and Rutter [87] employed gel chromatography on a Sephadex G-200 column in the final step of the isolation (Fig. 6.21).

Gel chromatography has also found wide application in the foodstuffs industry. The preparation of sugar-free milk has already been discussed in connection with group separations. Removal of the antinutritional factors from rapeseed is a similar task. With a 6-column gel chromatography pilot-plant set-up (sample volume: 3.8 l; elution rate 6 l/h) Janson [88] achieved practically the same separation as if this had been performed on the analytical scale (Figs. 6.22 and 6.23).

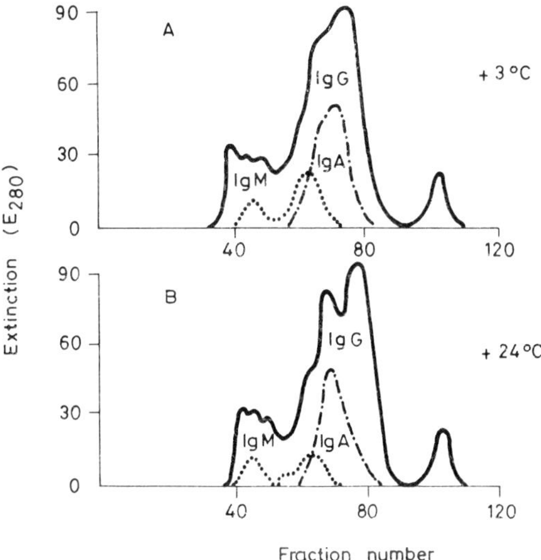

Fig. 6.18. Separation of human serum proteins on an Ultrogel AcA 34 column at 3 and at 24 °C [84]. (By kind permission of the authors and of LKB Produkter AB.)

In the fermentation industries, Raible and Engelhardt [89] have dealt with the components of beer, while Drews and Moeller [90] have studied the catalase in baker's yeast.

The gel chromatographic examination of the protein components of wine has also gained ground in the past few years. In addition to the subjective tasting tests, more and more importance is attached to instrumental and chromatographic investigations on these components. Proteins and their hydrolysis products, including peptides and amino-acids, play a determining role in the taste of various wines. Zakow [91] has studied the 'protein spectra' of wines, and Somers and Ziemelis [92, 93] the effect of clarification on wine proteins. They clarified samples of the wine 'Muscat Gordo Blanco' with various amounts of bentonite; the changes in the wine proteins were very

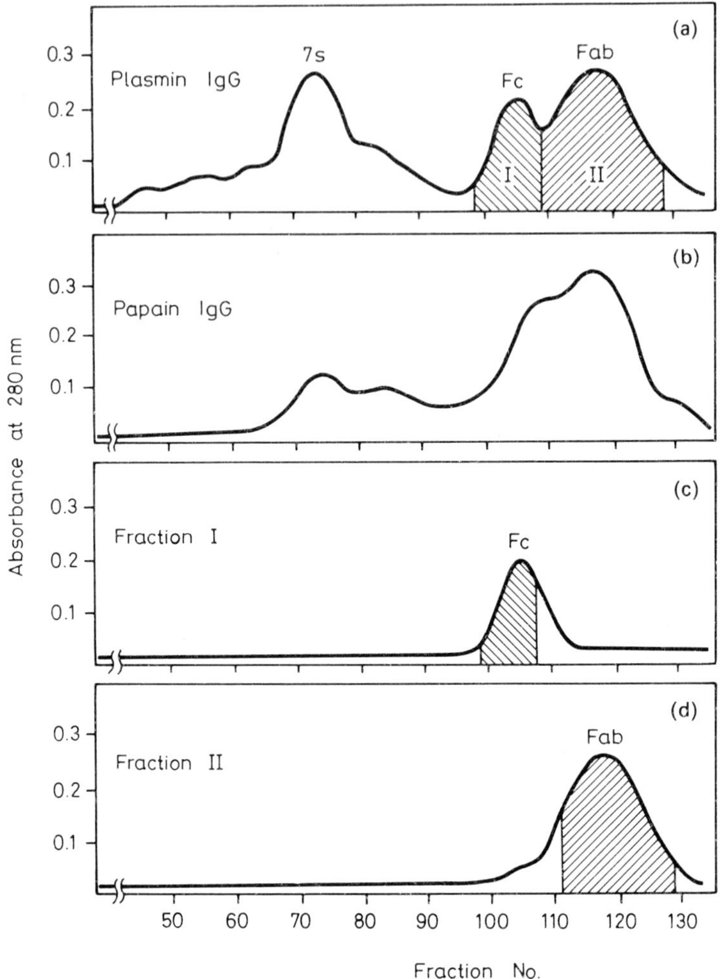

Fig. 6.19. Separation on Ultrogel AcA 44 of fragments F_{ab} and F_c cleaved with plasmin [85]. (By kind permission of the authors and of LKB Produkter AB.)

marked, even with a small quantity of bentonite (0.5 g/l). Wine proteins were studied by Avakjanc [94] on Sephadex G-25, G-50 and G-100. He found two main protein groups with Sephadex G-100, the first of which had a significant fructofuranosidase activity. Wucherpfennig and Franke [95] examined the constituents of must and wine by gel filtration, while Mesrob *et al.* [96] determined the molecular weights of soluble grape-proteins with Sephadex

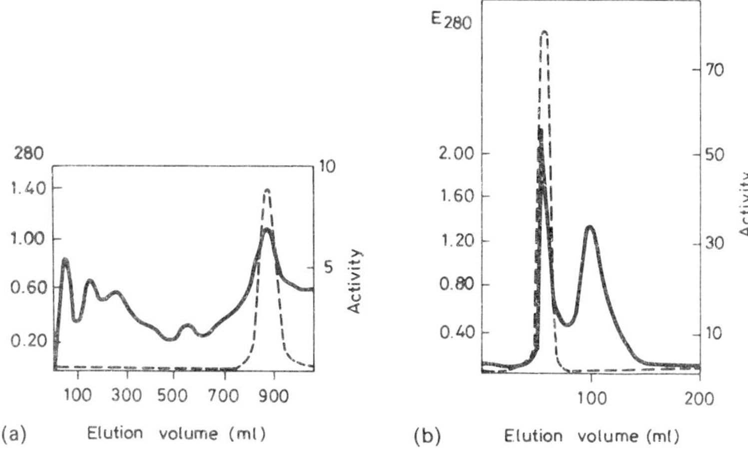

Fig. 6.20. Purification of NADP-dependent glutamate-dehydrogenase (a) on DEAE-Sephadex A-50, (b) by gel chromatography on Sephadex G-100 [86]. (By kind permission of the authors and of The American Society of Biological Chemists, Inc.)

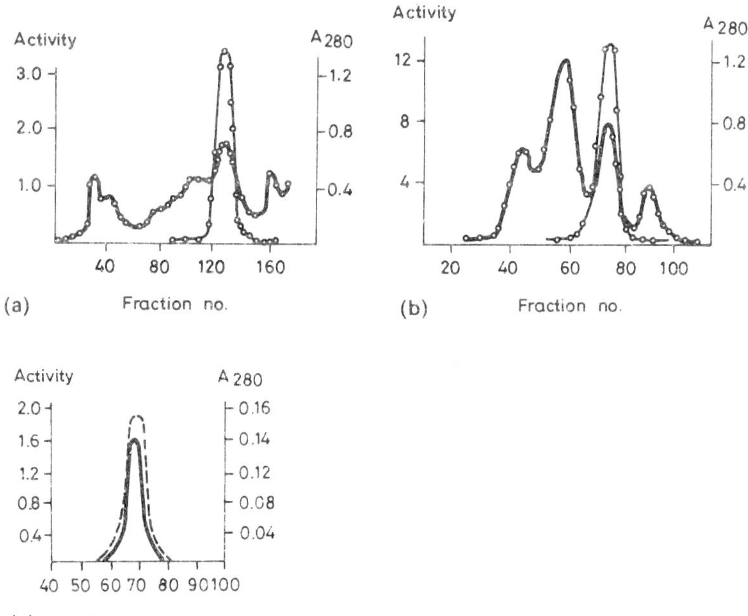

Fig. 6.21. Purification of aldolase obtained from *Micrococcus aerogenes*. The individual steps of the purification: (a) ion-exchange chromatography on DEAE-Sephadex A-50; (b) ion-exchange rechromatography on DEAE-Sephadex A-50; (c) gel chromatography on Sephadex G-200 [87]. (By kind permission of The American Society of Biological Chemists, Inc.)

G-75 and Sephadex G-100 columns, finding values of 1.75×10^4 and 9.5×10^4 by gel filtration. Radola [97] worked with Sephadex G-75 and Bio-Gel P-30. Of the four wine proteins he found, the main component comprised 73% of the total protein content, and had a molecular weight of 1.95×10^4. Feuillat and Bergeret [98] carried out preparative fractionation on Sephadex G-25, G-50 and G-100; the eluent was a 1:1 mixture of acetate buffer (pH 6) and 0.1 *M* sodium chloride. The fractions obtained on Sephadex G-25 were further separated on Sephadex G-100, and (after desalting) the homogeneities of the individual fractions were examined by PAGE.

The main components of the proteins of soya were separated by Okubo and Shibashaki [99] on Sephadex G-200 and a DEAE cellulose column.

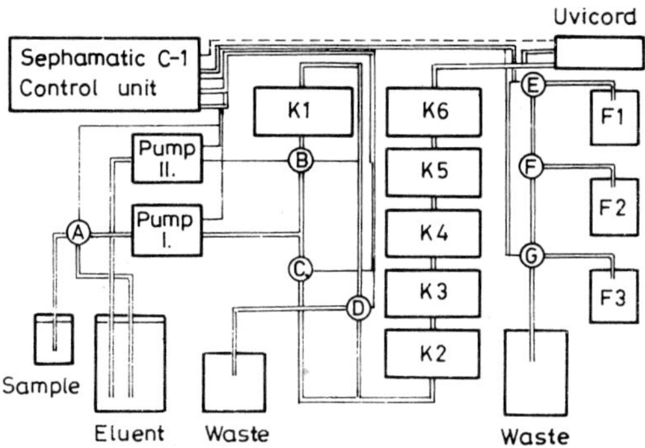

Fig. 6.22. Six-column gel-chromatographic set-up of Janson [88] for pilot-plant scale gel chromatography. (By kind permission of the author. Copyright 1971 American Chemical Soc.)

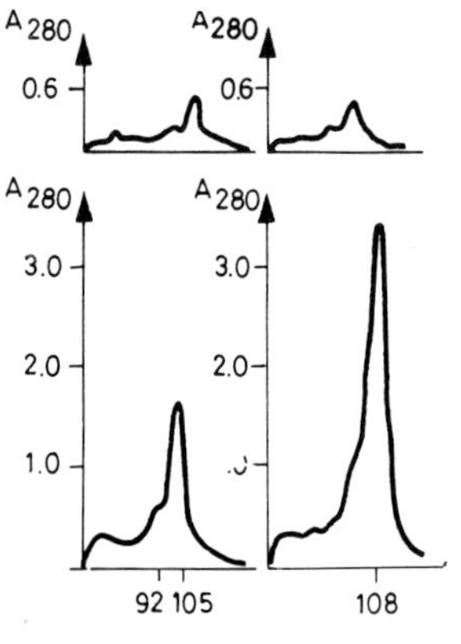

Fig. 6.23. Separation of antinutritional factors of rapeseed by Janson's method [88]. (By kind permisssion of the author. Copyright 1971 American Chemical Society)

Koshiyama [100] separated the 7S and 11S globulins of soya-bean, and investigated the optical characteristics of the purified, soluble proteins.

Arai [101] subjected the proteins of soya to decomposition with proteolytic enzymes, and examined the enzyme fragments. A similar study was reported by Yamashita [102], who utilized Sephadex G gels to 'map' the molecular weights of the fragments.

6.3.5 Gel chromatography of peptides

The gel chromatographic separation of peptides for preparative purposes is of importance for two reasons. It is necessary to discuss the products of partial hydrolysis of proteins, and also the purification of peptide hormones with biological activities. The hydrolysis products of proteins are usually analysed with a view to their structural investigation after partial or complete purification. One important aspect is that the hydrolysis products formed should be as different in size as possible; in this way, one gel chromatographic step permits the separation of 3–8 products. In the purification of peptides, gel chromatography is naturally combined with other methods involving ion-exchange, counter-current distribution, etc. On the basis of the work of Sajgó and Hajós [103, 104], we present the gel chromatography of the peptides obtained on cleavage with trypsin of the CBl fragment resulting from the splitting of rabbit muscle aldolase with cyanogen bromide: their labelling is also illustrated, together with the fitting of the resulting peptides into the amino-acid sequence of the CBl fragment (Figs. 6.24 and 6.25).

Isolation of the hypocalcaemic factor obtained from human saliva is presented as an example of the gel chromatography of peptides with

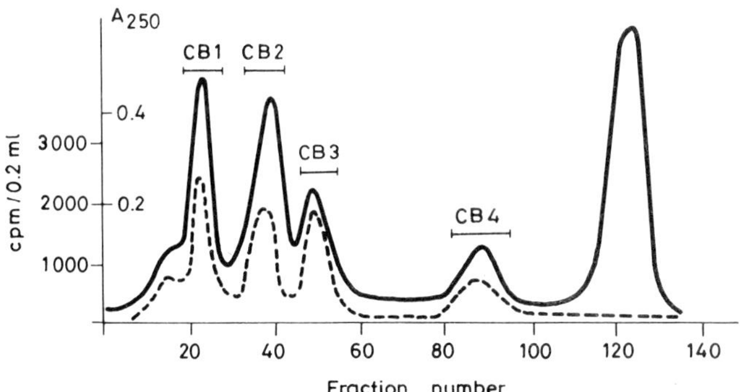

Fig. 6.24. Gel-chromatographic separation of CB1 fragments obtained by BrCN-cleavage on rabbit muscle aldolase [103]. (By kind permission of the author and of the Fed. Europ. Biochem. Soc.)

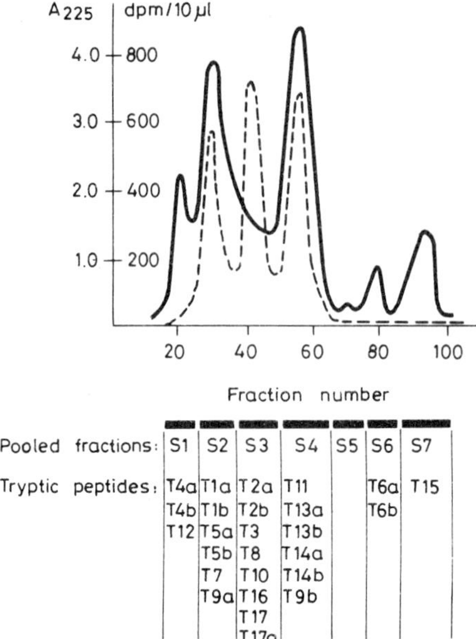

Fig. 6.25. Amino-acid sequence of CB1 fragments [104]. (By kind permission of the authors and of the Fed. Europ. Biochem. Soc.)

biological effects [105] (Fig. 6.26). The molecular weight of the pure polypeptide was examined on Sephadex G-75, and a value of 4260 was found.

A biologically important TRH (thyroid-releasing hormone) analogue has been isolated by Norwegian research workers [106]. The tripeptide was obtained from urine by gel chromatography, and could then be subjected to structural investigation.

The group of substances known as satietins was isolated by Knoll *et al.* [107]. One member of these polypeptides was produced in a pure state. The two gel chromatographic steps following ultrafiltration are aimed at purification, while the purpose of the gel chromatography on Bio-Gel P-30 after ion-exchange chromatography by stepwise elution on a DEAE cellulose column is merely desalting. The gel chromatograms are shown in Fig. 6.27

Nagy, Kenney and Singer purified the phenylhydrazine trimethylamine dehydrogenase adduct from the enzyme inhibitor reaction mixture by gel filtration on a Sephadex G-10 column. The gel filtration followed the centrifugation and precipitation steps and was applied before the separation on DEAE cellulose [108].

Nagy and Salach [109] have recently published the identity of the active site flavine-peptide fragments from the human 'A' form and the bovine 'B'

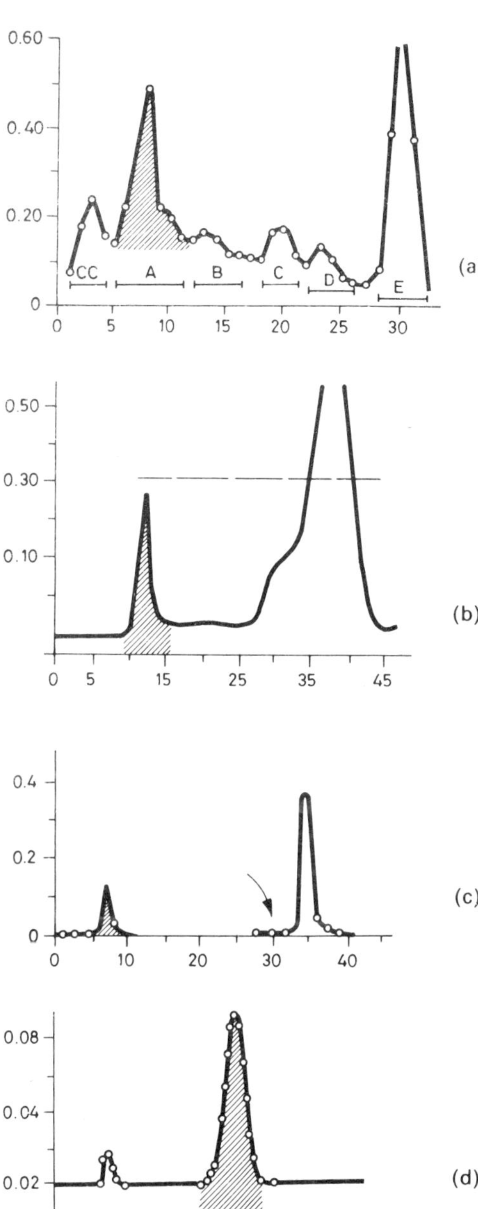

Fig. 6.26. Purification of hypocalcaemic factor (a) by preparative electrophoresis; (b) by gel-chromatography on a Sephadex G-100 column; (c) by ion-exchange chromatography on a DEAE-cellulose column; (d) by gel-chromatography on a Sephadex G-150 column. The shaded parts were purified further [105]. (By kind permission of M. Dekker, Inc.)

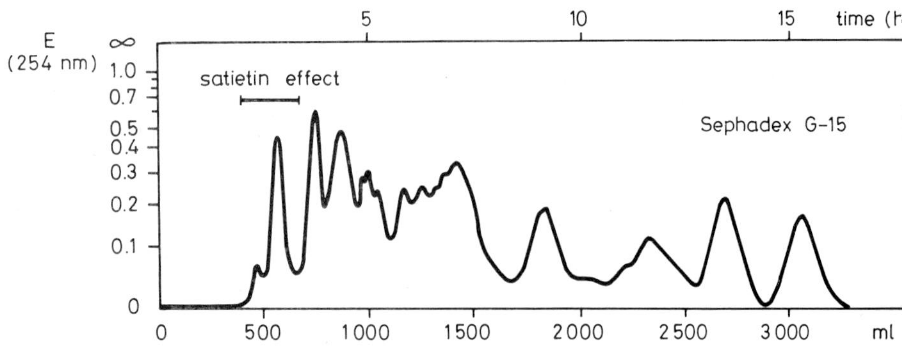

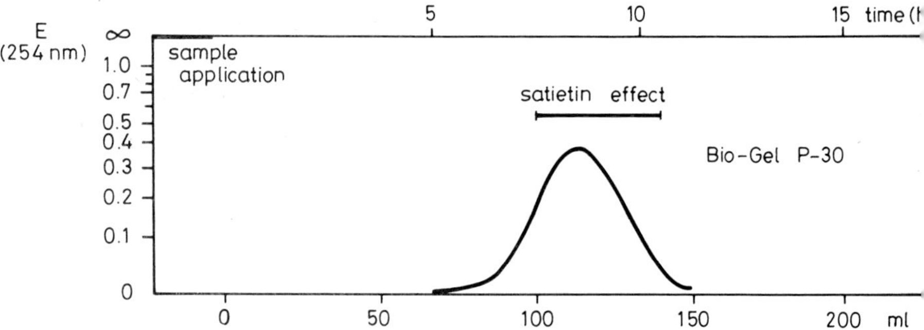

Fig. 6.27. The purification of satietin. The steps in the purification: (1) gel chromatography on Sephadex G-15; elution with physiological saline solution; (2) gel chromatography on Bio-Gel P-30; elution with distilled water. The purification step was preceded by ultrafiltration of the serum on an Amicon UM-10 membrane, and was followed in each case by ion-exchange chromatography on a DEAE-cellulose column [107]

form of monoamino oxidase. For the purification of flavin peptide from tryptic-chymotryptic hydrolysate, column gel chromatography was used on Bio-Gel P-2 column (2.4 × 90 cm, equilibrated and eluted by argon-saturated 0.5 *M* pyridinium-acetate buffer, pH 5.2) and a Sephadex G-15 (1.5 × 52 cm) column, using the same buffer. Gel chromatography on Bio-Gel P-2 and Sephadex G-15 columns was done after TCA treatment of the hydrolysate and adsorption on a Florisil column. The gel chromatographic fractions (in consequence of the volatile buffer) could be lyophilized. The yields of the gel chromatographic steps were about 33% and 50% on the Bio-Gel P-2 and Sephadex G-15 columns, respectively.

In following the gel chromatography of peptides it is particularly important to remember that utilization of the absorption at 280 nm assumes that the protein or peptide in question contains phenylalanine or tyrosine and tryptophan. If these are absent, the 280 nm absorption is of little value. In this respect, it is very helpful that proteins may also be detected by means of their absorption at 210 nm.

6.3.6 Gel chromatography of amino-acids

The theoretical and practical phenomena of the gel chromatographic separation of amino-acids fit into the picture of the gel chromatographic behaviour of small molecules, as described by Westernack [110], Janson [111], Gelotte [112] and others.

Compounds with very similar molecular dimensions are not eluted together, or in accordance with the order of their decreasing molecular size; rather the sequence of elution is determined by secondary gel chromatographic phenomena. The predominant phenomenon is gel filtration, i.e. on Bio-Gel P-2 the proteins are eluted together with the large polypeptides in the void volume V_o. However, for small molecules (with molecular weights less than 300), the position of elution is determined by the number of amino, hydroxyl or carboxyl groups, and aromatic rings: after V_∞, $K_d > 1$ [113].

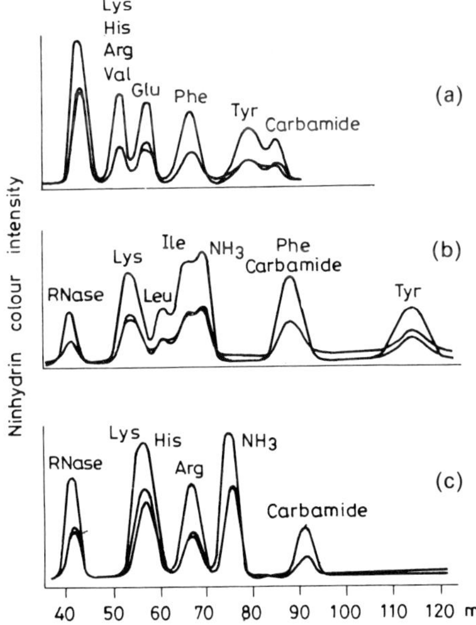

Fig. 6.28. Gel chromatography of amino-acids in solutions of various pH, on Sephadex G-10. (a) 0.2 *M* acetic acid, pH 2.7; (b) 0.2 *M* acetic acid + 1.0 *M* pyridine, pH 6.7; (c) 0.01 *M* NaOH, pH 12 [113]. (By kind permission of the authors and M. Dekker, Inc.)

In the separation of amino-acids, the pH, ionic strength and temperature, etc. of the eluent are of great importance. Variation of the pH of the eluent may change the whole elution pattern, and even the elution sequence. This latter phenomenon is illustrated by the example of phenylalanine and tyrosine in Fig. 6.28.

The gel chromatography of amino-acids is particularly important when a previously unknown compound causes a certain biological effect (the latter being known). Thus Kalász *et al.* [114] detected the presence of basic methylated amino-acids in cases where there were interesting aspects of cell proliferation. In similar work, it was assumed that amino-acids inhibiting cell proliferation were present (in bound form) among the organic constituents of the human tooth, and to confirm this hypothesis, the tooth protein components and the basic *N*-methylated amino-acids from the hydrolysate were purified. The tooth proteins were obtained by the method of Lezovic *et al.* [115]. Before detection of the amino-acids, the hydrolysate was separated by gel chromatography, and only the part containing the basic methylated amino-acids was analysed by gas chromatography.

6.3.7 Gel chromatography in organic solvents

In one type of gel chromatography with an organic solvent, the hydrophilic nature of the gel is utilized to bind an aqueous phase to the gel, the organic phase being situated outside the gel particles.

In this case, gel filtration based on steric hindrance plays only a negligible role; the fundamental mechanism is partition. Such a method has been employed to purify transfer RNA, to study native and synthetic oxytocin, and to separate actinomycins [116–120].

Another type of gel chromatographic procedure in an organic solvent is when the modified gel is used in organic solvent. The hydroxypropylated form of Sephadex G-25 also swells in organic solvents. It was employed by Neumann and Werner [119] to determine the free and bound tri-iodotyrosine contents of serum, and by Gregermann *et al.* [120] in an investigation of an octapeptide and its substituted derivatives.

6.4. HIGH-PERFORMANCE SIZE EXCLUSION CHROMATOGRAPHY (SE-HPLC)

6.4.1 High-performance liquid chromatography

According to the recent classifications [121], liquid column chromatography (LCC) can be carried out in the classical way or by using the modern high-pressure arrangements.

The classical LCC employs a column packed with 20–500 μm sorbents; in the case of gel chromatography the soft (or semi-rigid) gels are used for the separation of peptides and proteins. The mobile phase moves through the column under gravity, i.e. the difference in height between the free solvent surfaces in the reservoir and at the column outlet defines the flow rate (Fig. 6.2). The pump used in some cases gives a constant flow rate but does not serve to increase the speed of elution, which is limited by the mechanical stability and resistance of the gel applied. Usually, the time of each separation is not shorter than several hours but the loading capacity of the separation system both in size of sample and in amount of substances to be separated is much higher than in the case of high-performance methods.

The second type of LCC is called high-performance liquid chromatography (HPLC) but synonyms such as high-pressure liquid chromatography and high-speed liquid chromatography are also widely used. HPLC technique realizes forced flow of the eluent by pump (or pumps) through a small bore column packed with homodisperse microparticulate or pellicular particles [122–124] (particle size may be about 20, 10, 5 or 3 μm).

The whole HPLC set-up is arranged for

(a) high inlet pressure

(b) short separation time

(c) low dead volume

(d) sensitive detection of the separated components.

HPLC offers a unique possibility for the separation of amino-acids, peptides and proteins but mainly for an analytical purpose. Several recent reviews in this field demonstrate the advances made in the last decade and summarize the latest results [122–129].

Although the ion-exchange application of HPLC has been restricted to several fields (automatic amino-acid analysis, enzyme separations, etc.), the reverse phase (RP-) and size exclusion (SE-, an acronym of gel chromatography) complement each other and one of these methods can be used for any molecular size of amino-acids, peptides and proteins. The RP-HPLC of peptides was initiated by publications of Horváth's group as well as the recent results summarized in his books [124, 130, 131].

Unfortunately, the commercially available supports for RP-HPLC are generally produced with 80 or 100 Å pore size, the separation of peptides and proteins by HPLC can be irreproducible over a certain molecular size. At the same time, the preparative separation of peptides can be done by using the displacement mode of RP-HPLC, which is a recently published procedure of HPLC [132]. It provides a possible way of separation of several hundred milligrams with the standard (250 × 4.6 mm) columns generally employed in analytical HPLC.

6.4.2 Packing materials for size exclusion chromatography

Among the media for SE-HPLC are the so-called controlled-pore glasses (CPG) developed by Halled [133, 134]. CPGs are available as a series with pore-size between 4 and 150 nm and particle size 5 or 10 μm [135]. Unfortunately, the controlled-pore glasses are characterized by a high density of surface silanol groups (—Si—OH) which can behave as cation-exchangers and also result in adsorption and denaturation of some peptides and proteins. The irreversible interactions can sometimes be inhibited by coating of CPG, to give a product called glycophase-CPG [135–140]. The final solution of elimination of irreversible interactions seems to be this coating of covalently bonded organic modifiers onto both controlled pore glasses and micro-particulate silica gel. There are several basic types of 'silanol shielding' bonded phases, namely the diol-, carbohydrate-, polyethyleneglycol-, etc.

The bonded diol phase was described in detail by Roumeliotis and Unger [141], who attached the 1,2-dihydroxy-3-propoxypropyl groups covalently to the silica surface with Si—C binding and methylated the remaining two silanol groups. The structure of this modified support is as follows

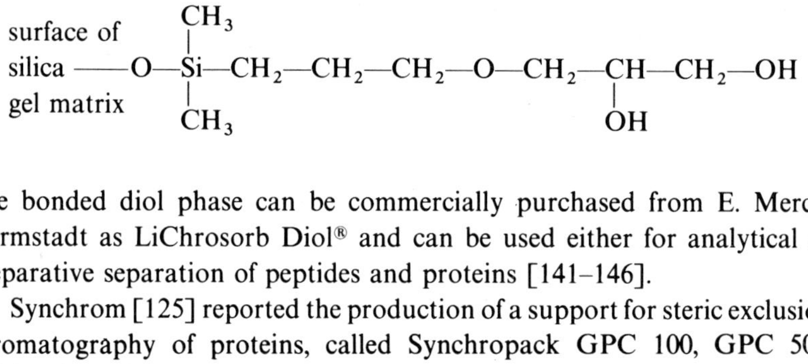

$$\text{surface of silica gel matrix}\text{—O—}\underset{\displaystyle CH_3}{\overset{\displaystyle CH_3}{\overset{|}{\underset{|}{Si}}}}\text{—}CH_2\text{—}CH_2\text{—}CH_2\text{—O—}CH_2\text{—}\underset{\displaystyle OH}{\underset{|}{CH}}\text{—}CH_2\text{—OH}$$

The bonded diol phase can be commercially purchased from E. Merck, Darmstadt as LiChrosorb Diol® and can be used either for analytical or preparative separation of peptides and proteins [141–146].

Synchrom [125] reported the production of a support for steric exclusion chromatography of proteins, called Synchropack GPC 100, GPC 500, GPC 1000 and GPC 4000. The support is silica gel with a covalently bonded carbohydrate layer. The particle size is 10 μm, pore size as shown in the name in Ångström, that is Synchropack GPC 500 has 50 nm pore size.

Poly(acryloyl morpholine) gels are commercially available as Enzacryl Gels, with different K numbers (K_1, K_2, etc.). They are widely used in the field of enzyme separation [147] and as a new support for preparing this-layers for gel chromatography of peptides and proteins. The Enzacryl Gels are commercially available from Koch-Light Laboratories (Bucks, England) [128, 129, 147–152].

Waters Associates (Milford, USA) commercialized the μ Bondapack-protein I-125 gel for hydrophilic molecular exclusion chromatography [153].

Toyo Soda Corporation of Japan has recently released TSK-GEL SW and PW type columns [126, 127]. Either gel can be used for aqueous high performance gel chromatography of proteins and enzymes. Although no real information is available on either the composition or the chemical structure of the gels, some investigations suggest the silica gel backbone to be covalently bonded with organic coating and the hydroxyl groups to be located on the organic layer. The TSK-GEL SW type of columns are made as G 2000 SW, G 3000 SW and G 4000 SW, and the TSK-GEL PW type of columns are available as G 1000 PW, G 2000 PW, G 3000 PW, G 4000 PW, G 5000 PW and G 6000 PW. These gels cover a wide molecular weight range, SW type from 5×10^2 to 6×10^5, PW type from 10^2 up to more than 10^7.

Although the TSK-GELs are designed for aqueous gel permeation chromatography, some components appear at $V_e > V_t$, which indicates that gel–solute interactions occur, in addition to the molecular sieving effects of the gel.

Two other rigid gels with solely organic composition have been commercialized. One of them is Shodex [154], available in Japan, the other is Spheron [155], which was developed in Czechoslovakia. Both gels are on methacrylate supports but only parts of their structure have been published [154, 155].

6.5 APPLICATION OF SE-HPLC

6.5.1 Controlled porosity glasses

Kearney and McGann [139] applied controlled porosity glass chromatography to milk proteins. CPG-10, 3000 Å (300 nm) was used after pretreatment with Carbowax 20 *M*. The relatively large pore size and a long separation time (350 min) were necessary, because the sample to be separated contained casein micelles and complexes.

6.5.2 Agarose and its octyl derivative

Hjerten and Kunquan [156] reported the use of agarose in HPLC. They demonstrated the chromatographic properties of a 12% agarose column in the separation of thyroglobulin, ferritin and phycoerythrin in 70 min. The hydrophobic interaction chromatography of the support was also studied after coupling an octyl ligand to the agarose.

6.5.3 TSK-GEL SW and TSK-GEL PW

A number of Japanese authors such as Kato, Hashimoto, Fukano *et al.* [126, 127, 157–176] and others [177, 178] published several papers on the application of TSK-GELs for high-performance steric exclusion chromatography (SE-HPLC) of water-soluble biopolymers, including peptides and proteins.

Hashimoto *et al.* [162] carried out molecular size measurement of proteins by means of high-speed gel permeation chromatography. They used TSK-GEL 3000 PW and 4000 PW columns, the exclusion limits of which (established with dextran) were 6×10^4 and 7×10^5. The pressure drops were 90 and 60 kg/cm^2 and the flow rates 0.9 and 1.0 ml/min, respectively. The total time of the experiments was 13–16 min per substance.

Fukano *et al.* [159] published the evaluation of the TSK-GEL SW type columns of three different grades—G 2000 SW, G 3000 SW and G 4000 SW. In their work, the separations of several non-enzymatic proteins, human serum, egg white, three commercial enzymes (catalase, alcohol dehydrogenase and peroxydase) as well as rat liver extract are presented. Their separations are really excellent examples of SE-HPLC of proteins although the highly prolonged elution times of the late-eluting components strongly suggest the existence of gel–solute interactions in addition to the exclusion processes. Some of their experiments raise the possibility of preparative application of SE-HPLC, as the sample size was 500 μl (separation of egg white proteins, rat liver extract) or 300 μl (separation of enzymes). The monitoring of the decomposition reaction of beta lactoglobulin by alpha chymotrypsin demonstrates the advantages of the rapid separation in process control. Fukano *et al.* also presented the molecular weight vs. elution volume calibration curves for TSK-GEL 2000 SW, 3000 SW and 4000 SW. Similar curves have been published by other authors [162, 172, 175, 176] and several papers have appeared which show the effect of eluent composition on

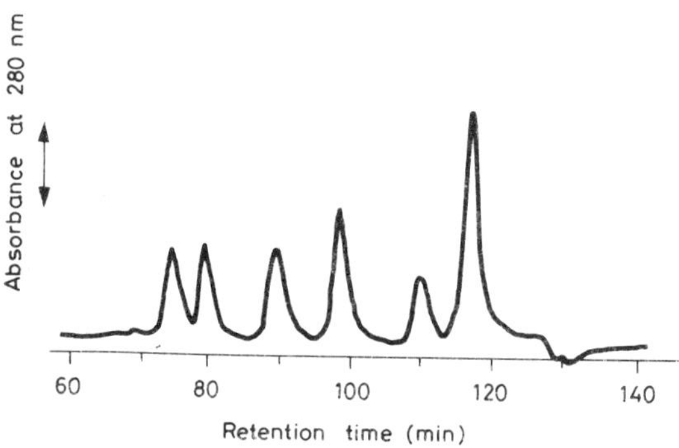

Fig. 6.29. Typical elution curves from two columns of TSK-GEL G3000 SW connected in series [175]. The substances (in order of their elution) are rabbit phosphorylase b, bovine serum albumin, ovalbumin, bovine carbonic anhydrase, soya-bean trypsin inhibitor and bovine lactalbumin. (Reproduced with kind permission of Dr. Tagaki and Elsevier Scientific Publishing Co.)

chromatographic properties and parameters of proteins and peptides [164, 168, 169]. The optimization of separation processes was suggested on the basis of specific resolution [159]:

$$R_s = 2(V_2 - V_1)/(w_2 + w_1)(\log M_1 - \log M_2)$$

The effect of the presence and concentration of salts, sodium dodecylsulphate and guanidine hydrochloride was also investigated.

Takagi [175] compared the advantages and disadvantages of SDS-PAGE and SE-HPLC, using TSK-GEL 3000 SW and found a high efficiency of SDS-PAGE in the resolution but difficulty in precise determination of mobility as well as poor recovery after separation of proteins. These shortcomings are inevitable in the column techniques. His excellent figures

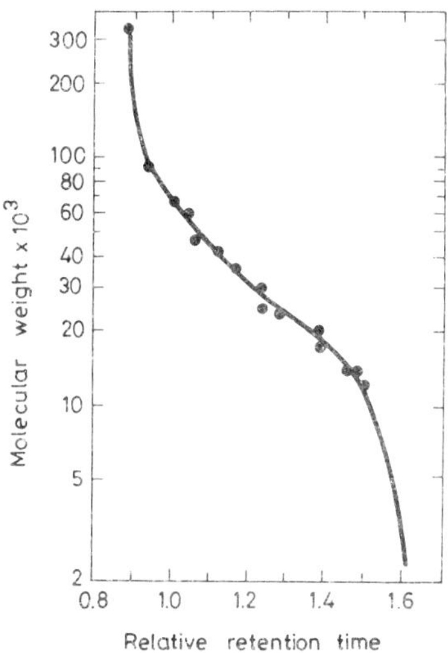

Fig. 6.30. Plots of molecular weights of proteins vs. their retention times relative to that of bovine serum albumin taken by SDS–HPLC system using TSK-GEL G3000 SW [175] (With the kind permission of Dr. Takagi and Elsevier Scientific Publishing Co.)

demonstrate the applicability of TSK-GEL 3000 SW for SE-HPLC in this field. Some of Takagi's results are shown in Figs. 6.29 and 6.30.

Okazaki *et al.* [174] described the rapid determination of cholesterol in lipoprotein fractions by SE-HPLC followed by the enzyme reaction, using reaction-type HPLC. Very small amounts of sample (20 μl of serum) were needed for a fast and complex process including separation, enzyme reaction and determination (in 70 min). The results demonstrate the cholesterol content of different lipoprotein fractions and albumin (no cholesterol), indicating the possible wide application of the SE-HPLC (Fig. 6.31).

The very fast analysis by TSK-GEL 3000 SW was also utilized by Murakami *et al.* [172] who studied the interactions between renin and renin-binding protein. The time- and temperature-dependent reaction and equilibrium cannot be studied by conventional gel filtration since it needs overnight elution. The separation was carried out, however, in 40 min with excellent resolution of the components to be separated by using a TSK-GEL 3000 SW (60 × 0.75 cm) column, on eluent containing sodium pyrophosphate, sodium chloride and sodium eluent. The chromatograms can be seen in Fig. 6.32.

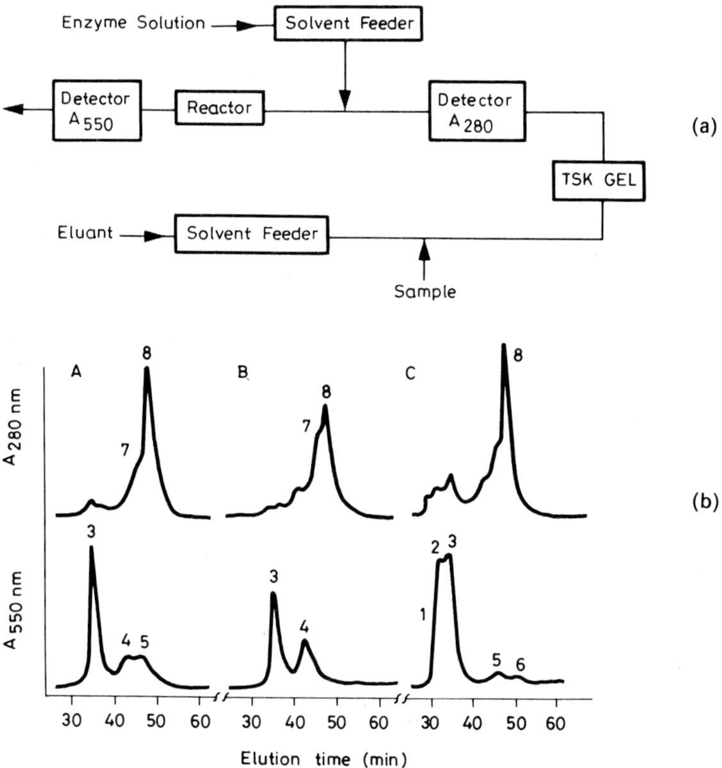

Fig. 6.31. (a) Flow diagram for the enzymatic detection of cholesterol by SE–HPLC. (b) Analysis of proteins and their cholesterol content from human sera. (A) normal female subject. (B) liver cirrhosis subject. (C) hyperlipidaemic subject. Loaded volume: 20 μl of the whole serum. Peaks: 1. VLDL; 2. IDL; 3. LDL; 4. HDL_2; 5. HDL_3; 6. VHDL; 7. γ-globulin; 8. albumin. Reproduced from [174] with the kind permission of the authors and Elsevier Scientific Publishing Co.)

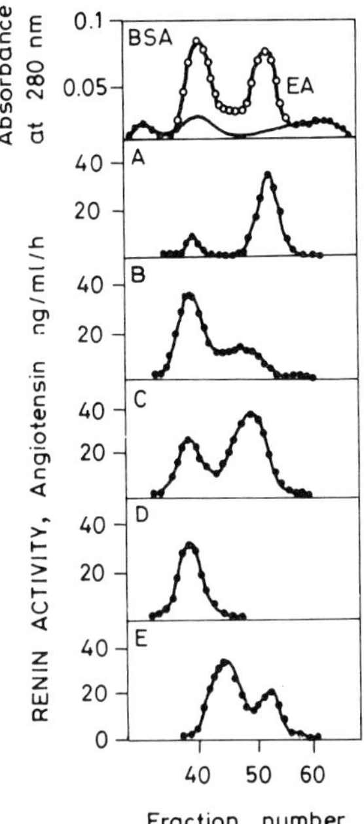

Fig. 6.32. HPLC separation of renin and renin-RBP complex formed in the following conditions: (A) 37 °C for 1 min, (B) 37 °C for 45 min, (C) 37 °C for 45 min followed by 4 °C for 16 h, (D) 37 °C for 45 min with 10 m*M* sodium tetrathionate, (E) 37 °C for 45 min with 4 *M* sodium chloride, bovine serum albumin (BSA) and egg albumin (EA) as standards together with extract (top chromatogram) and extracts alone (other chromatograms). [172]. (With kind permission of the authors and Elsevier Scientific Publishing Co.)

Ui [176] has recently reported the measurement of molecular size and hydrodynamic behaviour of several glycopeptides; he used a column packed with TSK-GEL for SE-HPLC in aqueous 6 *M* guanidine hydrochloride solution (Figs. 6.33 and 6.34).

Eleven reduced-alkylated glycoproteins were compared [176]; their molecular weights were calculated from their distribution coefficients as well as from the published data (taken from the amino-acid sequence and carbohydrate composition). The value of the calculated molecular sizes did not deviate generally from the actual molecular weights (Table 6.4) especially for glycoproteins with a carbohydrate content less than 10% (difference in calculated and apparent molecular weights less than 6%). The value of Ui's results is even more striking considering the very short time of SE-HPLC (about 50 min) and that Ui's findings were affirmed by Leach *et al.* [179, 180] who performed their experiments also in 6 *M* guanidine hydrochloride

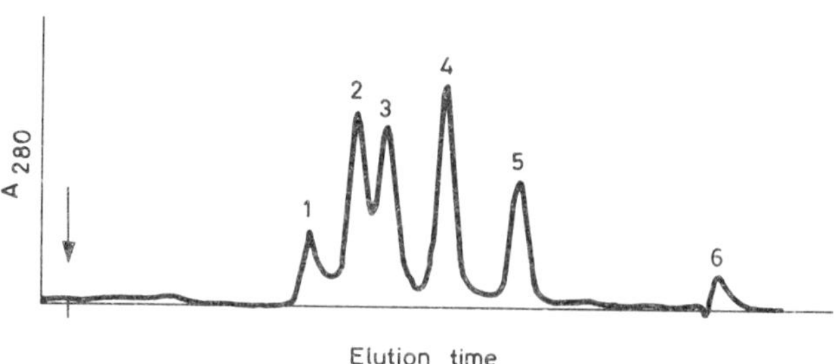

Fig. 6.33. SE-HPLC profile obtained with a glycopeptide mixture in the presence of 6 M guanidine hydrochloride [176]. 1 = Blue Dextran; 2 = transferrin, 3 = Taka-amylase A; 4 = Japanese quail ovoinhibitor; 5 = ribonuclease B; 6 = 2,4-dinitrophenylalanine. (Reproduced with kind permission of the author and Elsevier Scientific Publishing Co.)

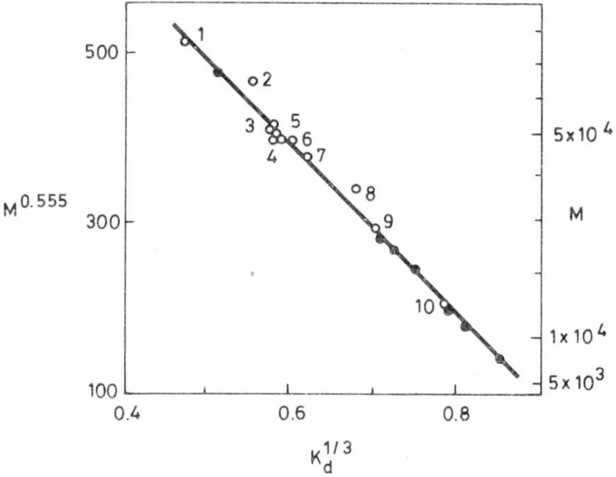

Fig. 6.34. Relationship between $M^{0.555}$ and $K_d^{1/3}$ obtained by SE-HPLC in 6 M guanidine hydrochloride [176]; ● represents simple polypeptides, ○ indicates glycopeptides as follows: 1 = transferrin; 2 = acid carboxypeptidase; 3 = Taka-amylase A; 4 = ovoinhibitor; 5 = globulin H chain; 6 = fetuin; 7 = ovalbumin; 8 = α_1 acid glycoprotein; 9 = ovomucoid; 10 = ribonuclease B. (With the kind permission of the author and Elsevier Scientific Publishing Co.)

solution but by using an agarose gel column at low elution rate. They also found a very small difference between the effective hydrodynamic volumes of glycoproteins and of linear polypeptides.

Calam and Davidson [173] dealt with the application of SE-HPLC to fractionation of glycoprotein hormones and human growth hormone, as allergen extracts of animal and plant origin. They used size exclusion chromatography on a TSK-GEL 3000 SW column (30 × 0.75 cm); the

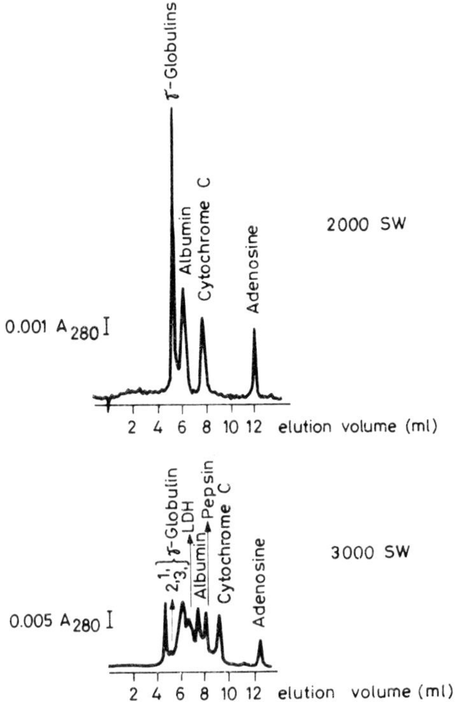

Fig. 6.35. Separation of several standard proteins on 2000 SW, 3000 SW columns [177]. (Reproduced with permission of the authors and Elsevier Scientific Publishing Co.)

separation time was less than 30 min. The fractions isolated were checked by SDS-PAGE.

Wehr and Abbot [177] demonstrated the difference in selectivity of TSK type 2000 SW and 3000 SW, and the value of a proper choice of gel for separation is shown in Fig. 6.35.

Somack, McKay and Giles [181] reported the biological application of Spherogel TSK-SW gel (Altex, Berkeley, USA). A 'standard' column size (600 × 7.5 mm) was used and the protein separation was completed in 20–25 min. Different samples such as serum proteins, milk proteins, thyroid hormone, nuclear binding protein from liver cell nuclei, and skeletal muscle contractile proteins were separated with excellent recovery and checked by multiple enzyme assay as well as other methods.

6.5.4 LiChrosorb Diol

Roumeliotis and Unger [141] investigated the preparative scale separation of proteins and enzymes on LiChrosorb Diol by SE-HPLC. By means of bovine albumin, chymotrypsinogen and lysozyme separations, the loading capacity could be determined. Under optimal conditions, on a large bore column (LiChrosorb Diol, 250 × 23.5 mm I.D., containing 50 g of packing), 30 mg of proteins can be separated with a flow rate of 20 ml/min. This means the preparative procedure takes less than 6 minutes with a 6×10^{-4} mass-load—phase ratio.

Zini *et al.* [145] applied the rarely used frontal analysis mode of SE-HPLC when they investigated the polymerization of human serum albumin. The LiChrosorb Diol (pore size 100 Å–10 nm) excludes the polymer but retains the monomor albumin.

Rubinstein [143] reported preparative SE-HPLC on LiChrosorb Diol, characterized by a dual retention mechanism. He studied the separation of components of foetal calf serum on a small bore LiChrosorb Diol column (4.6 × 250 mm). At the beginning of the procedure, the equilibration and elution were done with 80% propan-1-ol in 20% sodium acetate buffer, but after elution of several components, the solvent composition was changed to 50% propan-1-ol with a linear gradient. On this analytical type of column, 1 ml of dialysed serum could be separated in each run.

6.5.5 Synchropack GPC

The Synchropack GPCs have been designed for SE-HPLC of proteins. The 10 μm support coated with a bonded carbohydrate layer has been used extensively [125, 182–188]. Several comparative studies have been published on the separation of bovine serum albumin, fibrinogen, chymotrypsinogen and cytochrome C [183–187] and molecular weight calibration curves were taken on Synchropack GPC 100 and 500 [184, 187]. The separation of water soluble polymers [187], rapid analysis of bilirubin [185], determination of haemoglobin [186] and peptides of biological origin [183–186], etc. were also studied.

6.5.6 μ Bondapack-protein I-125 column

Micropreparative separation of ^{125}I labelled proteins was reported by von Stetten and Schlett [133]. They used a μ Bondapack-protein I-125 column (Waters, Milford, USA) [153, 189] for the separation of the iodinated sample components by hydrophilic SE-HPLC, with an iodide crystal as an on-line detector. The time of separation of the complex mixture of several iodinated proteins was 45 min. The authors also compared the resolution and separation time of Sephadex gel chromatography and Bondapack SE-HPLC, as well as optimization of pH, column size, etc.

6.5.7 Comparative investigations on several packings

Ulyashin *et al.* [148] reported the comparative investigations of Sephadex LH-20 and different silica gels in an organic solvent (dimethylformamide). They separated protected hydrophobic peptides by SE-HPLC. As some protected peptides are insoluble in water or aqueous-organic solvent systems, their separation could not be performed with other common methods (ion-exchange, partition, etc.). At the same time, the solubility of protected peptides in organic solvents suggested the application of SE-HPLC. The separation time was about 10 min in the case of microparticulate silica gel (Fig. 6.36) but up to 80 min using Sephadex LH-20. A UV detector and a refractive index monitor were used. At the same time, the Sephadex LH-20 column was found to be the most suitable for peptides with molecular weight of less than 2000. High performance columns packed with HPLC silica gel enable the separation of peptides with molecular sizes that differ by a factor of less than 1.1 and peptides with molecular sizes over 2000 can be analysed and separated adequately by HPLC. Even a mixture of approximately 30-membered peptides could be separated if the individual components differed in chain length by 2–3 amino-acid residues.

Valuable comparisons of the SE-HPLC packings can be found in several recent reviews of the separation of proteins and in papers about packings [122, 123, 128, 129, 149].

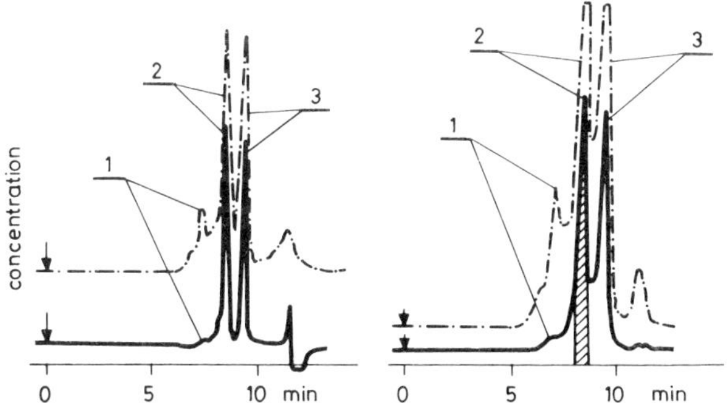

Fig. 6.36. Analytical (left) and preparative (right) scale separation of a synthetic peptide on a silica gel column (330 × 25 mm). The preparative procedure made possible the isolation of the substance to be purified (2) from impurities (1,3) with good recovery. The amount of sample was 50 mg, the separation was monitored by UV detector (broken line) and refractometer (solid line) [148]. (Reproduced with kind permission of Elsevier Scientific Publishing Co.)

6.6 THIN-LAYER GEL PERMEATION CHROMATOGRAPHY

Brough *et al.* published their study [190] on thin-layer gel permeation chromatography using Enzacryl gel. Dyed proteins, such as derivatives of insulin, ribonuclease A, chymotrypsinogen, ovalbumin, transferrin, serum albumin, fibrinogen, urease and thyroglobulin were prepared by Miller *et al.* [191]. Plots of log molecular weight vs. K_d were given for these dyed proteins on various Enzacryl gels.

REFERENCES

[1] Bombaugh, K. J.: *The Practice of Gel Permeation Chromatography.* in *Modern Practice of Liquid Chromatography*, Kirkland, J. J. (ed.), Wiley Interscience, New York, 1971.

[2] Blatt, W. F.: *J. Agr. Food Chem.*, **19,** 589 (1971).

[3] *Gel Filtration; Theory and Practice.* Pharmacia Fine Chemicals, Uppsala, Sweden, 1979.

[4] *Sephacryl® S-300 Superfine and Calibration Kits for Gel Filtration.* Pharmacia Fine Chemicals, Uppsala, Sweden, 1978.

[5] Kremmer, T., Boross, L.: *Gel Chromatography.* Akadémiai Kiadó, Budapest, 1979.

[6] *Materials, Equipment and Systems for Chromatography, Electrophoresis and HPLC.* Bio-Rad Laboratories, Richmond, Calif. U.S.A. Catalogue E, 1979.

[7] Boschetti, E., Tixier, R., Garelle, R.: *Science Tools*, **21,** 35 (1974).

[8] *Literature References 1959–72, 1973, 1974, 1975, 1976, 1977, 1978, 1979, 1980, 1981. Sephadex®, Sephadex® Ion Exchangers, CNBr-activated Sepharose®, Sepharose®, Dextran fractions, Dextran derivatives.* Pharmacia Fine Chemicals, Uppsala, Sweden.

[9] *Bio-Gel® Reference List. Gel Chromatography Applications.* Bio-Rad Laboratories, Richmond, Calif. U.S.A. January, 1973.

[10] Araki, C.: *Bull. Chem. Soc. Japan*, **29,** 543 (1956).

[11] Russell, B., Mead, T. H., Polson, A.: *Biochim. Biophys. Acta*, **86,** 169 (1964).

[12] Polson, A.: *Biochim. Biophys. Acta*, **50,** 565 (1961).

[13] *Ultragel®: Pre-swollen Polyacrylamide Agarose Gel Beads for High-resolution Gel Filtration.* LKB-Produkter, Sweden, Publ. No. 2204. 1976.

[14] *Sephadex LH-20. Chromatography in Organic Solvents.* Pharmacia Fine Chemicals, Uppsala, Sweden.

[15] *Sephasorb® HP Ultrafine*. Pharmacia Fine Chemicals, Uppsala, Sweden, 1978; *Separation News*, **2,** 1978.

[16] Byrne, S. H. Jr.: *Detectors*, in *Modern Practice in Liquid Chromatography*, Kirkland, J. J. (ed.), Wiley Interscience, New York, 1971.

[17] Morris, C. J. O. R., Morris, P.: *Separation Methods in Biochemistry*, 2nd Ed., Pitmans, London, 1976.

[18] Marsden, N. V. B.: *J. Chromatog.*, **58,** 304 (1971).

[19] *Sweetman, L., Nyhan, W. L: J. Chromatog.*, **59,** 349 (1971).

[20] Pecsok, R. L., Saunders, D.: *A Critical Evaluation of Gel Chromatography*, in *Separation Techniques in Chemistry and Biochemistry*, Keller, R. A. (ed.), Dekker, New York, 1967.

[21] Laurent, T. C., Killander, J.: *J. Chromatog.*, **14,** 317 (1964).

[22] Determann, H., Brewer, J. E.: *Gel Chromatography*, in *Chromatography*, Heftmann, E. (ed.), Van Nostrand Reinhold, New York, 1975.

[23] Fischer, L.: *Gel Filtration Chromatography*, in *Laboratory Techniques in Biochemistry and Molecular Biology*, Work, T. S., Burdon, R. H. (eds). Elsevier/North Holland Biomedical, Amsterdam, 1980.

[24] Kalász, H., Nagy, J., Kerecsen, L.: *Gel Chromatography of Phenylalkylamines* in *Monoamine Oxidases and their Selective Inhibition*, Magyar, K. (ed.). Pergamon Press–Akadémiai Kiadó, 1980, p. 57.

[25] Glueckauf, E.: *Ion Exchange and its Application*, Soc. Chem. Ind. London, 1955, p. 34.

[26] Kalász, H., Nagy, J., Knoll, J.: *J. Chromatog.*, **107,** 35 (1975).

[27] Giddings, J. C.: *Anal. Chem.*, **39,** 1027 (1967).

[28] Kalász, H., Nagy, J., Magyar, K., Knoll, J.: *Gel Chromatography of Catecholamines* in *Symposium on Pharmacology of Catecholaminergic and Serotoninergic Mechanisms*, Knoll. J. (ed.). Akadémiai Kiadó, Budapest, 1976, p. 91.

[29] Kalász, H.: *Proc. 17th Hung. Ann. Meet. Biochem.*, Kecskemét, Hungary, 1977, p. 161.

[30] Porath, J., Bennich, H.: *Arch. Biochem. Biophys.*, Suppl., **1,** 152 (1962).

[31] Bombaugh, K. J.: *J. Chromatog.*, **53,** 27 (1970).

[32] Kalász, H.: *J. Chromatog.*, **78,** 233 (1973).

[33] Kalász, H., Knoll, J.: *Science Tools*, **20,** 15 (1973).

[34] Margolis, S., Langdon, R. G.: *J. Biol. Chem.*, **241,** 469 (1966).

[35] Siegel, L. M., Monty, K. J.: *Biochim. Biophys. Acta*, **112,** 346 (1966).

[36] Killander, J.: *Biochim. Biophys. Acta*, **93,** 1 (1964).

[37] Determann, H., Michel, W.: *J. Chromatog.*, **25,** 303 (1966).

[38] Wieland, Th., Duesberg, P., Determann, H.: *Biochem. Z.*, **337,** 303 (1963).

[39] Whitaker, J. R.: *Anal. Chem.*, **35,** 1950 (1963).
[40] Andrews, P.: *Biochem. J.*, **91,** 222 (1964).
[41] Ackers, G. K.: *Biochemistry*, **3,** 723 (1964).
[42] Squire, P. G.: *Arch. Biochem. Biophys.*, **107,** 471 (1964).
[43] Carnegie, P. R.: *Nature*, **206,** 1128 (1965).
[44] Siegel, L. M., Monty, K. J.: *Biochim. Biophys. Res. Commun.*, **19,** 494 (1965).
[45] Anderson, D. M. W., Stoddard, J. F.: *Anal. Chim. Acta*, **34,** 401 (1966).
[46] Laurent, T. C., Killander, J.: *J. Chromatog.*, **14,** 317 (1964).
[47] Ackers, G. K., Thompson, I. E.: *Proc. Natl. Acad. Sci. U.S.A.*, **53,** 342 (1965).
[48] *Sephacryl® S-400, S-500, S-1000 Superfine.* Pharmacia Fine Chemicals AB. Uppsala, Sweden, 1981.
[48a] Determann, H.: *Gelchromatographie,* Springer-Verlag, Berlin (1967).
[49] Fish, W. W., Mann, K. G., Tanford, C.: *J. Biol. Chem.*, **244,** 4985 (1969).
[50] Pusztai, A., Watt, W. B.: *Biochim. Biophys. Acta*, **214,** 463 (1970).
[51] Pedersen, K. O.: *Arch. Biochem. Biophys., Suppl.*, **1,** 157 (1962).
[52] Kakiuchi, K., Kato, S., Imanishi, A., Isemura, T.: *J. Biochem.* (*Tokyo*), **55,** 102 (1964).
[53] Shepherd, G. R., Petersen, D. F.: *J. Chromatog.*, **9,** 445 (1962).
[54] Lionetti, F. J., Valeri, C. R., Bond, J. C.: *J. Lab. Clin. Med.*, **64,** 519 (1963).
[55] Ratcliff, A. P., Hardwicke, J.: *J. Clin. Pathol.*, **17,** 676 (1964).
[56] Hodgson, R., Sewell, P.: *J. Med. Lab. Tech.*, **22,** 130 (1965).
[57] Killander, J., Bengtsson, S., Philipson, L.: *Proc. Soc. Exptl. Biol. Med.*, **115,** 861 (1964).
[58] Manypol, W., Spitzy, H.: *Intern. J. Appl. Radiation Isotopes*, **13,** 647 (1962).
[59] Rivera, J. V., Toro-Goyco, E., Matos, M. L.: *Amer. J. Med. Sci.*, **249,** 371 (1965).
[60] Toro-Goyco, E., Martinez-Meldonado, M., Matos, M. L.: *Proc. Soc. Exptl. Biol. Med.*, **122,** 301 (1966).
[61] Ceska, M.: *Immunology*, **15,** 837 (1968).
[62] Gjedde, F.: *Acta Physiol. Scand.*, **70,** 69 (1967).
[63] Pearson, M. J., Martin, F. I. R.: *Diabetologia*, **6,** 581 (1970).
[64] Guenther, H. L., McDonald, H. J.: *Prep. Biochem.*, **2,** 397 (1972).
[65] Cuatrecases, P.: *J. Biol. Chem.*, **247,** 1980 (1972).
[66] Dixon, K., Exon, P. D., Hughes, H. R.: *Lancet*, 343 (1972: 1).
[67] Hummel, J. P., Dryer, W. J.: *Biochem. Biophys. Acta*, **63,** 530 (1962).
[68] Hocman, G., Knopp, J.: *J. Chromatog.*, **79,** 340 (1973).
[69] Fairclough, G. F. Jr., Fruton, J. S.: *Biochemistry*, **2,** 461 (1963).

[70] Spitzy, H., Skrube, H., Müller, K.: *Microchim. Acta*, 296 (1961).
[71] Lissitzky, S., Bismuth, J., Simon, C.: *Nature*, **199,** 1002 (1962).
[72] Shapiro, B., Rabinowitz, J. L.: *J. Nucl. Med.*, **3,** 417 (1962).
[73] Ceska, M., Pihl, P., Holmqvist, L., Grossmüller, F.: *J. Chromatog.*, **57,** 145 (1971).
[74] Samuelsson, E. G., Tibbling, P., Holm, S.: *Food Technol.*, **21,** 121 (1967).
[75] Philip, B. A., Herbert, P., Hollingworth, J. W.: *Proc. Soc. Exptl. Biol. Med.*, **123,** 576 (1966).
[76] Batchelor, F. R., Dewdney, J. M., Feinberg, J. G.: *Lancet*, 1175 (1967: 1).
[77] Tokomaru, T.: *J. Immunology*, **89,** 195 (1962).
[78] Wood, H. S., Lochmüller, H., Repertinger, C.: *Biochem. Z.*, **337,** 247 (1963).
[79] Numa, S., Ringelmann, E., Lynen, F.: *Biochem. Z.*, **340,** 228 (1964).
[80] Hunter, W. M., Greenwood, F. C.: *Nature*, **194,** 495 (1962).
[81] Greenwood, F. C., Hunter, W. M., Glover, J. S.: *Biochem. J.*, **89,** 114 (1963).
[82] Englander, S. W.: *Biochemistry*, **2,** 798 (1963).
[83] Vaerman, J. P.: *Studies on IgA in Man and Animals*, Université Catolique de Louvain, 1970.
[84] Ryley, H. C., Matthews, N.: *Science Tools*, **25,** 4 (1978).
[85] Skvaril, F., Theilkaes, L.: *Science Tools*, **25,** 10 (1977).
[86] Blumenthal, K. M., Smith, E. L.: *J. Biol. Chem.*, **248,** 6002 (1973).
[87] Lebherz, H. G., Rutter, W. J.: *J. Biol. Chem.*, **248,** 1650 (1973).
[88] Janson, J.-C.: *J. Agr. Food Chem.*, **19,** 581 (1971).
[89] Raible, K., Engelhardt, J.: *Brauwissenschaft*, **18,** 398 (1965).
[90] Drews, B., Moeller, P.: *Z. Physiol. Chem.*, **351,** 952 (1970).
[91] Zakow, D.: *Mitteilungen Rebe und Wein*, **19,** 437 (1970).
[92] Somers, T. C., Ziemelis, G.: *J. Sci. Food Agric.*, **23,** 441 (1972).
[93] Somers, T. C., Ziemelis, G.: *Amer. J. Enol. Viticult.*, **24,** 47 (1973).
[94] Avakjants, S. P.: *Vinodelie i Vinogradstvo SSSR*, **25,** 10 (1965).
[95] Wucherpfennig, K., Franke, I.: *Z. Lebensmittel Untersuch. Forschung*, **124,** 22 (1960).
[96] Mesrob, B., Iwanov, T., Zakow, D., Gorinowa, N.: *Mitteilungen Rebe und Wein*, **20,** 33 (1970).
[97] Radola, F. C.: *Rep. Nuclear Center, Seibersdorf*, 1961.
[98] Feuillat, M. Bergeret, S.: *Bull. Chim. Soc. France*, **5,** 2085 (1972).
[99] Okubo, K., Shibashaki, K.: *Agr. Biol. Chem.*, **31,** 1276 (1967).
[100] Koshiyama, I.: *Agr. Biol. Chem.*, **29,** 885 (1965).
[101] Arai, S.: *Agr. Biol. Chem.*, **34,** 1338 (1970).
[102] Yamashita, M.: *Agr. Biol. Chem.*, **34,** 1484 (1970).
[103] Sajgó, M.: *FEBS Letters*, **12,** 349 (1971).

[104] Sajgó, M., Hajós, Gy.: *FEBS Letters*, **38,** 341 (1974).
[105] Bonilla, C. A.: *Prep. Biochemistry*, **2,** 305 (1972).
[106] Reichelt, K. L., Foss, I., Trygstad, O., Edminson, P. D., Johansen, J. H., Boler, J. B.: *Neuroscience*, **3,** 1207 (1978).
[107] Knoll, J., Kalász, H., Knoll, B., Nagy, J.: *Hungarian Patent* RI-683 (1978).
[108] Nagy, J., Kenney, W. T., Singer, T. P.: *J. Biol. Chem.*, **254,** 2684 (1980).
[109] Nagy, J., Salach, J. I.: *Arch. Biochem. Biophys.*, **208,** 388 (1981).
[110] Wasternack, C.: *Pharmazie*, **27,** 67 (1972).
[111] Janson, J.-C.: *J. Chromatog.*, **28,** 12 (1967).
[112] Gelotte, B.: *J. Chromatog.*, **3,** 330 (1960).
[113] Eaker, D., Porath, J.: *Sep. Science*, **2,** 507 (1967).
[114] Kalász, H., Kovács, G. H., Nagy, J., Tyihák, E., Barnes, W. T.: *J. Dental Research*, **57,** 128 (1978).
[115] Lezovic, Y., Ziman, M., Hochman, G.: *J. Chromatog.*, **96,** 258 (1974).
[116] Drabarek, S., DuVigneaud, V.: *J. Amer. Chem. Soc.*, **86,** 4477 (1964); **87,** 3974 (1965).
[117] Klieger, E., Schröder, E.: *Tetrahedron Letters*, **25,** 2067 (1965).
[118] Schmidt-Kastner, G.: *Naturwiss.*, **51,** 38 (1964).
[119] Neumann, A., Werner, S. C.: *J. Clin. Invest.*, **46,** 1346 (1967).
[120] Gregermann, R. I., Weaver, T. Jr., Kowatch, M. A.: *J. Chromatog.*, **47,** 369 (1970).
[121] Ettre, L. S.: *J. Chromatog.*, **220,** 29 (1981).
[122] Majors, R. E.: *J. Chromatog. Sci.*, **15,** 334 (1977).
[123] Majors, R. E.: *J. Chromatog. Sci.*, **18,** 488 (1980).
[124] Horváth, Cs.: *High-Performance Liquid Chromatography–Advances and Perspectives.* Vols. 1, 2, 3. Academic Press, New York, 1980, 1981.
[125] *Steric Exclusion on Synchropack GPC.* SynChrom, Inc., Linden, IN. USA, 1980, pp. 3–4.
[126] *TSK-GEL SW Type Column. Technical Data.* Toyo Soda Manufacturing Co. Tokyo, Japan.
[127] *TSK-GEL PW Type Column. Technical Data.* Toyo Soda Manufacturing Co. Tokyo, Japan.
[128] Pryde, A., Gilbert, M. T.: *Applications of High Performance Liquid Chromatography.* Chapman and Hall, London, 1979, p. 213.
[129] Epton, R. (ed.): *Chromatography of Synthetic and Biological Polymers*, Vol. 1. *Column Packings, GPC, GF and Gradient Elution.* Ellis Horwood, Chichester, 1978.
[130] Molnár, I., Horváth, Cs.: *Clin. Chem.*, **22,** 129 (1976).
[131] Horváth, Cs., Melander, W., Molnár, I.: *Anal. Chem.*, **49,** 142 (1977).

[132] Kalász, H., Horváth, Cs.: *J. Chromatog.*, **215,** 295 (1981).

[133] Haller, W.: *Nature*, (London), **206,** 693 (1965).

[134] Haller, W.: *U.S.A. Patent*, 3, 549, 524 (1971).

[135] *Pierce Handbook and General Catalog*, 1979–1980 (Pierce Eurochemie B.V., Rotterdam, Holland): Pierce Controlled Pore Glass Support.

[136] Mizutani, T., Mizutani A.: *Anal. Biochem.*, **83,** 216 (1971).

[137] Hiatt, C. W., Sheloko, A., Rosenthal, E. J., Galimore, J. M.: *J. Chromatog.*, **56,** 32 (1971).

[138] Darling, T., Albert, J., Russel, P., Albert, D. M., Reid, T. W.: *J. Chromatog.*, **131,** 383 (1977).

[139] Kearney, R. D., McGann, T. C.: *Application of Controlled Pore Glass Chromatography to Milk Proteins* in, *Chromatography of Synthetic and Biological Polymers.* Vol. 1. Epton, R. (ed.) Ellis Horwood, Chichester, G. B., 1978, p. 269.

[140] Mizutani, T., Mizutani, A.: *J. Chromatog.*, **168,** 143 (1979).

[141] Roumeliotis, P., Unger, K. K.: *J. Chromatog.*, **185,** 445 (1979).

[142] Becker, N., Unger, K. K.: *Chromatographia,* **8,** 539 (1979).

[143] Rubinstein, M.: *Anal. Biochem.*, **98,** 1 (1979).

[144] Schmidt, D. E., Giese, R. W., Conron, B., Karger, B. L.: *Anal. Chem.*, **52,** 177 (1980).

[145] Zini, R., Barre, J., Bree, T., Tillement, J.-P., Sebille, B.: *J. Chromatog.*, **216,** 191 (1981).

[146] Roumeliotis, P., Kinkel, J., Unger, K.: *Proc. Vth. Int. Symp. Column Liquid Chromatog.* Abstr. No. **34.** Avignon, France, 1981.

[147] Epton, R., Holloway, C.: *An Introduction to Permeation Chromatography.* Koch-Light Laboratories Ltd., Bucks, G. B., 1973.

[148] Ulyashin, V. V., Deigin, V. I., Iwanov, V. T., Ovchinnikov, Yu. A.: *J. Chromatog.*, **215,** 263 (1981).

[149] Pfannkoch, E., Lu, K.-C., Regnier, F. E., Barth, H. G.: *J. Chromatog. Sci.*, **18,** 430 (1980).

[150] Epton, R., Holding, S. R., McLaren, J. V.: *J. Chromatog.*, **110,** 327 (1975).

[151] Galpin, I. J., Handa, B. K., Kenner, G. W., Moore, S., Ramage, R.: *Purification of Protected Peptides by GPC*, in: *Chromatography of Synthetic and Biological Polymers.* Vol. 1, (ed.: Epton, R.). Ellis Horwood Publ. Chichester, G. B., 1978, p. 331.

[152] Engelhardt, H., Mathes, D.: *J. Chromatog.*, **185,** 305 (1979).

[153] Rivier, J. E.: *J. Chromatog.*, **202,** 211 (1980).

[154] Nakamura, S., Ishiguro, S., Yumada, T., Moriizumi, S.: *J. Chromatog.*, **83,** 279 (1973).

[155] Borak, J., Cadersky, I., Kiss, F., Smrz, M., Viska, J.: *Poly(hydroxyethyl methacrylate) gels (Spheron®)*, in *Chromatography of Synthetic and Biological Polymers.* Vol. 1, Epton, R. (ed.), Ellis Horwood, Chichester, 1978, p. 91.
[156] Hjerten, S., Kunquan, Y. J.: *J. Chromatog.*, **215,** 317 (1981).
[157] Toyo Soda Company: *HLC-Report 801*, **1,** 36 (1974).
[158] Kato, Y., Mido, S., Yamamoto, M., Hashimoto, T.: *J. Polym. Sci.*, **12,** 1338 (1974).
[159] Fukano, K., Komiya, K., Sasaki, H., Hashimoto, T.: *J. Chromatog.*, **166,** 47 (1978).
[160] Rokushika, S., Ohkawa, T., Hatano, H.: *J. Chromatog.*, **176,** 456 (1978).
[161] Hashimoto, T., Sasaka, H., Aiura, M., Kato, Y.: *J. Polym. Sci. Polym. Phys. Ed.*, **16,** 1789 (1978).
[162] Kato, Y., Sasaki, H., Aiura, M., Hashimoto, T.: *J. Chromatog.*, **153,** 546 (1978).
[163] Hashimoto, T., Sasaki, H., Aiura, M., Kato, Y.: *J. Chromatog.*, **160,** 301 (1978).
[164] Kato, Y., Komiya, K., Sasaki, H., Hashimoto, T.: *J. Chromatog.*, **193,** 311 (1980).
[165] Hara, I., Okazaki, M., Ohno, Y.: *J. Biochem.*, **87,** 1863 (1980).
[166] Okazaki, M., Ohno, Y., Hara, I.: *J. Chromatog.*, **221,** 257 (1980).
[167] Kondo, H., Nakatani, H., Matsuno, R., Hiromi, K.: *J. Biochem.*, **87,** 1053 (1980).
[168] Kato, Y., Komiya, K., Sasaki, H., Hashimoto, T.: *J. Chromatog.*, **193,** 29 (1980).
[169] Kato, Y., Komiya, K., Sasaki, H., Hashimoto, T.: *J. Chromatog.*, **193,** 458 (1980).
[170] Imamura, T., Konishi, K., Yokayama, M., Konishi, K.: *J. Biochem.*, **86,** 639 (1979).
[171] Takagi, T., Takeda, K., Okuno, T.: *J. Chromatog.*, **208,** 201 (1981).
[172] Murakami, K., Ueno, N., Hirose, S.: *J. Chromatog.*, **225,** 329 (1981).
[173] Calam, D. C., Davidson, J.: *J. Chromatog.*, **218,** 581 (1981).
[174] Okazaki, M., Shiraishi, K., Ohno, Y., Hara, I.: *J. Chromatog.*, **223,** 285 (1981).
[175] Takagi, T.: *J. Chromatog.*, **219,** 123 (1981).
[176] Ui, N. J.: *J. Chromatog.*, **215,** 289 (1981).
[177] Wehr, C. T., Abbot, S. R.: *J. Chromatog.*, **185,** 453 (1979).
[178] Himmel, I.: *Intern. J. Protein Res.*, **17,** 365 (1981).
[179] Leach, B. S., Collawn, J. F., Fish, W. W.: *Biochemistry*, **19,** 5734 (1980).
[180] Leach, B. S., Collawn, J. F., Fish, W. W.: *Biochemistry*, **19,** 5741 (1980).

[181] Somack, R., McKay, V. S., Giles, J. W.: *Biological Applications on Spherogel TSK-SW Type Gel,* in *Size Exclusion Chromatography (GPC)*, Provder, T. (ed.), Am. Chem. Soc., Washington, D. C., USA., 1980.

[182] Chang, S. H., Gooding, K. M., Regnier, F. E.: *J. Chromatog.*, **125,** 103 (1976).

[183] Gooding, D. L., Chatfield, C., Coffin, B.: *American Laboratory*, **12,** No. 8, 48 (1979).

[184] Grubner, K. A., Whitaker, J. M., Morris, M.: *Anal. Biochem.*, **97,** 176 (1979).

[185] Lu, K.-C., Gooding, K. M., Regnier, F. E.: *Clin. Chem.*, **25,** 1608 (1979).

[186] Gooding, K. M., Lu, K.-C., Regnier, F. E.: *J. Chromatog.*, **164,** 506 (1979).

[187] Regnier, F. E., Gooding, K. M.: *Anal. Biochem.*, **103,** 1 (1980).

[188] Bostick, W. D., Denton, M. S., Dinsmore, S. R.: *Clin. Chem.*, **26,** 712 (1980).

[189] von Stetten, O., Schlett, R.: *J. Chromatog.*, **218,** 591 (1981).

[190] Brough, A. W., Epton, R., Marr, G., Shackley, A. T., Sniezko-Blocki, G. A.: *Poly(acryloyl morpholine) (Enzacryl®-Gel) Packings*, in *Chromatography of Synthetic and Biological Polymers*, Vol. 1, Epton, R. (ed.), Ellis Horwood, Chichester, 1978, p. 70.

[191] Miller, J. N., Erinle, O., Roberts, J. M., Thirkettle, C.: *J. Chromatog.*, **105,** 317 (1975).

CHAPTER 7

Experimental procedures

I. Kerese

It has been our intention that this book should serve a practical purpose, and this aspect has been taken into consideration in the compilation of the preceding chapters. However, details of the individual procedures would have disturbed the uniformity of those chapters, and so they are given separately here.

In selecting the methods to be given it was necessary to take into account the extremely rapid development of protein analysis during the past 20 years. The main trends in development suggested that (*i*) outdated analytical methods, such as the chromatography of proteins and peptides on ion-exchange resin columns, should be omitted and (*ii*) the methods given should be such that, even when they too become outdated, they will be of help in understanding the future methods.

For all the methods, the material examined is the same: hen egg-white. This permits the demonstration of how the various separation methods do not always separate a certain protein sample in the same way and with the same effectiveness; examples are the chomatography of egg-white on CM-Sepharose and on DEAE-Sepharose. This is not to say, of course, that the findings are generally applicable and that separations with similar resolution will be obtained with other protein samples: in the separation of the component proteins of a material examined by the various methods, characteristic effects are displayed by the physico-chemical properties of the invidual proteins concerned, such as their molecular weights, charges and steric structures.

The methods described have been carried out by experts who have great experience in laboratory work of this nature.

7.1 PREPARATION AND COMPOSITION OF THE MATERIAL FOR EXAMINATION

The whites of 10 eggs not older than one day are combined, thoroughly mixed, and forced through a plastic sieve. The sieved material is placed in Petri dishes to a depth of about 5 mm, and freeze-dried. The residue after freeze-drying is ground in a porcelain mortar.

By subjecting egg-white to chromatography on a CM-cellulose column, Rhodes *et al.* [1] isolated the following proteins:

	p*I*	Amount of protein from N content, mg	Distribution of N in fractions, %
Ovomucoid + flavoprotein	3.9–4.3	304	12.3
Ovomucoid	3.9–4.3	6	0.3
Ovalbumin A_1	4.58	1106	50.8
Ovalbumin A_2	4.65	237	10.8
Ovalbumin A_3	4.75	70	3.2
Globulin		22	1.5
Conalbumin	6.5–6.8	256	12.4
Conalbumin	6.5–6.8	97	4.7
Globulin		46	2.2
Globulin		15	0.7
Globulin		7	0.3
Avidin	10	6	0.3
Globulin		8	0.4
Lysozyme	10.7–11.3	11	0.16
		2191	100.0

7.2 DETERMINATION OF NITROGEN CONTENT

Through determination of the nitrogen content and knowledge of the nitrogen conversion factor for the protein being examined, reliable data are obtained on the protein content of a sample. If the appropriate nitrogen conversion factor is not known, it may be calculated from the amino-acid composition of the proteins.

Of the numerous methods for determination of the protein contents of samples, the following procedures have been selected on the basis that they are reliable and can be carried out in any laboratory.

7.2.1 Kjeldahl determination by titration

Weigh a sample containing 200–300 mg of protein into a 100-ml Kjeldahl flask, add 10 ml of concentrated sulphuric acid, 6 g of potassium sulphate and 0.2 g of copper sulphate and heat at 320–350 °C for about 90 min. When the solution has cooled, dilute it to about 50 ml, cool it again, transfer it quantitatively to a 100-ml standard flask and make up to volume with distilled water. Pipette 25 ml of this solution into a Parnas–Wagner distillation apparatus [2], add 10 ml of 40% sodium hydroxide solution and steam-distil the ammonia, collecting it in either 25 ml of 0.5 *M* hydrochloric acid (accurately measured) or 25 ml of approximately 4% boric acid solution, together with 4 drops of Tashiro indicator (0.2 g of Methyl Red and 0.19 g of Methylene Blue dissolved in 100 ml of ethanol). In the Parnas–Wagner method [2], the excess of hydrochloric acid is back-titrated with 0.05 *M* sodium hydroxide, and in the Winkler procedure [3] the ammonium borate formed is titrated with 0.05 *M* hydrochloric acid.

7.2.2 Photometric determination [4, 5]

This method is suitable for the determination of the nitrogen content of large numbers of samples. There are several advantages to the method: (*i*) in the case of a homogeneous substance with a particle size smaller than 0.2 mm, very small samples are sufficient and a large number of samples may be decomposed simultaneously in a thermal block; (*ii*) the photometry is fast and accurate; (*iii*) if duplicate samples are decomposed and each resulting solution analysed in duplicate, the mean of the four results gives a satisfactory measure of the nitrogen content of the material.

Reagents

Sulphuric acid for decomposition: 2 g of selenium are heated with 250 ml of sulphuric acid until a colourless solution is obtained; after cooling, this solution is carefully poured into 250 ml of saturated potassium sulphate solution (27.4 of potassium sulphate dissolved in 245 ml of warm distilled water) with cooling.

Nessler reagent [6]

Iodine (22.5 g) is dissolved in a solution of 30 g of potassium iodide in 20 ml of water, and 30 g of metallic mercury are added. The vessel is shaken vigorously while being cooled under running water, until the solution is practically colourless. A few drops of the solution are then tested with 1% starch solution to check that there is not an appreciable amount of unreacted iodine left. The solution is then decanted. If the starch test does not reveal the presence of iodine, a solution of 2.25 g of iodine and 3 of potassium iodide in 2.0 ml of water is added dropwise until a pale yellow colour is restored. The decanted solution is made up to 200 ml with water, and finally 975 ml of 10% sodium hydroxide solution are added with stirring.

Standard solutions

Nitrogen concentration 0.1 mg/ml: 47.16 mg of $(NH_4)_2SO_4$ (dried over phosphorus pentoxide) dissolved in 0.2*N* H_2SO_4 and diluted to 100 ml.

Procedure

The material should have a particle size of at most 0.2 mm. Duplicate samples containing 0.1–0.5 mg of nitrogen are weighed into normal test-tubes and decomposed by heating with 4 drops of H_2O_2 and 0.4 ml of sulphuric acid mixture, the test-tubes being inserted in holes drilled in an aluminium block, which is heated to 320 °C. The decomposition requires 1.5–2.5 h (the solutions should become colourless). The operation is best performed in a fume-cupboard. The standards (0.1, 0.2, 0.3 . . . 0.8 ml of 0.1 mg/ml nitrogen standard) are treated similarly.

After decomposition the solution is cooled and made up to 10 ml with distilled water. Then, depending on the expected nitrogen content, two 0.2–4.0 ml portions are taken in graduated 10-ml test-tubes and made up to 4 ml with distilled water if necessary, then 2.0 ml of Nessler reagent are added with continuous shaking. The absorbances are then measured in 1-cm cells at 440 nm against water. The calibration curve is drawn from the results for the standards, with 4-ml portions of the diluted solution from the decomposition step (the absorbances should be approximately the same as the number of ml of 0.1 mg/ml standard originally taken).

7.2.3 Evaluation of results

The choice of titration or spectrophotometry is governed by the purpose of the analysis. The titration procedure must be regarded as the fundamental method, but its use is frequently hindered by sufficient sample not being available; in such a case, spectrophotometry may be of assistance. The larger measurement errors in the spectrophotometric procedure necessitate the use of replicate determination. Incorrect results due to faulty sampling or pipetting can be distinguished among the four results obtained. The spectrophotometric method is particularly useful for the routine analysis of large numbers of well-homogenized samples.

7.3 DETERMINATION OF AMINO-ACID COMPOSITION

Determination of its amino-acid composition is indispensable for the general characterization of a protein, elucidation of its primary structure, and nutritional and biological evaluation.

The amino-acid composition cannot generally be found by chromatography of a hydrolysate, as the individual amino-acids undergo destruction to differing extents during the different hydrolyses. The hydrolyses used must therefore permit the exact establishment of the amino-acid composition.

The following hydrolyses are usually performed: (1) with 6 *M* hydrochloric acid for 22 and 48 h from the results of which the value for zero time can be extrapolated [7];

(2) with 5 *M* sodium hydroxide, for determination of Trp [8];

(3) performic acid oxidation, and then hydrolysis with 6 *M* hydrochloric acid, for accurate determination of Cys and Met [9].

7.3.1 Hydrolysis with 6 *M* HCl

About 13 mg of sample is weighed into the hydrolysis vessel (Fig. 7.1), which is then fitted together and immersed in a "dry ice"–ethanol cooling mixture; when the solution has frozen, the vessel is evacuated to 7.9 Pa (0.06 mmHg)

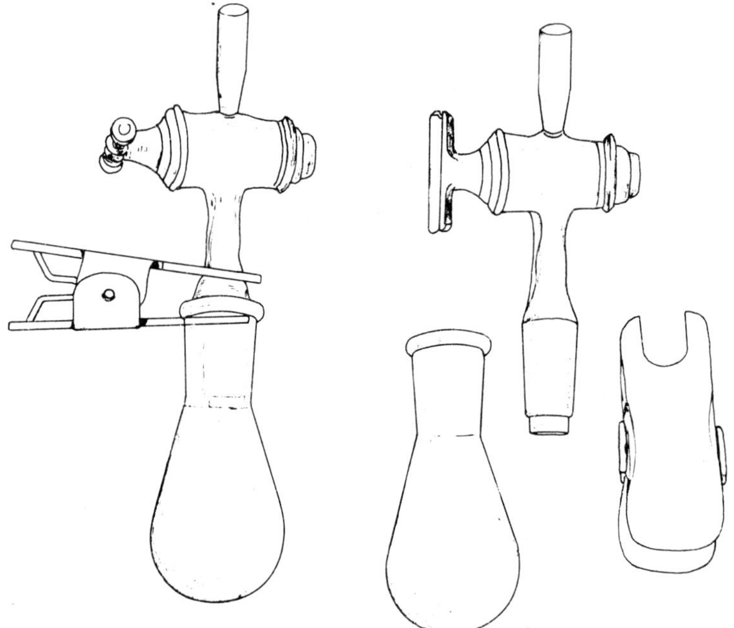

Fig. 7.1. Evacuable vessel for hydrolysis with 6 *M* HCl

(measured with a McLeod gauge). To remove traces of air dissolved in the acid, the hydrolysis vessel is taken from the cooling mixture and left, after the pump has been switched off, for the frozen solution to melt slowly. As soon as bubbles begin to emerge from the viscous solution, the hydrolysis vessel is immediately reimmersed in the cooling mixture and evacuation continued until 7.9 Pa (0.06 mmHg) is reached (ca. 10–15 min), but it is not necessary to

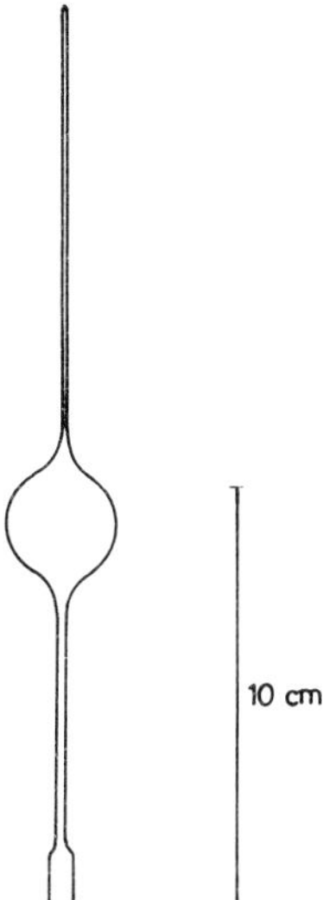

Fig. 7.2. "Ballonki" for filtration of hydrolysates. The bulb is warmed above a low flame, and the open end (packed with scraps of Seitz filter fibre) is immersed to the bottom of the solution to be filtered. As the bulb cools, the solution is filtered upwards; the filter fragments are removed, the end of the capillary is broken off, and the filtered solution can be poured out

refreeze the solution. The hydrolysis vessel is then isolated by means of the tap, and disconnected from the vacuum pump. The evacuated vessel is placed in an oven at 110 °C for 22 or 48 h. After the hydrolysis, the vessel is cooled to room temperature and connected to a rotary still inclined at an angle of 30°, and the hydrochloric acid is distilled off by heating to not more than 40 °C.

The hydrolysate is dissolved in 1 ml of 0.2 *N* sodium citrate buffer (pH 2.2), and any insoluble matter is filtered off with a "ballonki" (Fig. 7.2). (Solutions not utilized on the same day are stored frozen until required.) The optically clear solution (30-μl sample) is chromatographed on a 4 × 23 mm column of Durrum DC 4-A resin with (*a*) 0.2 *N* sodium citrate buffer (pH 3.25) containing 2.5% ethanol; (*b*) 0.7 *N* sodium citrate buffer (pH 3.50); (*c*) 1.6 *N* sodium citrate (pH 3.65) at a flow rate of 20 ml/h. The chromatography takes 75 min and the elution temperature is 35–75 °C. The eluate is mixed with ninhydrin solution added at a rate of 10 ml/h. The ninhydrin solution [7] consists of 20.0 g of ninhydrin, 750 ml of methylcellosolve, 250 ml of 3.7 *M* sodium citrate (pH 5.5) and 0.4 g of $SnCl_2 \cdot 2\,H_2O$. The absorbance of the ninhydrin/eluate mixture is monitored at 440 and 570 nm.

7.3.2 Hydrolysis with 5 *M* NaOH for determination of Trp

A quartz tube containing as starting materials about 13 mg of egg-white, 25 mg of hydrolysed starch and 5 ml of 5 *M* sodium hydroxide is immersed in a "dry ice"–ethanol cooling mixture until the materials are frozen, and transferred to a similarly cooled Oehlshlegel hydrolysis vessel [10] (Fig. 7.3). This is then connected to a vacuum pump and evacuated to 23.9 Pa (0.18 mmHg) with cooling, and the glass tap is closed. After being allowed to warm up to room temperature, the hydrolysis vessel is placed in an oven at 110 °C for 16 h. After cooling to room temperature, the quartz tube is taken out of the hydrolysis vessel, 4.6 ml of 6 *M* hydrochloric acid are added to the hydrolysate, and the volume is made up to 10 ml with distilled water.

The solution (30 μl) is then chromatographed in the same way as the 6 *M* hydrochloric acid hydrolysates, and Leu and Trp are determined.

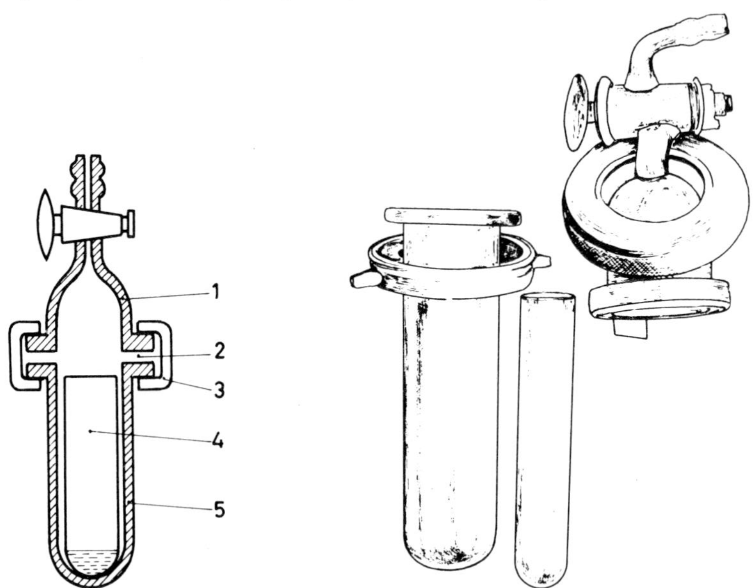

Fig. 7.3. Evacuable vessel for alkaline hydrolysis according to Oehlshlegel *et al.* [10]: 1, upper glass part with tap; 2, position of rubber ring; 3, clamp; 4, quartz tube containing hydrolysate; 5, lower glass vessel

7.3.3 HCl hydrolysis of protein oxidized with performic acid

The performic acid is prepared by adding 0.5 ml of 30% H_2O_2 to 4.5 ml of 80% formic acid, and heating in a water-bath at 50 °C for 3 min.

A mixture of 0.4 ml of performic acid and about 12 mg of sample is heated in a water-bath at 50 °C for 15 min, and then evaporated to dryness in a rotary

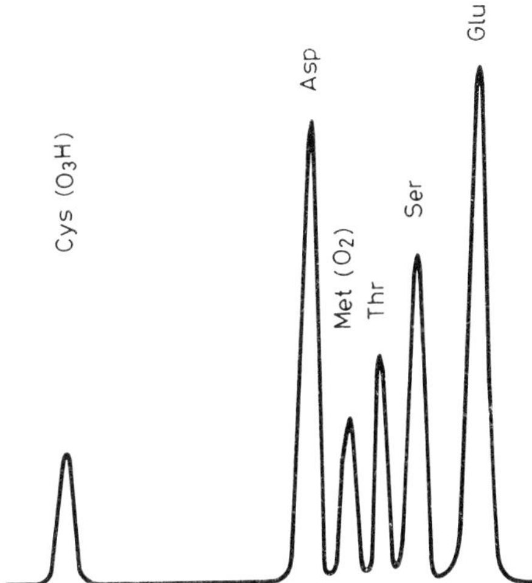

Fig. 7.4. Cys(O_3H) and Met(O_2) chromatographed after performic acid oxidation and 6 *M* HCl hydrolysis

vacuum still. The residue is then hydrolysed with 5 ml of 6 *M* hydrochloric acid for 22 h, by the procedure already described.

The chromatography is done only with the buffer of pH 3.25. A typical chromatogram is shown in Fig. 7.4.

7.3.4 Collation of chromatographic results

In the first step, the quantities of amino-acids found by the chromatography of the four hydrolysates are recalculated on the basis of 100 μg of egg-white. Typical quantities found, in μg per 100 μg of freeze-dried egg-white, are given below (Table 7.1).

The results obtained from the 22-h and 48-h hydrolyses are next extrapolated to zero time to obtain the values before the destruction caused during the acid hydrolysis, except for Cys, Val, Met, Ile and Trp. (During the

Table 7.1

Hydrolysis	6 *M* HCl 22 h	6 *M* HCl 48 h	5 *M* NaOH	6 *M* HCl after performic acid oxidation
Asp	8.92	8.85		
Thr	3.26	3.00		
Ser	5.45	4.96		
Glu	12.34	12.34		
Pro	2.75	2.76		
Gly	2.84	2.82		
Ala	3.06	3.02		
Cys	0.85	0.80		1.67
Val	4.54	4.58		
Met	0.74	0.27		2.03
Ile	3.41	3.56		
Leu	5.85	5.84		
Tyr	2.10	1.95		
Phe	5.07	4.90		
Lys	5.36	5.31		
His	3.36	3.36		
Trp	0	0	1.00	
Arg	3.76	3.75		
NH_2	2.08	2.20		

22-h hydrolysis, the Val and Ile are not liberated quantitatively; the correct values for these are obtained from the 48-h hydrolysis.) These calculations give the following values for the amino-acid composition (% w/w) of the freeze-dried egg-white:

Asp	9.0%	Ile	3.55%
Thr	3.5%	Leu	5.85%
Ser	5.75%	Tyr	2.2%
Glu	12.35%	Phe	5.15%
Pro	2.75%	Lys	5.4%
Gly	2.85%	His	3.35%
Ala	3.1%	Trp	1.0%
Cys	1.65%	Arg	3.75%
Val	4.5%	NH_2	1.95%
Met	2.05%		

Total amino-acid content: 79.7%

7.4 ELECTROPHORETIC METHODS

In contrast to the other methods, the description of the electrophoretic separation methods not only contains data and information on their practical application, but also gives an insight into the decisive extent to which the nature of the support influences the electrophoretic migration of protein molecules with identical charges.

7.4.1 Paper electrophoresis

Support: Schleicher & Schüll 2043 b, 2 cm wide, 8 cm long (between the fixed points).
Sample applied: 3 μl.
Electrode buffer: veronal–sodium acetate, pH 8.6, $\mu = 0.1$.

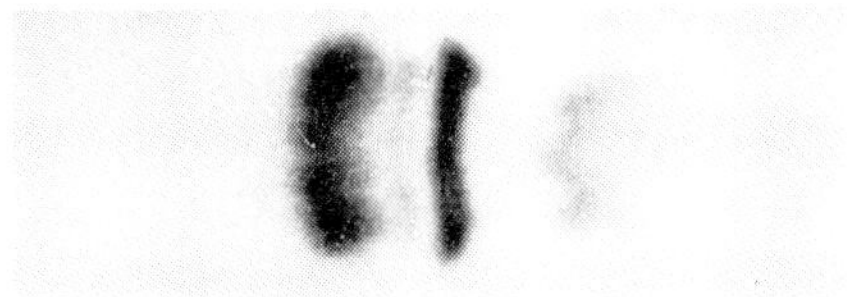

Fig. 7.5. Paper electropherogram of egg-white

Electrophoresis: 130 V, 4 mA, 160 min.
Staining: 0.4% Ponceau in 7.5% TCA.
Destaining: with 5% acetic acid.
Evaluation: 4 distinguishable bands are found on the paper (Fig. 7.5).
(Work by J. Máday, Labor MIM Laboratory, Budapest.)

7.4.2 Starch gel electrophoresis

Support: 35 × 100 × 6 mm 13% starch gel, arranged vertically in triple parallels in a gel holder according to Smithies [12].
Sample applied: 0.05 ml of 1% freeze-dried egg-white solution.
Buffer: 5.4 g of TRIS, 0.4 g of EDTA, 2.75 g of H_3BO_3, 12.5 ml of 1*M* NaOH, H_2O to 1000 ml, pH 8.6.
Electrophoresis: 320 V, 18 mA, for 17 h.
Staining: with a 0.2% solution of Amido Black in acetic acid–methanol–water (1:4:5) for 30 min.
Destaining: with acetic acid–methanol–water (1:4:5) for 24 h.
Evaluation: Starch gel electrophoresis separation of the proteins of egg-white yields only the main components, in fairly diffuse bands (Fig. 7.6).
(Work by Mrs. I. Péter, Central Laboratory, Agricultural College, Kaposvár.)

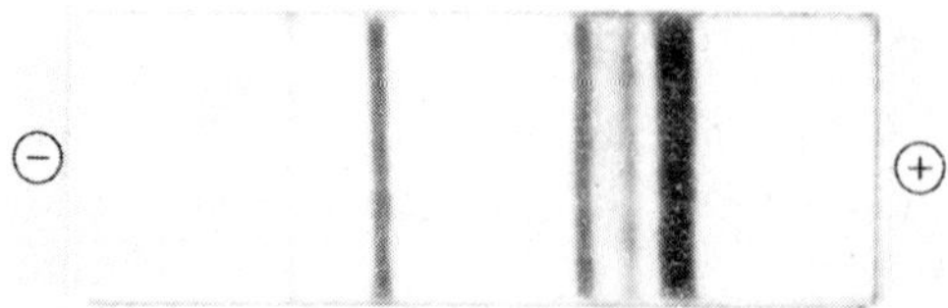

Fig. 7.6. Starch gel electropherogram of egg-white

7.4.3 Agarose gel electrophoresis

Support: 2 ml of 0.8% Koch-Light agarose-containing gel, 26 × 76 × 1 mm.
Sample applied: 7 μl of 5% egg-white solution.
Buffer: 50 m*M* veronal–veronal-Na, pH 8.6.
Electrophoresis: 140 V, 5 mA, 20 min.
Fixation: saturated picric acid solution–20% acetic acid (3 : 1 v/v).
Staining: 0.1% Coomassie BB R-250 in water–ethanol acetic–acid (8 : 1 : 1).
Destaining: water–ethanol–acetic acid (8 : 1 : 1).
Evaluation: in the anodic direction from the position of application of the sample to the gel, strongly diffuse bands can be seen, the number of which is difficult to determine; in the cathodic direction, a scarcely visible band is formed by the lysozyme (p*I* 10.7–11.3) comprising 0.16% of the egg-white. (Work by Gy. Kocsis, Phylaxia State Serum Institute, Budapest.)

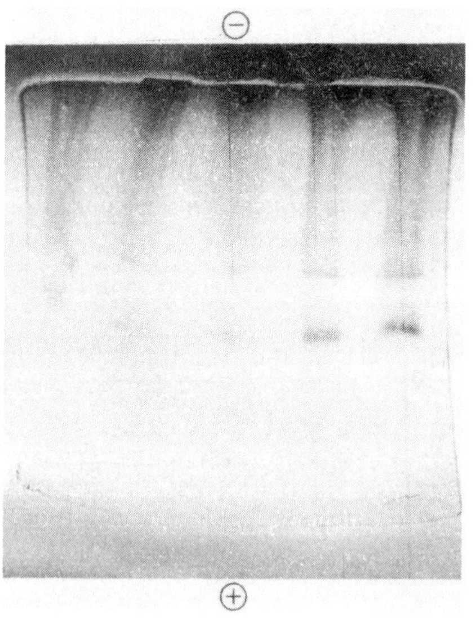

Fig. 7.7. Electropherogram of egg-white in homogeneous polyacrylamide gel

7.4.4 Polyacrylamide homogeneous gel electrophoresis

Support: 7.5% ($C = 2.6\%$) gel (see Table 3.5, gel 3, p. 113), 78 × 78 × 2.7 mm, positioned vertically.
Sample applied: 5 and 10 μl of 2% egg-white solution.
Electrode buffer: TRIS–glycine buffer, pH 8.3 (see Table 3.4, No. 36, p. 107)
Electrophoresis: 100 V, 80 mA, for 40 min; 200 V, 120 mA, for 180 min.
Fixation: in 7% acetic acid for 60 min.
Staining: with 0.1% Amido Black solution in 7% acetic acid.
Destaining: with 7% acetic acid.
Evaluation: the large number of bands of various widths resulting from the component proteins gives a good reflection of the advantages of PAGE over the previous electrophoretic methods (Fig. 7.7).
(Work by Mrs. I. Péter, Central Laboratory, Agricultural College, Kaposvár.)

7.4.5 Polyacrylamide linear-gradient gel electrophoresis

Support: Pharmacia Polyacrylamide Gradient Gel, PAA 4/30.
Apparatus: Pharmacia GE-4 II Gel Electrophoresis Apparatus.
Sample applied: 5, 10, 15 and 20 μl of 1% egg-white solution.
Reference substances (ascending from the bottom on the two edges of the gel): ovalbumin (M.W. 43,000), bovine serum albumin (M.W. 67,000), phosphorylase b (M.W. 94,000).
Electrode buffer: TRIS–glycine buffer, pH 8.3 (see Table 3.4, No. 36, p. 107)
Electrophoresis: stable potential 125 V, for 17 h.
Fixation: in 7% acetic acid, for 60 min.
Staining: with 0.1% Amido Black solution in 7% acetic acid.
Destaining: with 7% acetic acid.
Evaluation: the bands obtained in the linear-gradient gel support very well the multiplicity of component proteins in egg-white, as reported by Rhodes *et al.* [1] (Fig. 7.8.) (Work by T. Låås and I. Olsson, Pharmacia Fine Chemicals AB, Sweden.)

7.4.6 Electrophoresis on SDS-containing, staged-gradient polyacrylamide gel

Support: gels prepared with 40% acrylamide solution ($C = 5\%$) [13].
Composition of lower gel (4 cm high):

40% acrylamide solution	10 ml
gel buffer, pH 8.2	20 ml
DMAPN	0.1 ml
10% $(NH_4)_2S_2O_8$ solution	0.1 ml
H_2O	to 40 ml

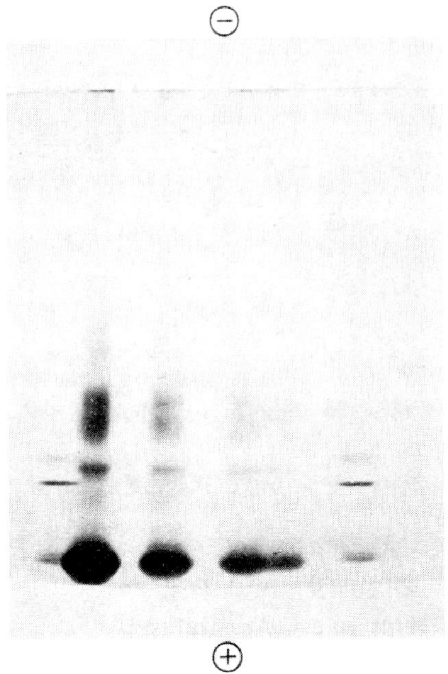

Fig. 7.8. Electropherogram of egg-white in linear gradient polyacrylamide gel

Composition of middle gel (5 cm high):

40% acrylamide solution	7.5 ml
gel buffer, pH 8.2	20 ml
DMAPN	0.1 ml
10% $(NH_4)_2S_2O_8$ solution	0.1 ml
H_2O	to 40 ml

Composition of upper gel (5 cm high):

40% acrylamide solution	5 ml
gel buffer, pH 8.2	20 ml
DMAPN	0.1 ml
10% $(NH_4)_2S_2O_8$ solution	0.1 ml
H_2O	to 40 ml

Gel buffer: 0.1 *M* TRIS–acetic acid, pH 8.2, 0.1% SDS.
Electrode buffer: the gel buffer was used.
Samples applied: reference proteins (bovine serum albumin, serum albumin, egg albumin, catalase, equine myoglobin, cytochrome C), A: 50 μg; B: 100 μg; freeze-dried egg-white: C: 50 μg; D: 100 μg.
Electrophoresis: 125 V, 80 mA, for 3.5 h.

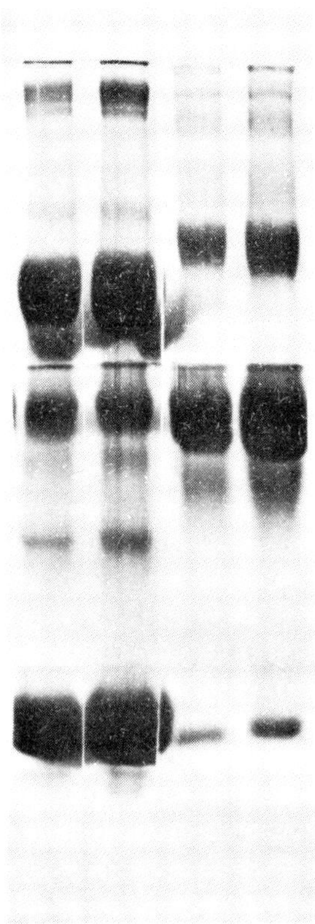

Fig. 7.9. Electropherogram of egg-white in stepwise gradient polyacrylamide gel containing SDS

Fixation: in methanol–acetic acid–water (5 : 5 : 1), for 60 min.
Staining: 0.25% Coomassie BB R-250 solution in methanol–acetic acid–water (5 : 5 : 1), for 2 h.
Destaining: a mixture of 250 ml of methanol, 375 ml of acetic acid and 4375 ml of water is allowed to flow through overnight.
Evaluation: the separations obtained in cases C and D very clearly illustrate not only the quantitative proportions of the egg-white component proteins, but also their distribution according to molecular weight (Fig. 7.9).
(Work by Ö. Takács, Dept. of Biochem., Med. Univ., Szeged.)

7.4.7 Rocket immunoelectrophoresis [14]

Support: 120 × 150 × 1.8 mm gel slab containing 32 ml of 1% Indubiose A 37 agarose + 0.7 ml of antiovalbumin rabbit serum.
Buffer: Svendsen buffer (barbiturate–TRIS–Gly), pH 8.6, $\mu = 0.08$.

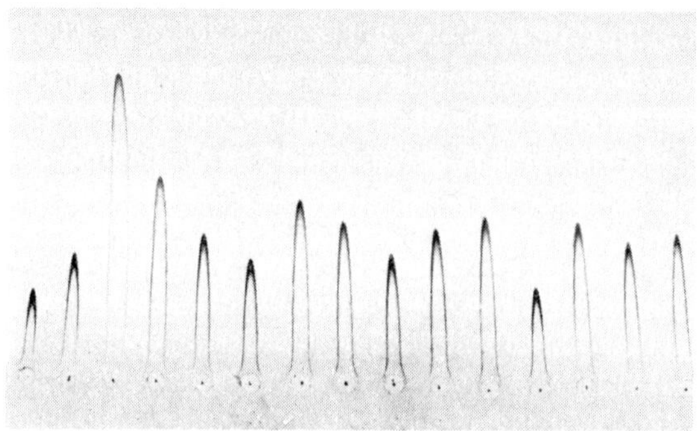

Fig. 7.10. Rocket immunoelectropherogram of egg-white

Sites of application of samples: holes 3 mm in diameter, separated by distances of 8 mm.
Volume of sample applied: 5 μl. Egg-white solutions were applied in a standard dilution series (the first 5 peaks from the left in Fig. 7.10). with concentrations of 0.0125, 0.025, 0.2, 0.1 and 0.05 μg/ml; the other peaks originate from 0.9% NaCl extracts of various pastry products containing egg-white.
Electrophoresis: 4 V/cm standard potential, for 16 h, with the gel-supporting glass plate situated on a cooling plate.
Staining: with a 0.1% solution of Coomassie BB R-250.
Destaining: according to Laurell [15].
Evaluation: The egg-white quantities in the NaCl solution extracts of the various pastry products can be well evaluated on the basis of the lengths of the electroendo-osmotic movements; these are compared with the standard dilution series containing known amounts. (Work by F. Péterfy, "Human" Inst. for Serobact. Prod. and Research, Gödöllő, Hungary.)

7.5 ISOELECTRIC FOCUSING (IEF)

IEF in an ampholyte-containing polyacrylamide gel slab in the pH interval 3–10 provides valuable information for the selection of subsequent separation methods in which the fractionation is based on the charges of the substances to be separated (IEF in a narrower pH interval, various electrophoresis techniques, ion-exchange chromatography).

If IEF in the pH interval 3–10 is used to establish the limits within which the p*I* values of the components of the substance under examination are to be found, the p*I* values of the individual components or of one particularly important protein may be determined accurately by means of a second IEF. For this, an ampholyte with a narrower pH interval is employed. In addition, the focusing may first be performed with the electrodes a large distance apart; the electrodes are then brought nearer to each other, care being taken that the bands of importance always lie between the electrodes.

7.5.1 IEF focusing of egg-white in ampholyte-containing polyacrylamide gel slab in the pH interval 3–10

The purpose of the IEF is to obtain information on the p*I* values of the egg white proteins, and also to demonstrate that during the IEF the identical proteins migrate from any position to the position with pH corresponding to the p*I* value.

A Pharmacia FBE 3000 flat bed apparatus and ECPS 2000/300 constant power supply are used.

The acrylamide–BIS solution ($T=10\%$, $C=3\%$) has the composition:

acrylamide	9.7 g
BIS	0.3 g
H_2O	to 100 ml

and when mixed is shaken with 1.5 g of Elgalite or Amberlite MB 1, and stored with this at 4 °C.

The TCA–sulphosalicylic acid solution is made from:

trichloroacetic acid	20 g
sulphosalicylic acid	10 g
H_2O	to 200 ml

A 0.2% Coomassie BB R-250 solution in methanol–water–acetic acid (9 : 9 : 2) is used for staining. The destaining solution is methanol–water–acetic acid (9 : 9 : 2).

The sample is 10 mg of freeze-dried egg-white in 1 ml of water.

The gel slab is prepared with:

acrylamide solution	18 ml
glycerol	4 ml
Pharmalyte, pH 3–10	1.8 ml
H_2O	6 ml

This solution is placed in a 100-ml suction flask, which is then closed with a rubber stopper and connected to a water-pump, and the solution is degassed. Ammonium persulphate solution (2.28%, 0.2 ml) is added to the degassed solution, and after mixing the solution is poured into the 230 × 115 × 1 mm glass plate chamber of the FBE 3000. Gelation occurs about 30 min later. The plastic plate and the glass plate (previously treated with a 0.4% solution of

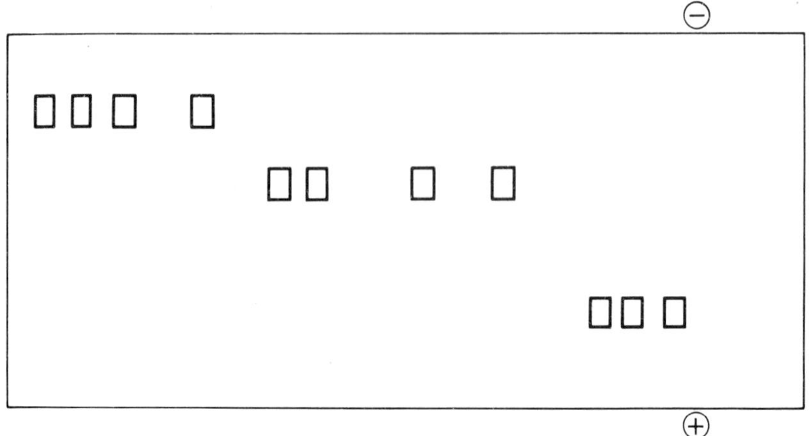

Fig. 7.11. Sites of sample application on gel slab for isoelectric focusing of egg-white

Silane A-174, Union Carbide Co.) are carefully removed from the gel layer adhering to the glass plate, and the 1 mm thick gel slab adhering to the glass plate is placed in the FBE 3000 apparatus. The sample applicator is placed on

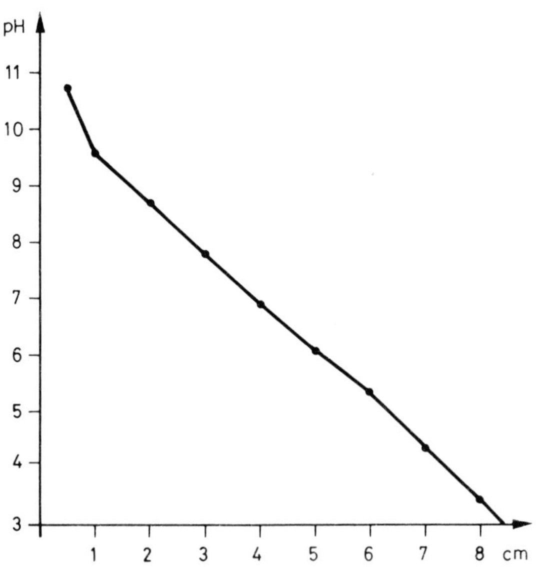

Fig. 7.12. pH gradient produced on gel slab during isoelectric focusing of egg-white

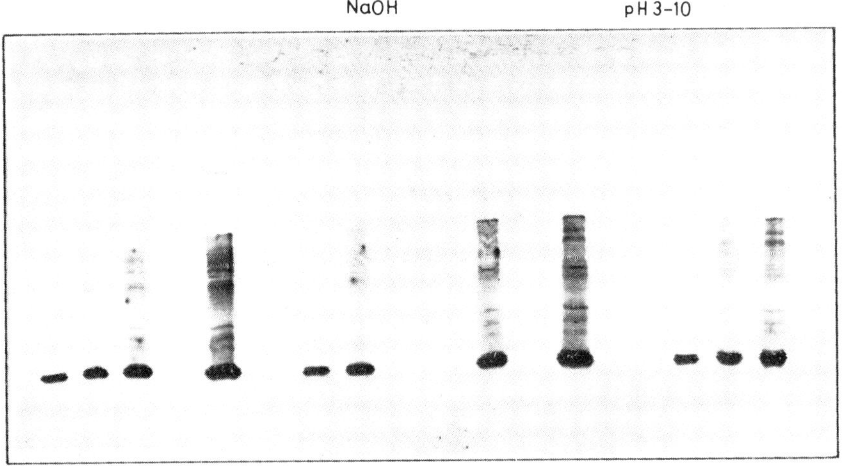

Fig. 7.13. Focused proteins of egg-white in PAG containing ampholyte of pH 3–10

the gel in three positions, with three excisions at each position, into which 5, 10 and 15 μl samples are added side by side (Fig. 7.11).

During IEF the current strength is maintained at almost the same level: initially, a potential of 900 V is necessary for 33 mA, and at the end of the IEF 2000 V is necessary for 35 mA. The IEF lasts for 90 min; the pH distributions (Fig. 7.12) developing in the gel slab during this time are measured with a surface electrode.

The gel slab adhering to the glass is fixed in TCA–sulphosalicylic acid solution for 1 h, then stained for 1 h, and destained overnight.

Evaluation: depending on their p*I* values, the proteins of the sample solutions applied in the various positions migrate to the gel band of corresponding pH. The egg-white proteins are predominantly situated in the pH interval 4–5. A characteristic deviation from this is displayed by the migration of lysozyme; after IEF, this is found on the cathodic edge of the gel. The experimental results depicted in Fig. 7.13 show that the pH distribution of the gel was uniform.

(Work by T. Låås and I. Olsson, Pharmacia Fine Chemicals AB, Sweden.)

7.5.2 IEF of egg-white in a PAG slab containing an ampholyte of pH 4–6.5

The purpose of the IEF is to determine the p*I* values of the proteins occurring in larger amounts in egg-white.

The apparatus and reagents are the same as for IEF in the pH interval 3–10, plus Pharmalyte (pH 4–6.5), 0.03 *M* glutamic acid to wet the anode strip, and 0.1 *M* histidine to wet the cathode strip.

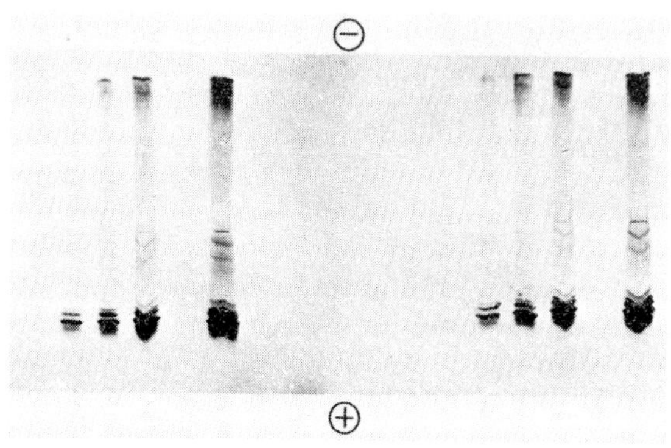

Fig. 7.14. Focused proteins of egg-white in PAG containing ampholyte of pH 4–6.5

The IEF technique is the same as for IEF at pH 3–10. A standard current strength of 36 mA is applied for 2 h.

Evaluation: the separations achieved with this IEF (Fig. 7.14) clearly demonstrate the importance of the quantity of sample applied, insofar as the bands of the ovalbumins (present in predominant amounts) can be well observed in the cases of the 5 and 10 μl samples. The many types of egg-white proteins are very well separated on the basis of their p*I* values in the pH interval 4–6.5.

(Work by T. Låås and I. Olsson, Pharmacia Fine Chemicals AB, Sweden.)

7.5.3 IEF of egg-white in PAG containing an ampholyte of pH 4–6.5, in a narrowed electric field

The aim is to attain very high resolution in the separation of the proteins with isoelectric points lying in the pH interval 4.5–4.8. Accordingly, the IEF is begun with the electrodes 220 mm apart, the distance of separation being subsequently gradually decreased.

The apparatus is the same as for IEF at pH 3–10, and the chemicals those for IEF ar pH 4–6.5.

The gel slab is 230 × 115 × 1 mm, 5 and 10 μl samples of 10-mg/ml freeze-dried egg-white solution are applied.

Before application of the samples, with the electrodes 220 mm apart, the gel slab is subjected to prefocusing in the longitudinal direction for 1 h at 2000 V and 30 W. Next, the samples are applied midway between the

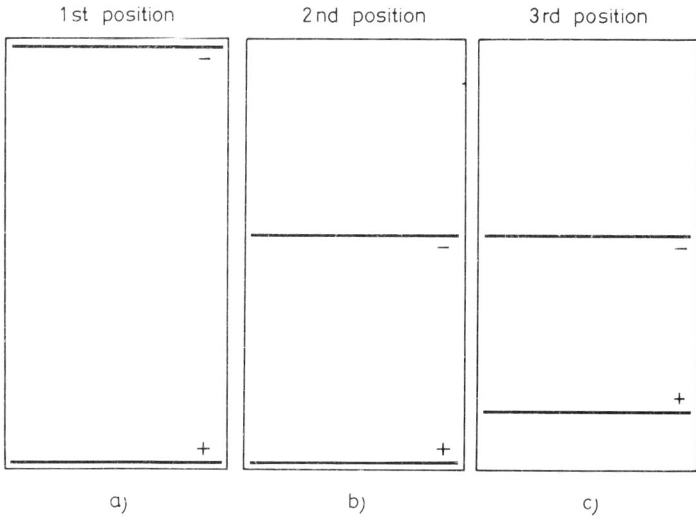

Fig. 7.15. Positions of electrodes during IEF

electrodes (still 220 mm apart), and the IEF is begun at 2000 V. After 100 min, the cathode is brought 80 mm closer to the anode, and the IEF continued for a further 120 min. Following this stage, both the anode and the cathode are moved 36 mm towards each other, so that the distance of separation is 50 mm (Fig. 7.15). The IEF is continued under these conditions at 2000 V for a further 45 min. The field of the IEF is then 400 V/cm, and pH interval between the electrodes is 1.

The fixation, staining and destaining are done as already described above. Evaluation: IEF performed with electrodes separated by a decreasing distance very clearly demonstrates that the bands of the proteins with p*I* values in the interval pH 4.5–5.5 are separated with very high resolution (Fig. 7.16). This method appeared to be more sensitive than all the others attempted. (Work by T. Låås and I. Olsson, Pharmacia Fine Chemicals AB, Sweden.)

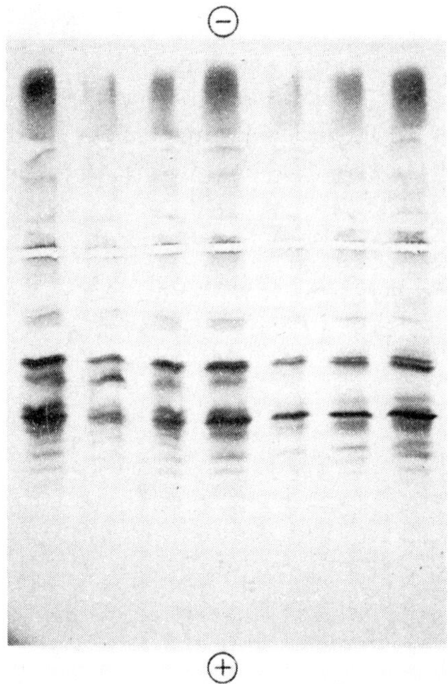

Fig. 7.16. High-resolution separation of egg-white proteins of p*I* 4.5–4.8 by means of IEF with decreasing electrode separation

7.5.4 Preparative IEF of egg-white in a Sephadex G-200 flat bed containing an ampholyte of pH 4–6.5

The aim was to obtain preparative quantities of the fractionated proteins present in greatest amount in egg-white and focused in the pH interval 4–6.5, for possible subsequent analyses.

The apparatus was the same as before, used with Whatman 1 MM filter paper, 217 × 217 mm.

Preparation of gel bed and application of sample

Sephadex IEF (8 g) is swollen with 225 ml of water, and then is degassed in a suction flask. The supernatant liquid is poured off, 20 ml of Pharmalyte (pH 4–6.5) are mixed with the gel, and the slurry is poured into the FBE 3000 preparative tray. Next, to bind the excess of liquid in the gel, sufficient Sephadex IEF must be sieved uniformly onto the surface of the gel bed so that this no longer moves when the bed is inclined at an angle of 45°. The tray containing the gel is placed on the cooling plate of the FBE 3000 apparatus, the preparative sample applicator is positioned on roughly the middle of the gel slab, and the gel enclosed by the two discs of the applicator is lifted out with a spatula. The freeze-dried egg-white solution (13 mg/ml) and 9 ml of aqueous 5% Pharmalyte solution (pH 4–6.5) are mixed into the gel removed. This sample-containing gel is next replaced in the empty site of the preparative sample applicator, and the two discs of the applicator are removed. The electrodes are positioned on the two edges of the gel bed, parallel with the line of the sample (anode strip wetted with 0.1 *M* phosphoric acid, cathode strip with 0.1 *M* sodium hydroxide), and the IEF is carried out at 650 V and 100 mA for 3 h, and then at 1150–1180 V and 50 mA for 75 min. After completion of the IEF, the gel bed is covered with rolled-up Whatman 1 MM filter paper, without air bubbles. Two min later, the paper (containing absorbed protein) is placed in 10% TCA + 5% sulphosalicylic acid solution for fixation, the fixation solution is washed out of the paper (with the following destaining solution), and staining is performed with 0.2% Coomassie BB R-250 solution in methanol–acetic acid–water (9:2:9), and destaining with methanol–acetic acid–water (9:2:9).

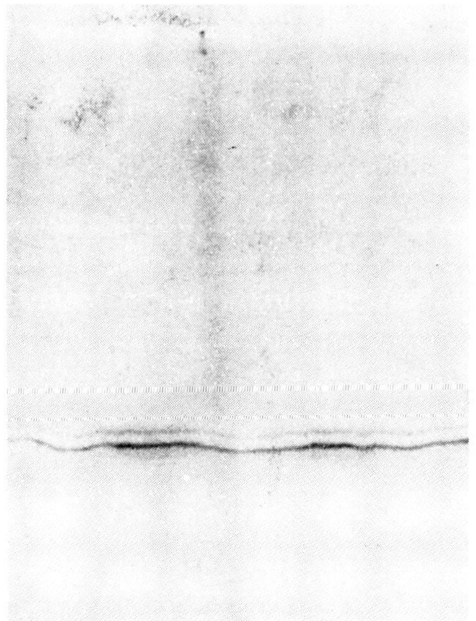

Fig. 7.17. Focused proteins of egg-white in preparative gel bed containing ampholyte of pH 4–6.5

Evaluation: on the basis of the paper fingerprint (Fig. 7.17), by appropriate positioning of the discs used with the preparative sample applicator, the gel band containing the desired fraction may be lifted out with a spatula, the gel packed into a column, and the protein fraction eluted. (Work by T. Låås and I. Olsson, Pharmacia Fine Chemicals AB, Sweden.)

7.6 COLUMN-CHROMATOGRAPHIC METHODS WITH ION-EXCHANGE SORBENTS

Although of great value in certain special cases, adsorption and partition column-chromatography and the column-chromatographic methods performed with ion-exchange resins have been virtually completely replaced by the use of cellulose-, dextran- and agarose-based ion-exchange gels, which have a higher capacity and give a better resolution. Accordingly, we have not carried out separations by the three above-mentioned chromatographic procedures.

The various types of ion-exchange chromatography actually performed do not always give uniformly good separations, but for just this reason they are important for illustrative purposes.

7.6.1 Ion-exchange chromatography on a CM-cellulose column

The CM-cellulose (0.7 meq/g) was prepared for chromatography as described in Chapter 4 (p. 184) and packed in a column 16 mm in diameter and 120 mm long. The sample applied was 100 mg of freeze-dried egg-white, dissolved in

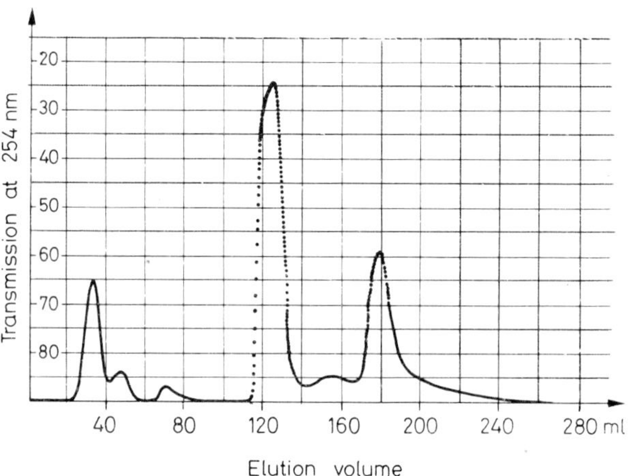

Fig. 7.18. Chromatogram of egg-white, chromatographed with linear gradient elution on CM-cellulose

4 ml of the starting buffer (0.04 *M* sodium acetate, pH 4.4). The limiting buffer was 0.5 *M* sodium acetate, pH 9.5, and the flow rate was 40 ml/h. The eluate was monitored at 254 nm (Fig. 7.18).

The elution started from the acidic region; the sodium ion concentration and pH were increased with a linear gradient.

Evaluation: On the chromatogram there are six well separated peaks. (Work by Mrs. M. Fritz, "Human" Natl. Inst. for Serobacteriological Production and Research, Gödöllő.)

7.6.2 Ion-exchange chromatography on a CM-Sepharose CL-6B column

The CM-Sepharose CL-6B column was 16 mm in diameter and 100 mm long and the sample was 100 mg of freeze-dried egg-white, dissolved in 4 ml of starting buffer. The linear gradient elution started with 0.04 *M* sodium acetate

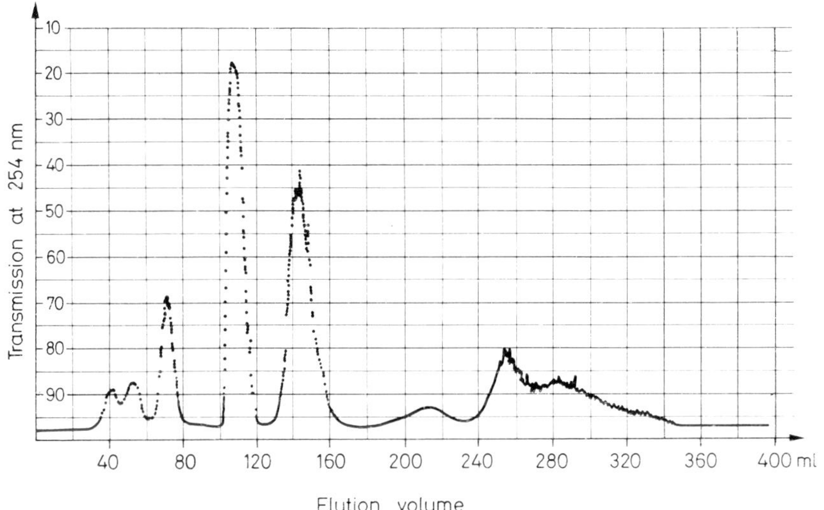

Fig. 7.19. Chromatogram of egg-white, chromatographed with linear gradient elution on CM-Sepharose CL-6B

buffer, pH 4.4, and ended with 0.5 *M* sodium acetate buffer, pH 9.5, the flow rate being 40 ml/h, and the eluate was monitored at 254 nm (Fig. 7.19). Evaluation: Chromatography led to nine fractions here, in contrast to six fractions obtained with the CM-cellulose column, due to the fact that CM-Sepharose CL-6B is a better sorbent. (Work by Mrs. M. Fritz, "Human" Natl. Inst. for Serobacteriological Production and Research, Gödöllő.)

7.6.3 Ion-exchange chromatography on a DEAE-Sephacel column

Elution with a buffer of constant TRIS concentration and decreasing pH

The DEAE-Sephacel column was 16 mm in diameter, and 100 mm long and the sample was 100 mg of freeze-dried egg-white, dissolved in 4 ml of starting buffer. The linear gradient elution started with 0.04 *M* TRIS buffer, pH 8.6, and ended with 0.04 *M* TRIS buffer, pH 3.4, the flow rate being 40 ml/h, and the eluate was monitored at 254 nm (Fig. 7.20).

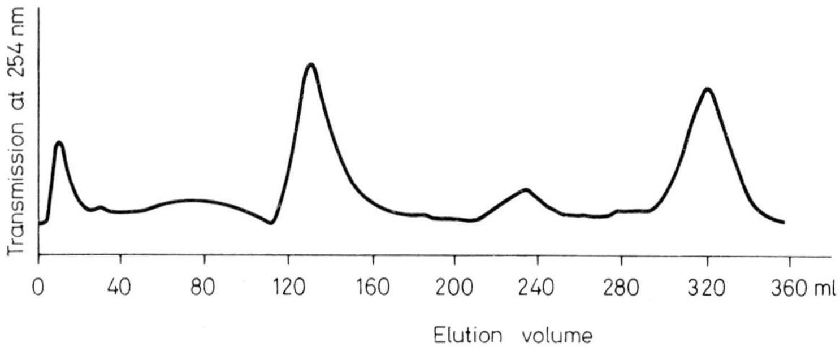

Fig. 7.20. Chromatogram of egg-white, chromatographed with a buffer of constant TRIS concentration and decreasing pH on DEAE-Sephacel

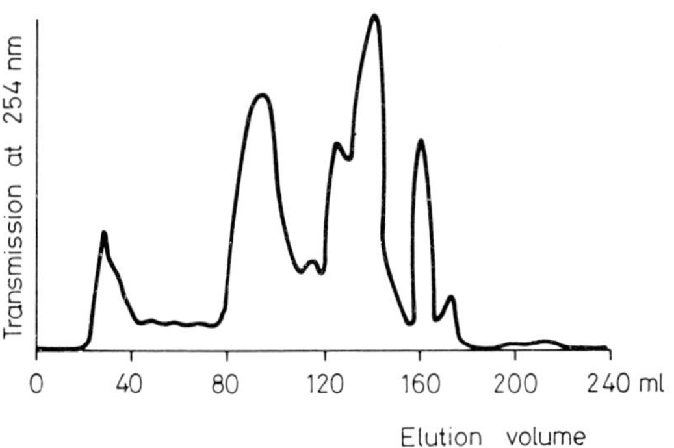

Fig. 7.21. Chromatogram of egg-white, chromatographed with a buffer of increasing TRIS concentration and decreasing pH on DEAE-Sephacel

Elution with a buffer of increasing TRIS concentration and decreasing pH

The DEAE-Sephacel column was 16 mm in diameter and 100 mm long and the sample was 100 mg of freeze-dried egg-white, dissolved in 4 ml of starting buffer. The linear gradient elution started with 0.04 *M* TRIS buffer, pH 8.6, and ended with 0.5 *M* TRIS buffer, pH 3.4, the flow rate being 40 ml/h, and the eluate was monitored at 254 nm (Fig. 7.21).
Evaluation: The two chromatographic separations demonstrate clearly the importance of the ion-concentration of the buffer in ion-exchange chromatography: the elution with a buffer of increasing ion-concentration required a smaller volume of buffer, but yielded more protein fractions. (Work by Mrs. M. Fritz, "Human" Natl. Inst. for Serobacteriological Production and Research, Gödöllő.)

7.6.4 Ion-exchange chromatography on a DEAE-Sepharose CL-6B column

Elution by linear gradient elution, with increasing TRIS concentration

The DEAE-Sepharose CL-6B column was 16 mm in diameter and 100 mm long and the sample was 100 mg of freeze-dried egg-white, dissolved in 4 ml of starting buffer. The linear gradient elution started with 0.04 M TRIS buffer, pH 8.6, and ended with 0.5 M TRIS buffer, pH 3.4, the flow rate being 40 ml/h,

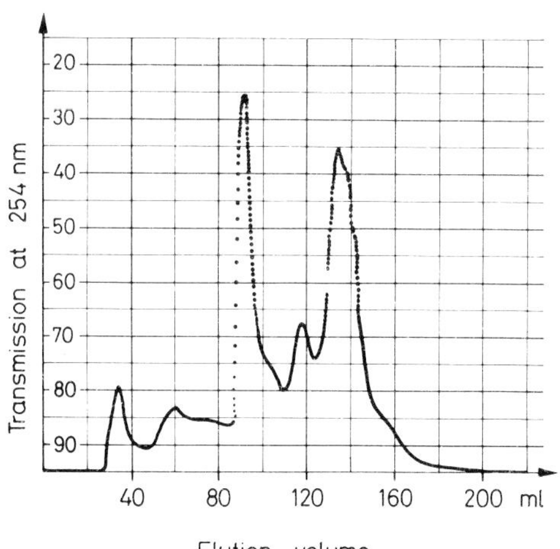

Fig. 7.22. Chromatogram of egg-white, chromatographed by linear gradient elution, with increasing TRIS concentration on DEAE-Sepharose CL-6B

and the eluate was monitored at 254 nm (Fig. 7.22). (Work by Mrs. M. Fritz, Nat. "Human" Inst. for Serobacteriological Production and Research, Gödöllő.)

Elution by stepwise gradient elution with increasing sodium ion concentration

The intention here was to compare the effects of linear gradient and stepwise gradient elution.

The DEAE-Sepharose CL-6B column was 25 mm in diameter and 100 mm long and the sample was 100 mg of freeze-dried egg-white, dissolved in 4 ml of starting buffer. The stepwise gradient elution started with 0.01 *M* TRIS-HCl, pH 8.4, the flow rate being 150 ml/h and the eluate was monitored at 254 nm. The stepwise concentrations of the second gradient component (NaCl) were:

1. 0
2. 0.09 *M*
3. 0.16 *M*
4. 0.185 *M*
5. 0.21 *M*
6. 0.265 *M*
7. 0.35 *M*
8. 0.5 *M*

Evaluation: Stepwise increase of the ion-concentration of the buffer gave a better separation then elution with a linear gradient. The differences between the chromatograms given by the two elution methods, especially at the beginning of elution, are caused by the different concentrations of the starting buffers.
(Work by L. Szabó, Department of Biochemistry, József Attila University, Szeged.)

7.6.5 Conclusion on the results of the chromatographic methods above

The significance of the quality of the ion-exchanger is shown well in the separations on CM-derivatives, where better separations are achieved with better ion-exchanger (CM-Sepharose CL-6B). As regards the three basic factors in ion-exchange chromatography (temperature of the column, ion-concentration and pH of the eluent), the importance of the ion-concentration is demonstrated well by the chromatography on the DEAE-Sephacel column (Figs. 7.21 and 7.22).

With respect to the chromatography carried out by linear and stepwise gradient elution with increasing ion-concentration, the stepwise procedure gave more fractions. This is caused mainly by the increase in the ion-concentration, and to a smaller degree by the different chromatographic dimensions.

From the negatively charged egg-white proteins, more fractions may be obtained on a DEAE-Sephacel column, eluted with a buffer with ion-concentration increased by linear gradient, than under the same chromatographic conditions on DEAE-Sepharose CL-6B.

In order to establish definitely whether the sequences of the protein fractions eluted from CM-Sepharose CL-6B and DEAE-Sephacel columns are mirror images of each other, it is necessary to make control examination of the separated fractions by IEF.

7.7 GEL FILTRATION BY A COLUMN-CHROMATOGRAPHIC TECHNIQUE

The column form of gel filtration provides a possibility for separation of various proteins into fractions of proteins with the same molecular weights, by elution of a solution containing them. Since the basis of the fractionation is separation according to molecular weight, if the solution applied to the gel contains different proteins of nearly the same molecular weight, these will be contained in one eluted fraction. In such cases, therefore, gel filtration yields group separation. The composition or homogeneity of each fraction can only be assessed after further treatment of the fractions resulting from preparative-scale gel filtration (see Fig. 1.1, p. 24).

The gel filtration of the freeze-dried egg-white was facilitated by the fact that the molecular weights of the component proteins were known from the literature data. On this basis, Sephadex G-75 and G-100, Sepharose 6-B and Sephacryl S-200 were used for the gel filtration.

7.7.1 Gel filtration on Sephadex G-75 and G-100 columns

Gel preparation: 12 g of Sephadex G-75 or G-100 Fine were hydrated with 200 ml of boiling 0.9% sodium chloride solution.

Gel column dimensions: 25 mm diameter, 400 mm long. Sample applied: 200 mg of freeze-dried egg-white, dissolved in 5 ml of 0.9% sodium chloride solution. (The membranes remaining undissolved after 6 h, comprising 4–6% of the total proteins, were filtered off with a densely-woven nylon sieve.)

Eluent: 0.9% sodium chloride solution.

Flow rate: 1 ml/min.

Detection: continuously at 254 nm.

7.7.2 Gel filtration on Sepharose 6B and Sephacryl S-200 columns

Gel column: 25 mm diameter, 350 mm long, Sepharose 6B or Sephacryl S-200 column. The other conditions were as in Section 7.7.1. (All work by L. Szabó, Department of Biochemistry, József Attila University, Szeged.)

7.7.3 Evaluation of the gel filtrations

The gel filtrations performed gave very good illustrations of the molecular-sieve properties of the various gel beds (Figs. 7.23–7.26).

With the Sephadex G-75 gel column, which is recommended for the separation of protein molecules in the M.W. range 3000–80,000, the ovalbumin and the proteins with M.W. above 45,000(the first two peaks) were

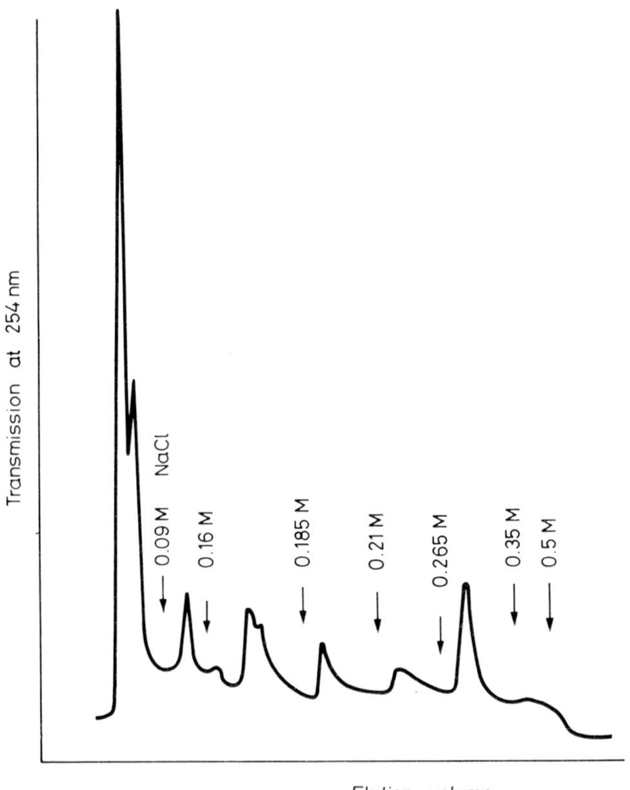

Fig. 7.23. Chromatogram of egg-white, chromatographed by stepwise gradient elution, with increasing sodium ion concentration on DEAE-Sepharose CL-6B

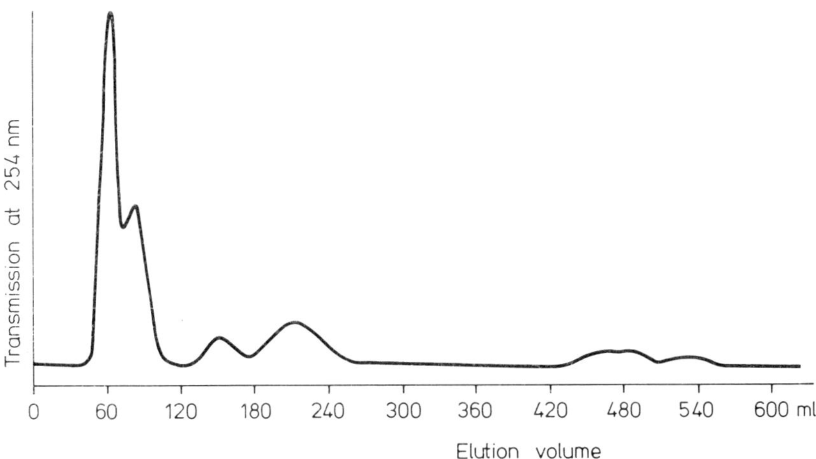

Fig. 7.24. Gel filtration of egg-white on a Sephadex G-75 column

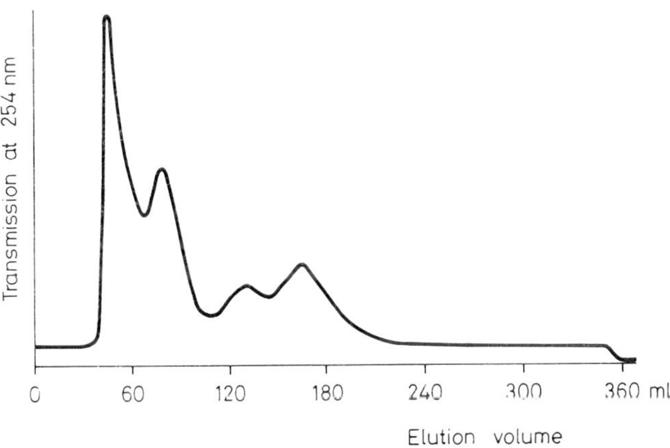

Fig. 7.25. Gel filtration of egg-white on a Sephadex G-100 column

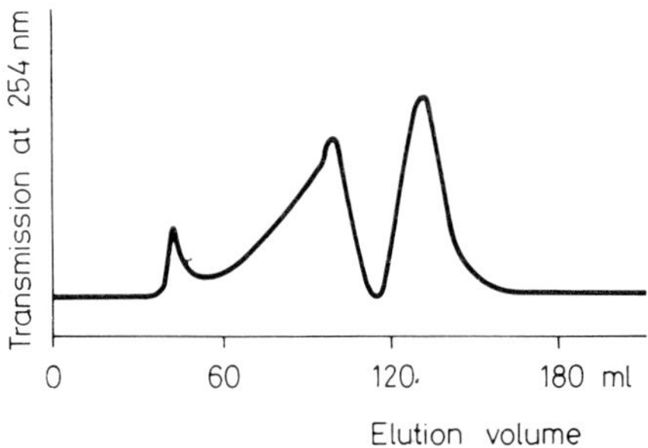

Fig. 7.26. Gel filtration of egg-white on a Sepharose 6B column

well separated. In the chromatogram obtained with Sephadex G-100, this separation was even more marked, in contrast to the proteins with lower M.W. (peaks 3 and 4). Proteins with M.W. higher than those of the ovalbumins were separated very well with the Sephacryl S-200 gel column. Understandably, much poorer separations were achieved with the Sepharose 6B column, which is recommended for the gel filtration of proteins with M.W. in the range 10^4–4×10^6.

7.8 THIN-LAYER GEL FILTRATION

In thin-layer gel filtration, separation in the gel layer also occurs on the basis of the sizes of the sample substances; accordingly, this technique may be employed for the determination of the approximate molecular weights of the dissolved proteins. Pharmacia TLG apparatus is used with Whatman 3 MM filter paper, 200 × 200 mm.

Reference solutions: cytochrome-C solution, 20 mg/ml, M.W. 12400; soya-bean trypsin inhibitor solution, 20 mg/ml, M.W. 21,500; ovalbumin solution, 20 mg/ml, M.W. 45,000. Sample solution: freeze-dried egg-white, 20 mg/ml. (The reference substances and the sample are dissolved in 0.9% sodium chloride solution.)

Preparation of gel: with gentle stirring, 150 ml of boiling 0.9% sodium chloride solution are poured onto 8 g of Sephadex G-100 superfine in a 300 ml beaker. Swelling of the Sephadex is complete within 5–6 h. Preparation of gel layer: with the TLG spreader, a 0.6 mm thick layer is spread on a 200 × 200 mm glass plate. The gel layer is washed overnight with 0.9% sodium chloride solution, with the apparatus inclined at an angle of 15°

Sample application and gel filtration: the sample and reference substances are applied in quantities of 5 and 10 μl to the gel layer, about 15 mm apart. With the plate inclined at an angle of 30°, the gel filtration is carried out with 0.9% sodium chloride solution for 4 h. The gel layer is lifted out of the apparatus and positioned horizontally, and a Whatman 3 MM filter paper of the same size is rolled onto it. After 1 min, the moist filter paper is removed and immersed for 1 min in 0.1% Bromophenol Blue solution (in 9:1 v/v methanol–acetic acid mixture) and the excess of dye is then dissolved out by treating the filter paper with 5% acetic acid solution as in the procedure of Radola [16].

Results and calculations: the reference basis for the calculations was taken as the distance covered by the ovalbumin (96 mm, $R = 1$) (Fig. 7.27).

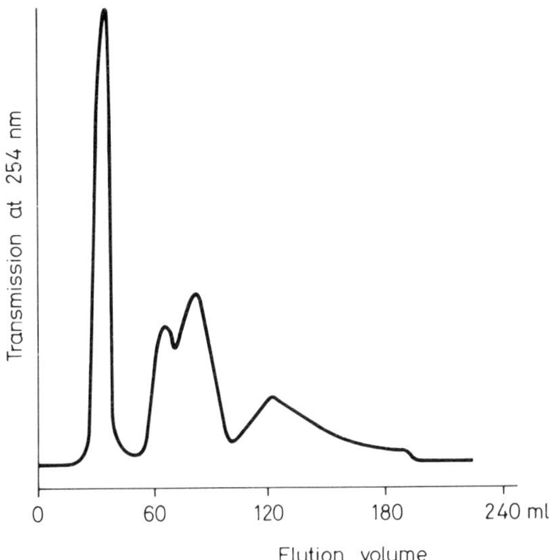

Fig. 7.27. Gel filtration of egg-white on a Sephacryl S-200 column

	Distance covered, mm	R ovalb.	$1/R$ ovalb.
1. cytochrome C	63	63/96 = 0.65	1.53
2. soya-bean trypsin inhibitor	74	74/96 = 0.77	1.3
3. ovalbumin	96	96/96 = 1	1
egg-white			
4. lower edge of band	103	103/96 = 1.07	0.93
5. lower focus of band	97	97/96 = 1.01	0.99
6. middle focus of band	79	79/96 = 0.82	1.21
7. upper focus of band	58	58/96 = 0.60	1.67

From the linear plot of 1/R vs log (M.W.), the molecular weights of the egg-white protein fractions separable by gel filtration are obtained: 9300, 26,500, 45,500 and 53,500. Evaluation: the results of the thin-layer gel filtration show the distribution of the component proteins according to molecular weight, as in the molecular sieving on a gel column. It must be emphasized, however, that with thin-layer gel filtration the separation of individual component proteins has much lower resolution than that in molecular weight determinations by means of electrophoretic procedures.

7.9 GENERAL CONCLUSIONS ON THE APPLICABILITY OF THE METHODS PRESENTED IN THIS CHAPTER

The procedures described are not all the best within the groups of methods in question, but even so, in our view they permit various comparisons.

A guide towards the selection of the most appropriate procedure is given by analytical isoelectric focusing (see Fig. 1.1, p. 24), in particular to the choice of pH for the eluents and the choice of the sorbents in ion-exchange chromatography.

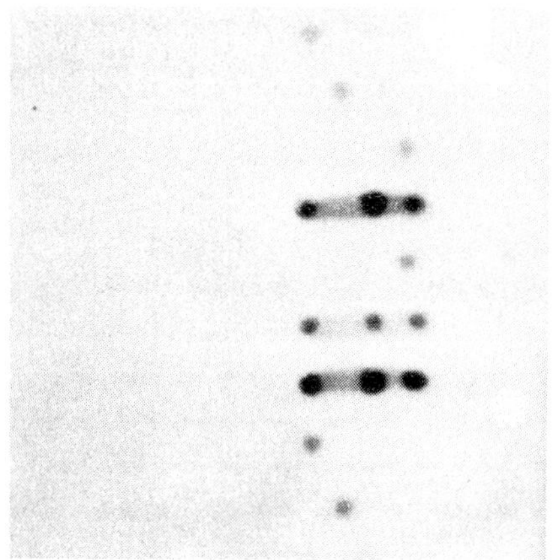

Fig. 7.28. Gel filtration of egg-white on a Sephadex G-100 Superfine thin-layer

When our model protein preparation, freeze-dried egg-white, was separated by means of the different analytical methods, very good agreement was found with regard to the distribution of the component proteins. Although the resolutions proved different, the characteristic fractions were obtained by the electrophoretic methods, the liquid-chromatographic procedures, and with column and thin-layer gel filtration alike. Depending on the methods used, these fractions showed up as a large number of bands, or sometimes only as tailing (e.g. in thin-layer gel filtration).

REFERENCES

[1] Rhodes, M. B., Azari, P. R., Feeney, R. E.: *J. Biol. Chem.*, **230,** 399 (1958).
[2] Parnas, J. K.: *Z. Anal. Chem.*, **114,** 261 (1938).
[3] Winkler, L. W.: *Z. Angew. Chem.*, **26,** 231 (1913).

[4] Lang, C. A.: *Anal. Chem.*, **30,** 1692 (1958).
[5] Strauch, L.: *Z. Klin. Chem.*, **3,** 165 (1965).
[6] Koch, F. C., McMeekin, T. L.: *J. Am. Chem. Soc.*, **46,** 3066 (1924).
[7] Moore, S., Stein, W. H.: *Methods in Enzymology*, **VI,** 819 (1963).
[8] Hugli, T. E., Moore, S.: *J. Biol. Chem.*, **247,** 2828 (1972).
[9] Hirs, C. H. W.: *J. Biol. Chem.*, **219,** 611 (1956).
[10] Oelshlegel, F. J., Schroeder, J. R., Stahmann, M. A.: *Anal. Biochem.*, **34,** 331 (1970).
[11] Matsui, K., Yaeno, K.: *Anal. Biochem.*, **6,** 491 (1963).
[12] Smithies, O.: *Adv. Protein Chem.*, **14,** 65 (1959).
[13] Talbot, D. N., Yphantis, D. A.: *Anal. Biochem.*, **44,** 246 (1971).
[14] Axelsen, N. H., Krøll, J., Weeke, B.: *A Manual of Quantitative Immunoelectrophoresis*, Universitetsforlaget, Oslo, 1973.
[15] Laurell, C. B.: *Anal. Biochem.*, **15,** 45 (1966).
[16] Radola, B. J.: *J. Chromatog.*, **38,** 61 (1968).

Index